Unix/Linux系统编程

[美] K. C. 王（K. C. Wang）著
肖堃 译

Systems Programming in Unix/Linux

机械工业出版社
China Machine Press

图书在版编目（CIP）数据

Unix/Linux 系统编程 /（美）K. C. 王（K. C. Wang）著；肖堃译．—北京：机械工业出版社，2020.5（2023.11 重印）
（计算机科学丛书）
书名原文：Systems Programming in Unix/Linux

ISBN 978-7-111-65671-5

I. U…　II. ① K…　② 肖…　III. ① UNIX 操作系统 - 程序设计　② Linux 操作系统 - 程序设计　IV. TP316.8

中国版本图书馆 CIP 数据核字（2020）第 092039 号

北京市版权局著作权合同登记　图字：01-2019-0940 号。

Translation from the English language edition:
Systems Programming in Unix/Linux
by K. C. Wang

本书讲授 Unix/Linux 系统编程的理论和实践，提供了广泛的计算机系统软件知识和高级编程技巧，使读者能够与操作系统内核进行交互，并有效利用系统资源来开发应用软件。本书还为读者提供了学习计算机科学 / 工程高级课程（如操作系统、嵌入式系统、数据库系统、数据挖掘、人工智能、计算机网络、网络安全、分布式计算和并行计算）所需的背景知识。

本书可作为面向技术的系统编程课程的教科书。因为本书包含带有完整源代码的详细示例程序，所以也适合高级程序员自学使用。

出版发行：机械工业出版社（北京市西城区百万庄大街 22 号　邮政编码：100037）
责任编辑：冯秀泳　　　　责任校对：马荣敏
印　　刷：固安县铭成印刷有限公司　　　　版　　次：2023 年 11 月第 1 版第 4 次印刷
开　　本：185mm×260mm　1/16　　　　印　　张：24.25
书　　号：ISBN 978-7-111-65671-5　　　　定　　价：139.00 元

客服电话：（010）88361066　68326294

译者序

我在联想集团工作期间参与了多款掌上电脑、智能手机产品的研发，在其中承担操作系统移植和定制、设备驱动开发、产测工具开发等工作。当移动智能终端的主流操作平台向Android切换时，我第一次接触到Linux操作系统及其编程。当时感觉需要学习的知识非常多，GCC、GDB、makefile、系统调用、C库、shell命令、shell脚本、文件I/O、字符设备、块设备、套接字等各种名词扑面而来，让人手足无措。在这种情况下，我非常渴望有一本专著能够系统、深入地介绍Linux系统编程的知识，能够帮助我和我的团队迅速掌握基本原理和开发技能。遗憾的是，直到我离开联想集团都没有找到这样一本书。

在电子科技大学任教以后，我先后承担了"Linux操作系统编程""嵌入式系统导论"以及"操作系统基础"等本科课程的教学工作。在教学过程中发现当前操作系统类课程的教材大多侧重于讲解经典的理论，并没有和实际的操作系统关联起来，甚至部分理论在当前的操作系统设计中根本没有实际的应用。另外，操作系统课程和操作系统编程课程的独立开设也让学生很难将操作系统的设计理念、实现机制与程序开发联系起来。为此，电子科技大学信息与软件工程学院启动了核心课程平台建设，尝试将"操作系统基础"与"Linux操作系统编程"课程融合，希望以Linux系统为蓝本，让学生能够理解内核的实际代码设计中蕴含的操作系统理论，通过编程项目进一步要求学生熟练掌握程序接口，帮助他们领会操作系统的工作机制和设计艺术。而要想达到理想的教学效果，就需要有Linux系统编程方面的教材或参考书。

感谢机械工业出版社的刘锋先生让我拜读了K. C. Wang的这本*Systems Programming in Unix/Linux*，本书完美地契合了我心中对于工程与教学两个方面的需求，因此决定将本书翻译出来，希望能够让更多国内计算机相关专业的学生和从事Linux开发工作的工程师看到。在这里我要感谢我的妻子和父母，正是他们的默默付出才让我顺利完成了翻译工作。我还要感谢为本书编辑、校对付出辛勤努力的编辑。本书成稿之际正值新冠肺炎疫情全球肆虐之时，在这里祝福世界各国人民能够免受病毒的困扰，早日恢复生产、生活秩序。

由于Unix/Linux系统编程涉及的知识较多，加之译者的能力和水平有限，错误、疏漏及不足之处在所难免，敬请广大读者批评指正。

肖　堃

2020年3月于成都

前 言

系统编程是计算机科学和计算机工程教育中不可或缺的一部分。计算机科学 / 工程课程中的系统编程课程起着两个重要作用。首先，它为学生提供了有关计算机系统软件和高级编程技能的广泛知识，使他们能够与操作系统内核进行交互，执行文件操作和网络编程，并有效利用系统资源来开发应用程序。其次，它为准备学习计算机科学 / 工程高级课程的学生提供了背景知识，这些高级课程包括操作系统、嵌入式系统、数据库系统、数据挖掘、人工智能、计算机网络以及分布式计算和并行计算。由于其重要性，Unix/Linux 系统编程已成为计算机科学 / 工程教育中的热门主题，并且也是高级程序员自学的主题，因此该领域有大量书籍和在线文章。尽管如此，我仍然发现很难为我在华盛顿州立大学讲授的系统编程课程选择一本合适的教材或参考书。多年以来，我在课程中不得不使用自己的课堂笔记和编程作业。经过深思熟虑，我决定将这些材料整理成书。

本书是为讲授和学习系统编程的理论和实践而服务的。与大多数其他书籍不同，这本书更深入地介绍了系统编程主题，并强调了编程实践。书中引入了一系列编程项目，让学生运用所学知识和编程技能来开发实用的程序。本书的目标是作为面向技术的系统编程课程的教科书。因为本书包含带有完整源代码的详细示例程序，所以也适合高级程序员自学使用。

事实证明，本书的编写是另一项非常艰巨且耗时的工作。在准备书稿期间，我得到了许多人的帮助和鼓励，这是我的荣幸。我想借此机会感谢他们。在此，我要特别感谢 Yan Zhang 为本书的插图和校对所提供的帮助。

还要特别感谢 Cindy 一如既往的支持和启发，没有她就没有本书。最重要的是，我要感谢家人的支持，他们接受我无时无刻不在忙碌的所有借口。

本书中编程项目的示例解决方案可从 http://wang.eecs.wsu.edu/~kcw 下载。如需获取源代码，请通过电子邮件（kwang@eecs.wsu.edu）与作者联系。

K. C. Wang
美国华盛顿州普尔曼市
2018 年 4 月

目　录

Systems Programming in Unix/Linux

第1章
Systems Programming in Unix/Linux

引　言

摘要

本章是书的引言部分。其中描述了本书的范围、目标读者以及将本书作为计算机科学/工程专业教材的适宜性。本章简单介绍了Unix的历史，包括贝尔实验室开发的Unix早期版本、AT&T System V以及Unix的其他版本，如BSD、HP UX、IBM AIX和Sun/Solaris Unix。此外，还介绍了Linux的开发及其各种发行版（包括Debian、Ubuntu、Mint、Red Hat和Slackware），列出了适用于Linux的各种硬件平台和虚拟机，并展示了如何将Ubuntu Linux同时安装到Microsoft Windows中的VirtualBox虚拟机和VMware虚拟机上。本章还解释了Linux的启动过程（从Linux内核启动到用户登录和命令执行），描述了Unix/Linux文件系统组织、文件类型和常用的Unix/Linux命令。最后，本章介绍了用户管理和维护Linux系统需执行的一些系统管理任务。

1.1　关于本书

本书是一部研究Unix/Linux系统编程的专著（Thompson和Ritchie 1974，1978；Bach 1986；Linux 2017）。其中涵盖Unix/Linux的所有基本组件，包括进程管理、并发编程、定时器和时钟服务、文件系统、网络编程和MySQL数据库系统。除介绍Unix/Linux的功能之外，还着重探讨了编程实践。书中包含编程练习和实际编程项目，让学生通过实践来练习系统编程。

1.2　系统编程的作用

系统编程是计算机科学和计算机工程教育不可或缺的一部分。计算机科学/工程专业中的系统编程课程主要有两个目的：1）教授学生计算机系统软件方面的广博知识以及高级编程技巧，使其能够与操作系统内核交互，从而有效利用系统资源来开发应用软件；2）为学生打下扎实的专业基础，以便在操作系统、嵌入式系统、数据库系统、数据挖掘、人工智能、计算机网络、网络安全、分布式和并行计算等计算机科学/工程领域继续深造。

1.3　本书的目标

本书旨在实现以下目标。

1.3.1　强化学生的编程背景知识

计算机科学/工程专业的学生一般是在一年级或二年级学习编程入门课程。这些课程通常涵盖C、C++或Java的基本编程。由于时间有限，这些课程只涉及编程语言的基本语法、数据类型、简单的程序结构以及I/O库函数的使用。实际上，由于缺乏足够的实践，刚修完这些课程的学生在编程技能方面相当薄弱。例如，他们通常不太了解软件工具、程序开发步骤和程序执行的运行时环境。本书的第一个目标是为学生提供高级编程所需的背景知识和技能。书中详细介绍了程序开发步骤，包括汇编器、编译器、链接器、链接库、可执行文件内容、程序执行映像、函数调用约定、参数传递方案、局部变量、栈帧、将C程序与汇编代码链接、程序终止和异常处理。此外，还说明了如何使用makefile管理大型编程项目，以及如何使用GDB调试程序的执行。

1.3.2　动态数据结构的应用

大多数基础数据结构课程都介绍了链表、链队列、链栈和链树等概念，但却很少说明这些动态数据结构在实践中的用处和使用方式。本书将详细讨论这些主题，包括C结构、指针、链表和链树。书中包含编程练习和实际编程项目，让学生能够在现实生活中应用动态数据结构。例如，有一个编程练习是让学生通过多个层次的函数调用来显示一个程序的栈内容，加深对函数调用序列、所传递的参数、局部变量和栈帧的理解，另一个编程练习是让学生实现参数可变的类printf函数，这需要使用参数传递方案来访问栈上的可变参数。还有一个编程练习是让学生打印磁盘的分区情况，其中，扩展分区会在磁盘中形成一个“链表”，但不是像在编程语言中学习的那样通过传统指针来形成。有一个编程项目是实现一个二叉树来模拟Unix/Linux文件系统树，该二叉树支持pwd、ls、cd、mkdir、rmdir、creat、rm操作，就像真正的文件系统一样。除使用动态数据结构之外，该编程项目还需要学生运用各种编程技术，如字符串标记化、搜索树节点、插入和删除树节点等。此外，学生需要使用二叉树遍历算法将二叉树保存为文件，并根据保存的文件重构二叉树。

1.3.3　进程概念和进程管理

系统编程的重点是各种进程的抽象概念，这往往很难领会。本书将使用一个简单的C程序（附有一段汇编代码），向学生展示运行中的真实进程。该程序实现了一个多任务环境来模拟操作系统内核中的进程操作，让学生能够创建进程、按优先级调度进程、通过上下文切换运行不同进程、通过二叉树维护进程关系、使用睡眠（sleep）和唤醒（wakeup）原语实现等待（wait for）子进程终止等。这一具体示例让学生更容易理解各种进程和进程管理功能。然后，本书详细介绍了Unix/Linux中的进程管理，包括fork()、exit()、wait()、通过exec()更改进程执行映像。另外，还介绍了I/O重定向和管道。其中提供了一个编程项目，让学生实现一个sh模拟器来执行命令。该sh模拟器的功能需要与Linux的bash完全相同。它可以执行所有Unix/Linux命令和I/O重定向，以及通过管道连接的复合命令。

1.3.4　并发编程

并行计算代表着计算的未来。因此，当务之急是在计算机科学/工程专业学生的早期学习阶段引入并行计算和并发编程。本书详细介绍了并行计算和并发编程，解释了线程的概念及其相对于进程的优势。其中详细描述了Pthreads编程（Pthreads 2017），解释了各种线程同步工具，如线程连接（threads join）、互斥量（mutex lock）、条件变量（condition variable）、信号量（semaphore）和屏障（barrier）。书中通过实例阐述了线程的应用，包括矩阵计算、快速排序和并发线程求解线性方程组等，还介绍了竞态条件、临界区、死锁和死锁预防方案等概念。除使用Pthreads之外，书中还提供了一个编程项目，让学生尝试使用用户级线程，实现线程同步工具，并通过用户级线程练习并发编程。

1.3.5 定时器和定时功能

定时器和定时功能对于进程管理和文件系统而言必不可少。本书详细讨论了 Unix/Linux 中硬件定时器的原理、定时器中断和时钟服务功能，展示了如何在系统编程中使用间隔定时器和定时器产生的信号，并提供了一个编程项目，让学生实现支持并发任务的间隔定时器。

1.3.6 信号、信号处理和进程间通信

信号和信号处理是理解程序异常和进程间通信（IPC）的关键。本书介绍了 Unix/Linux 环境下的信号、信号处理和进程间通信，解释了操作系统内核中的信号源、信号传递和处理以及信号与异常之间的关系，并展示了如何安装信号捕捉程序来处理用户模式下的程序异常。书中还提供了一个编程项目，让学生使用 Linux 信号和管道来实现一个进程间通信机制，允许并发任务之间相互交换消息。

1.3.7 文件系统

除进程管理之外，文件系统是操作系统的第二大组成部分。本书对文件系统进行了深入介绍。其中描述了不同层次的文件操作，并解释了它们之间的关系，包括存储设备、Unix/Linux 内核中的文件系统支持、文件操作的系统调用、库 I/O 函数、用户命令和 sh 脚本。书中包含一系列编程练习，让学生实现一个与 Linux 完全兼容的完整 EXT2 文件系统。这些练习不仅能让学生全面了解各种文件操作，还让其拥有了编写完整文件系统的独特经验。书中还提供了一个编程项目，让学生实现并比较不同 I/O 缓冲区管理算法的性能。其中，模拟系统支持进程和磁盘控制器之间的 I/O 操作。该项目为学生提供了练习 I/O 编程、中断处理、进程与 I/O 设备之间同步的机会。

1.3.8 TCP/IP 和网络编程

网络编程是操作系统的第三大组成部分。本书介绍了 TCP/IP 协议、套接字 API、UDP 和 TCP 套接字编程，以及网络计算中的服务器 - 客户机模型。其中提供了一个编程项目，让学生实现一个基于服务器 - 客户机的系统，用于通过互联网执行远程文件操作。如今，WWW 和互联网服务已成为人们日常生活中不可或缺的一部分。因此，有必要向计算机科学 / 工程专业的学生介绍这门技术。本书详细探讨了 HTTP 和 Web 编程，使学生能够在自己的 Linux 机器上配置 HTTPD 服务器来支持 HTTP 编程。书中说明了如何使用 HTML 创建静态网页以及如何使用 PHP 创建动态网页。另外，还介绍了 MySQL 数据库系统，展示了如何在命令和批处理模式下创建和管理数据库，以及如何同时实现 MySQL 与 C 和 PHP 之间的连接。其中提供了一个编程项目，让学生使用 CGI 编程来实现一个网络服务器，用于通过 HTTP 表单和动态网页支持各种文件操作。

1.4 目标读者

本书将作为一本系统编程课程教材，供以技术为导向、理论与编程实践并重的计算机科学 / 工程专业学生使用。书中包含许多详细的工作示例程序，并附有完整源代码。适合高级程序员和计算机爱好者自学。

1.5　本书的独特之处

本书有许多不同于其他书籍的独特之处。

（1）本书自成一体。其中包含学习系统编程所需的所有基础和背景信息，包括程序开发步骤、程序执行和终止、函数调用约定、运行时栈的使用以及如何将 C 程序和汇编代码链接。书中还介绍了 C 结构、指针和动态数据结构（如链表、队列和二叉树），并说明了动态数据结构在系统编程中的应用。

（2）本书使用一个简单的多任务系统来介绍和阐释了进程的抽象概念。该系统允许学生创建进程、通过上下文切换调度和运行不同进程、实现进程同步机制、控制和观察进程的执行。这种独特方法可让学生具体地感受并更好地理解真实操作系统中的各种进程操作。

（3）本书详细描述了 Unix/Linux 中各种进程的来源和作用，包括系统启动、初始化进程、守护进程、登录进程以及用户 sh 进程等。此外，还说明了用户进程的执行环境（如标准输入、标准输出、标准错误文件流和环境变量），解释了 sh 进程如何执行用户命令。

（4）本书详细描述了 Unix/Linux 中的各种进程管理函数，包括 fork()、exit()、wait()、通过 exec() 更改进程执行映像、I/O 重定向和管道。此外，书中提供了一个编程项目，让学生能够应用这些概念和编程技术，实现一个 sh 模拟器来执行命令。该 sh 模拟器的功能需要与 Linux 的标准 bash 完全相同。它支持执行简单命令和 I/O 重定向，以及通过管道连接的复合命令。

（5）本书介绍了并行计算和并发编程。书中通过示例详细讨论了 Pthreads 编程，包括矩阵计算、快速排序和并发线程求解线性方程组，并演示了连接、互斥量、条件变量和屏障等线程同步工具的使用。还介绍了并发编程中的重要概念和问题，如竞态条件、临界区、死锁和死锁预防方案。此外，它还让学生有机会尝试使用用户级线程，设计和实现线程同步工具（如线程连接、互斥量、信号量），并通过并发程序演示用户级线程的功能。这可以加深学生对 Linux 中线程工作原理的理解。

（6）本书详细介绍了定时器和时钟服务功能，还让读者有机会实现间隔定时器，以支持多任务系统中的并发任务。

（7）本书介绍了中断和信号的统一处理，即：将信号作为进程中断来处理，类似于 CPU 中断。其中详细解释了信号源、信号类型、信号产生和信号处理，展示了如何安装信号捕捉程序来让进程处理用户模式下的各种信号，并讨论了如何将信号应用到进程间通信和 Linux 中的其他进程间通信机制中。书中还提供了一个编程项目，让学生使用 Linux 信号和管道来实现一个进程间通信机制，允许多任务系统中的各项任务之间交换消息。

（8）本书将各种文件操作组织成具有多个不同层次的层次结构，包括存储设备、Unix/Linux 内核中的文件系统支持、文件操作的系统调用、库 I/O 函数、用户命令和 sh 脚本。该层次结构向读者展示了文件操作的全貌。其中详细介绍了不同层次的文件操作，解释了它们之间的关系，并说明了如何使用系统调用来开发文件操作命令，例如用于列出目录和文件信息的 ls、用于显示文件内容的 cat 以及用于复制文件的 cp。

（9）本书提供了一个编程项目，引导学生实现一个完整的 EXT2 文件系统。该编程项目已在华盛顿州立大学 CS360 系统编程课程中使用多年，并取得了良好效果。它为学生提供了编写一个完整文件系统的独特经验，该系统不仅能正常工作，还与 Linux 完全兼容。这或许是本书最独特之处。

（10）本书介绍了 TCP/IP 协议、套接字 API、UDP 和 TCP 套接字编程，以及网络编程中的服务器 – 客户机模型。其中提供了一个编程项目，让学生实现一个基于服务器 – 客户机的系统，用于通过互联网执行远程文件操作。书中还详细探讨了 HTTP 和 Web 编程，说明了如何配置 Linux HTTPD 服务器来支持 HTTP 编程，以及如何使用 PHP 创建动态网页。其中提供了一个编程项目，让学生使用 CGI 编程来实现一个 Web 服务器，用于通过 HTTP 表单和动态网页支持各种文件操作。

（11）本书着重介绍了编程实践。整本书都在通过完整和具体的示例程序来说明系统编程技术。尤其是，它提供了一系列编程项目，让学生练习系统编程。这些编程项目包括：

1）实现一个二叉树来模拟支持各种文件操作的 Unix/Linux 文件系统树（第 2 章）。

2）实现一个 sh 模拟器来执行命令，包括 I/O 重定向和管道（第 3 章）。

3）实现用户级线程、线程同步和并发编程（第 4 章）。

4）实现间隔定时器，以支持多任务系统中的并发任务（第 5 章）。

5）使用 Linux 信号和管道来实现一个进程间通信机制，允许并发任务之间交换消息（第 6 章）。

6）在 EXT2 文件系统中将文件路径名转换为文件的索引节点（INODE)(第 7 章)。

7）通过系统调用，递归复制文件（第 8 章）。

8）实现一个类 printf 函数，只使用 putchar() 的基本操作来进行格式化打印。

9）使用 sh 函数和 sh 脚本，递归复制文件（第 10 章）。

10）实现一个与 Linux 兼容的完整 EXT2 文件系统（第 11 章）。

11）在 I/O 系统模拟器中实现和比较 I/O 缓冲区管理算法（第 12 章）。

12）实现互联网文件服务器和客户机（第 13 章）。

13）通过 CGI 编程，实现一个使用 HTTP 表单和动态网页的文件服务器（第 13 章）。

在这些编程项目中，除 4）、5）、10）、11）外，其余都是华盛顿州立大学 CS360 课程的标准编程作业。10）已在 CS360 课程中使用多年。其中许多项目，如 12）和 13)，尤其是文件系统项目 10）都需要学生组成两人团队来完成，学生可在以团队为导向的工作环境中，习得并练习与同龄人沟通与合作的技巧。

1.6　将本书用作系统编程课程的教材

本书适合作为系统编程课程教材，供以技术为导向、理论与实践并重的计算机科学 / 工程专业学生使用。将本书用作一学期课程教材时，可包含以下主题：

（1）Unix/Linux 介绍（第 1 章）。

（2）编程背景（第 2 章）。

（3）Unix/Linux 进程管理（第 3 章）。

（4）并发编程（第 4 章部分内容）。

（5）定时器和时钟服务功能（第 5 章部分内容）。

（6）信号与信号处理（第 6 章）。

（7）文件操作级别（第 7 章）。

（8）操作系统内核中的文件系统支持和文件操作的系统调用（第 8 章）。

（9）I/O 库函数和文件操作命令（第 9 章）。

（10）使用 sh 编程和 sh 脚本进行文件操作（第 10 章）。

（11）EXT2 文件系统的实现（第 11 章）。

（12）TCP/IP 和网络编程、HTTP 和 Web 编程（第 13 章）。

（13）MySQL 数据库系统介绍（第 14 章部分内容）。

部分章节的习题部分都包含一些特定习题，旨在回顾该章所介绍的各种概念和原理。其中很多习题都适合作为编程项目，让学生获得更多系统编程方面的经验和技能。

本书的大部分资料已在华盛顿州立大学（WSU）电气工程与计算机科学系（EECS）的 CS360 系统编程课程中使用多年，并取得了良好效果。该课程一直是 WSU EECS 计算机科学和计算机工程专业课程的主干。请访问网址 http://www.eecs.wsu.edu/~cs360 查看现行 CS360 课程大纲、课堂讲稿及编程作业。

该编程项目旨在引导学生实现一个与 Linux 完全兼容的完整 EXT2 文件系统。事实证明，它是 CS360 课程中最成功的部分，该行业雇主和学术机构都反馈良好。

1.7　其他参考书

由于其重要性，Unix/Linux 系统编程已经成为计算机科学 / 工程教育和高级程序员自学的热门学科。因此，该领域拥有大量可供参考的书籍和网络文章。参考文献节列出了其中一些书籍（Curry 1996；Haviland 等 1998；Kerrisk 2010；Love 2013；Robbins 和 Robbins 2003；Rochkind 2008；Stevens 和 Rago 2013）。这些书籍中有很多都是绝佳的程序员参考指南，适合作为额外参考书目阅读。

1.8　关于 Unix

Unix（Thompson 和 Ritchie 1974，1978）是一种通用操作系统。该系统诞生于 20 世纪 70 年代早期，由肯·汤普森（Ken Thompson）和丹尼斯·里奇（Dennis Ritchie）采用贝尔实验室的 PDP-11 微型计算机开发。1975 年，贝尔实验室向公众发布了 Unix，称为 V6 Unix。该 Unix 系统的最初使用者以各大高校和非营利性机构为主。这一早期版本，连同有关 C 编程语言的经典著作（Kernighan 和 Ritchie 1988），在操作系统领域掀起了 Unix 革命，其影响一直持续至今。

1.8.1　AT&T Unix

在整个 20 世纪 80 年代，美国电话电报公司（AT&T）都在持续开发 Unix，最终发布了 AT&T System V Unix（Unix System V 2017），这是 AT&T 发布的代表性 Unix。System V Unix 是一个单处理器（单 CPU）系统。在 20 世纪 80 年代后期，该系统被扩展为多处理器版本（Bach 1986）。

1.8.2　Berkeley Unix

Berkeley Unix（Leffler 等 1989，1996）是指，加州大学伯克利分校的伯克利软件发行中心（BSD）在 1977 年至 1985 年间开发的一组 Unix 操作系统变体。BSD Unix 最重要的贡献是实现了 TCP/IP 协议族和套接字接口，作为一种标准的组网方式，几乎所有其他操作系统都会纳入 TCP/IP 协议族和套接字接口，这促进了互联网在 20 世纪 90 年代的飞速发展。此外，BSD Unix 从一开始就提倡开源，这刺激了 BSD Unix 的进一步移植和开发。后续的 BSD Unix 版本为多个开源软件开发项目提供了基础，包括 FreeBSD（McKusick 等 2004）、

OpenBSD 和 NetBSD 等，且一直沿用至今。

1.8.3 HP Unix

HP-UX（HP-UX 2017）是惠普公司开发的专有 Unix 操作系统，第一版发布于 1984 年。最新版本的 HP-UX 支持基于 PA-RISC 处理器体系结构的 HP 9000 系列计算机系统，以及基于英特尔安腾处理器的 HP Integrity 系统。HP-UX 的独特功能包括可代替 Unix 标准 rwx 文件权限的用于大型文件系统和访问控制表的内置逻辑卷管理器。

1.8.4 IBM Unix

AIX（IBM AIX 2017）是 IBM 为其多个计算机平台开发的一系列专有 Unix 操作系统。AIX 最初是为 IBM 6150 RISC 工作站而开发，现可支持各种硬件平台，包括 IBM RS/6000 系列以及基于 POWER 和 PowerPC 的更高版本系统、IBM System I、System/370 大型机、PS/2 个人计算机和苹果网络服务器（Apple Network Server）。AIX 基于 UNIX System V，具有兼容 BSD4.3 的扩展包。

1.8.5 Sun Unix

Solaris（Solaris 2017）最初是 Sun 公司（Sun OS 2017）开发的一款 Unix 操作系统。自 2010 年 1 月起，该系统更名为 Oracle Solaris。Solaris 以其可扩展性而闻名，特别是在 SPARC 系统上。它支持由 Oracle 和其他供应商提供的各种基于 SPARC 和基于 x86 的工作站和服务器。

可以看出，大多数 Unix 系统都是专有的，并且与特定的硬件平台相关联。普通人可能无法访问这些系统。对于希望在 Unix 环境下练习系统编程的读者来说，这是一个不小的挑战。因此，我们将使用 Linux 作为编程练习和实践的平台。

1.9 关于 Linux

Linux（Linux 2017）是一个类 Unix 系统。它最初是林纳斯·托瓦兹（Linus Torvalds）在 1991 年为基于 Intel x86 的个人计算机开发的一个实验性内核。后来，世界各地的人都开始加入 Linux 的研发队伍。Linux 的一个重要里程碑发生在 20 世纪 90 年代末，当时，它与 GNU（Stallman 2017）相结合，纳入了许多 GNU 软件，如 GCC 编译器、GNU emacs 编辑器和 bash 等，极大地促进了 Linux 的进一步发展。不久之后，Linux 实现了访问互联网的 TCP/IP 协议族，并移植了支持 GUI 的 X11（X-window），成为一个完整的操作系统。

Linux 包含其他 Unix 系统的许多特性。在某种意义上，它是由各种最为流行的 Unix 系统组合而成。在很大程度上，Linux 是兼容 POSIX 标准的。Linux 已被移植到许多硬件体系结构中，如摩托罗拉、SPARC 和 ARM 等。主要的 Linux 平台仍然是基于 Intel x86 的个人计算机，包括广泛可用的台式机和笔记本电脑。此外，Linux 可免费使用，且易于安装，因此，颇受计算机科学专业的学生欢迎。

1.10 Linux 版本

Linux 内核的开发是在 Linux 内核开发小组的严格监督下进行的。除不同的发行版之外，所有 Linux 内核都完全相同。但根据发行版的不同，Linux 有许多不同的版本，这些版本可能在发行版软件包、用户界面和服务功能上有所不同。下面列出了一些最流行的 Linux 发行版。

1.10.1 Debian Linux

Debian 是专注于免费软件的 Linux 发行版。可支持很多软件平台。Debian 发行版采用 .deb 包格式和 dpkg 包管理器及其前端。

1.10.2 Ubuntu Linux

Ubuntu 是基于 Debian 的 Linux 发行版。它支持定期发布新版本，能够在台式机和服务器上提供一致的用户体验和商业支持。Ubuntu Linux 有多个官方发行版。这些 Ubuntu 变体只是安装了一套不同于原始 Ubuntu 的软件包。由于这些变体可以从与 Ubuntu 相同的存储库中获取更多软件包和更新，因此，同一套软件可用于所有变体。

1.10.3 Linux Mint

Linux Mint 是基于 Debian 和 Ubuntu 的社区主导型 Linux 发行版。Linux Mint 致力于创建一个“既强大又易于使用的现代、优雅和舒适的操作系统”。Linux Mint 通过纳入一些专有软件来提供完全开箱即用的多媒体支持，并与各种免费和开源应用程序绑定在一起。因此，它受到了许多 Linux 初学者的欢迎。

1.10.4 基于 RPM 的 Linux

Red Hat Linux 和 SUSE Linux 是最早使用 RPM 文件格式的主要发行版，目前仍有一些软件包管理系统在使用这种格式。这两种发行版都分为商业和社区支持版本。例如，Red Hat Linux 提供了一个由 Red Hat 赞助的社区支持发行版（称为 Fedora）和一个商业支持发行版（称为 Red Hat Enterprise Linux），而 SUSE 分为 openSUSE 和 SUSE Linux Enterprise。

1.10.5 Slackware Linux

Slackware Linux 发行版以高度可定制而著称，专注于通过尖端的软件和自动化工具来提供维护便捷性和可靠性。该发行版适合 Linux 高级用户使用。它允许用户选择 Linux 系统组件来安装系统和配置已安装的系统，使其能够了解 Linux 操作系统的内部工作原理。

1.11 Linux 硬件平台

Linux 最初是为基于 Intel x86 的个人计算机而设计。早期的 Linux 版本是以 32 位保护模式在基于 Intel x86 的个人计算机上运行。现在，它可同时在 32 位和 64 位模式下使用。除 Intel x86 之外，Linux 已移植到许多其他的计算机体系结构上，包括摩托罗拉 MC6800、MIP、SPARC、PowerPC 以及最近的 ARM。但 Linux 的主要硬件平台仍然是基于 Intel x86 的个人计算机，尽管基于 ARM 的 Linux 嵌入式系统正在迅速普及。

1.12 虚拟机上的 Linux

目前，大多数基于 Intel x86 的个人计算机都以 Microsoft Windows 为默认操作系统，例如 Windows 7、8 或 10。在同一台个人计算机上安装 Linux 和 Windows 相当容易，并且可在开机时使用双启动来启动 Windows 或 Linux。然而，大多数用户都不愿意这样做，原因要么是存在技术上的困难，要么是更偏向于 Windows 环境。通常的做法是在 Windows 主机内的虚拟机上安装和运行 Linux。下面将说明如何在 Microsoft Windows 10 内的虚拟机上安装和运行 Linux。

1.12.1 VirtualBox

VirtualBox（VirtualBox 2017）是 Oracle 推出的一款 x86 和 AMD64/Intel64 虚拟化产品，具有强大功能。可在 Windows、Linux、Macintosh 和 Solaris 主机上运行。它可支持大量的客户机操作系统，包括 Windows（NT 4.0、2000、XP、Vista、Windows 7、Windows 8、Windows 10）和 Linux（2.4、2.6、3.x 和 4.x)，Solaris 和 OpenSolaris，OS/2 及 OpenBSD。请按照以下步骤在 Windows 10 上安装 VirtualBox。

1. 下载 VirtualBox

VirtualBox 网站（http://download.virtualbox.org）上有 VirtualBox 二进制文件及其源代码的链接。对于 Windows 主机，VirtualBox 二进制文件为

VirtualBox-5.1.12-112440-win.exe

此外，还需要下载

VirtualBox 5.1.12 Oracle VM VirtualBox Extension Pack

支持英特尔卡的 USB 2.0 和 USB 3.0 设备、VirtualBox RDP、磁盘加密、NVMe 和 PXE 启动。

2. 安装 VirtualBox

下载完 VirtualBox-5.1.12-112440-win.exe 文件后，双击文件名运行，在 Windows 10 上安装 VirtualBox VM，并在桌面上创建 Oracle VM VirtualBox 图标。

3. 创建 VirtualBox 虚拟机

启动 VirtualBox。弹出一个初始 VM 窗口，如图 1.1 所示。

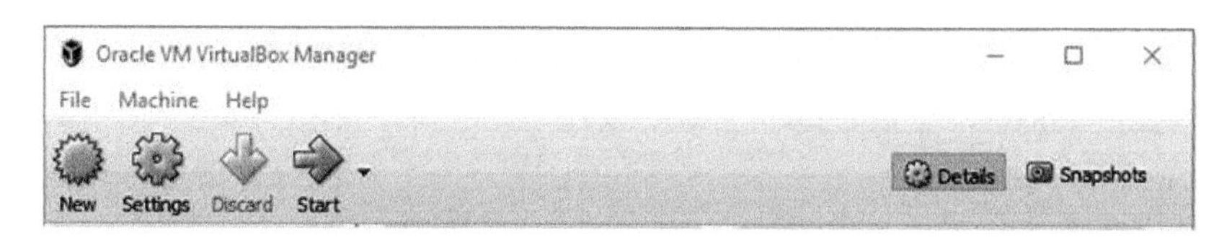

图 1.1　VirtualBox 虚拟机窗口

点击“New”（新建）按钮创建一个新的 VM，其具有 1024MB 的内存和 12GB 的虚拟磁盘。

4. 将 Ubuntu 14.04 安装到 VirtualBox VM 上

将“Ubuntu 14.04 install DVD”插入光盘驱动器。从 DVD 启动 VM，将 Ubuntu 14.04 安装到 VM 上。

5. 调整屏幕分辨率

出于某种未知原因，当 Ubuntu 首次启动时，屏幕分辨率将停留在 640×480。如需更改屏幕分辨率，请打开终端并输入命令行“xdiagnose”。在“X 诊断”设置窗口中，启用“Debug”(调试) 下的所有选项，包括

- 附加图形调试消息
- 显示启动消息

- 启用自动崩溃错误报告

尽管这些选项似乎都与屏幕分辨率无关，但确实可将普通 Ubuntu 显示屏的分辨率更改为 1024×768。图 1.2 显示了 VirtualBox 虚拟机上的 Ubuntu 屏幕。

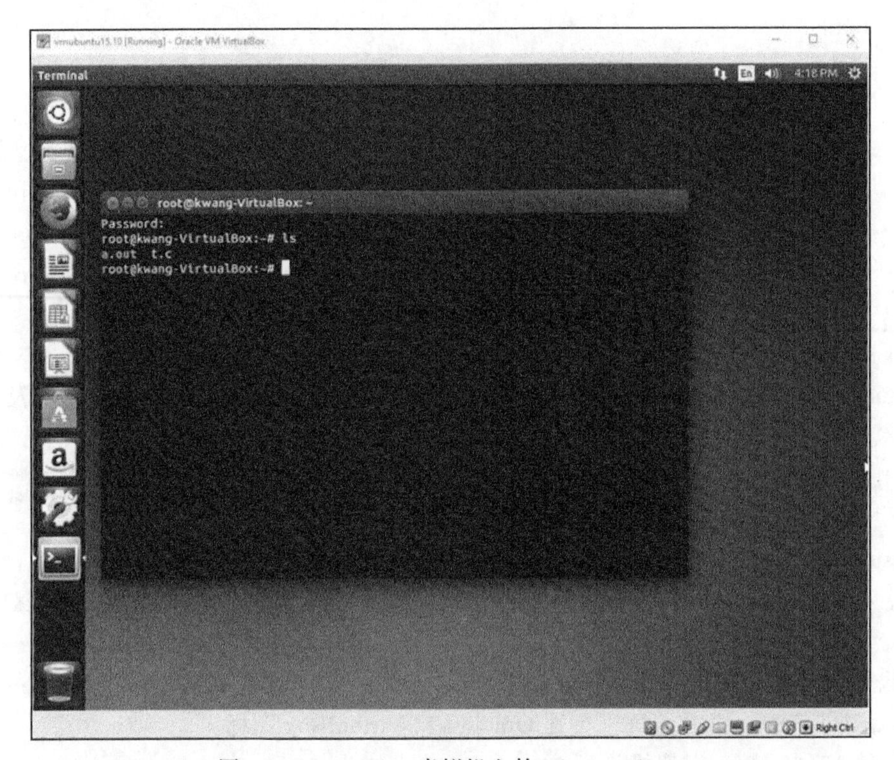

图 1.2　VirtualBox 虚拟机上的 Ubuntu Linux

6. 在 Ubuntu 下测试 C 语言编程

Ubuntu 14.04 装有 gcc 包。安装 Ubuntu 后，用户可以创建和编译 C 源文件，并运行 C 程序。若用户需要 emacs，可通过以下工具安装：

```
sudo apt-get install emacs
```

emacs 为在 GDB 环境下进行文本编辑、编译 C 程序和运行生成的二进制可执行文件提供了一个集成开发环境（IDE）。我们将在第 2 章中介绍和讨论 emacs IDE。

1.12.2　VMware

另一款较为流行的 VM 为 VMware，适用于基于 x86 的个人计算机。所有版本的 VMware（包括 VM 服务器）都要收费，但 VMware Workstation Player 是免费的，它足以执行简单的 VM 任务。

1. 在 Windows 10 上安装 VMware Player

VMware Workstation Player 可从 VMware 的下载站点得到。下载完成后，双击文件名运行，安装 VMware，并在桌面上创建 VMware Workstation 图标。点击图标，启动 VMware VM，弹出 VMware 虚拟机窗口，如图 1.3 所示。

2. 将 Ubuntu 15.10 安装到 VMware 虚拟机上

请按照以下步骤将 Ubuntu 15.10 安装到 VMware 虚拟机上。

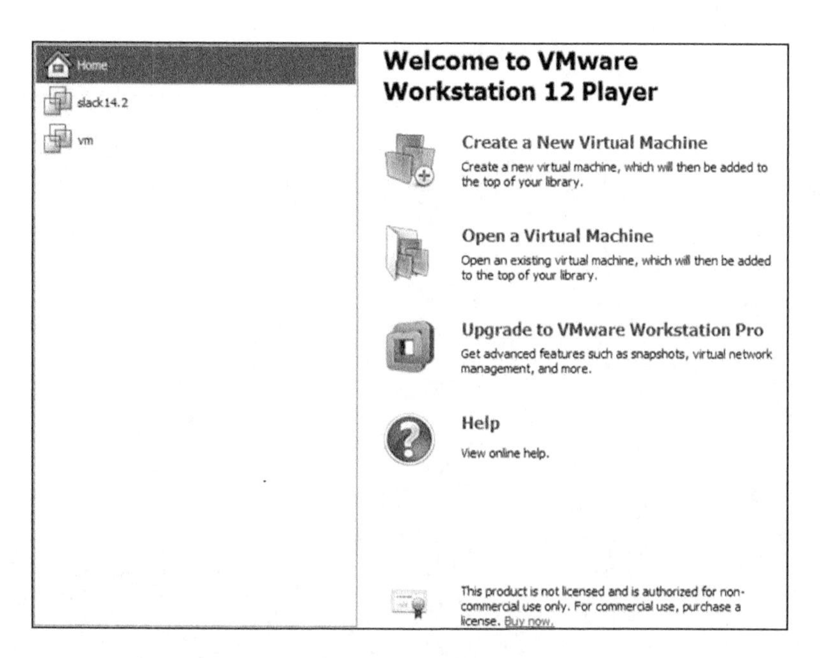

图 1.3　VMware 虚拟机窗口

（1）下载“Ubuntu 15.10 install DVD”映像；将其刻录成 DVD 光盘。

（2）下载适用于 Windows 10 的 VMware Workstation Player 12 exe 文件。

（3）安装 VMware Player。

（4）启动 VMware Player。

- 选择：创建一个新的虚拟机。
- 选择：安装光盘：DVD-RW 驱动器（D:）
  => 插入安装程序光盘，直到准备好安装
  然后，输入“Next”(下一步)
- 选择：Linux
  版本：ubuntu
- 虚拟机名称：更改为合适的名称，例如 Ubuntu。
- VMware 将创建一个虚拟机，其具有 20GB 的磁盘、1GB 的内存等。
- 选择“Finish”(完成)，完成新虚拟机的创建。
- 下一屏：选择：“play virtual machine”(启动虚拟机）来启动虚拟机。
- 虚拟机将从“Ubuntu install DVD”启动，以安装 Ubuntu。

（5）在 Ubuntu Linux 下运行 C 程序。

图 1.4 显示了 Ubuntu 的启动界面以及在 Ubuntu 下运行 C 程序。

1.12.3　双启动 Slackware 和 Ubuntu Linux

免费的 VMware Player 只支持 32 位的虚拟机。除安装程序光盘为“Slackware14.2 install DVD”之外，安装步骤与上述相同。安装 Slackware 14.2 和 Ubuntu 15.10 之后，设置 LILO，以使这两个系统中的任一个都能在虚拟机启动时启动。若先安装 Slackware，再安装 Ubuntu，则 Ubuntu 将识别 Slackware，并将 GRUB 配置为双启动。图 1.5 显示了 LILO 的双启动菜单。

```
Password:
root@kwang-virtual-machine:~# ls
a.out  t.c
root@kwang-virtual-machine:~# cat t.c
#include <stdio.h>
int main()
{
   int i;
   for (i=0; i<10; i++)
     printf("print a test line %d times\n", i);
}
root@kwang-virtual-machine:~# cc t.c
root@kwang-virtual-machine:~# a.out
print a test line 0 times
print a test line 1 times
print a test line 2 times
print a test line 3 times
print a test line 4 times
print a test line 5 times
print a test line 6 times
print a test line 7 times
print a test line 8 times
print a test line 9 times
root@kwang-virtual-machine:~#
```

图 1.4　VMware 虚拟机上的 Ubuntu

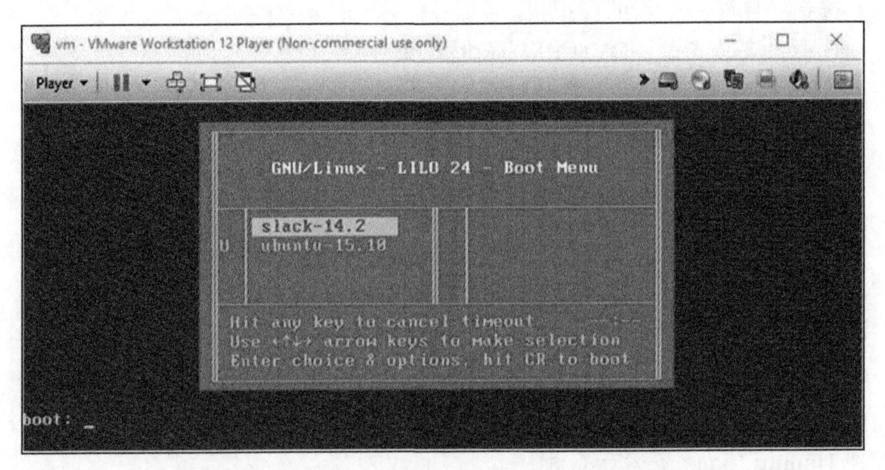

图 1.5　Slackware 和 Ubuntu Linux 的双启动窗口

1.13　使用 Linux

1.13.1　Linux 内核映像

在典型的 Linux 系统中，Linux 内核映像位于 /boot 目录中。可启动的 Linux 内核映像名为

vmlinuz-generic-VERSION_NUMBER

initrd 是 Linux 内核的初始 ramdisk 映像。

一个可启动的 Linux 内核映像由三部分组成：

| BOOT | SETUP | linux kernel |

BOOT 是一个 512 字节的启动程序，用于从软盘映像启动早期的 Linux 版本。但现在，它不再用于 Linux 启动，而是包含一些参数，以供“SETUP”使用。SETUP 是一段 16 位和 32 位的汇编代码，用于在启动期间将 16 位模式转换为 32 位保护模式。Linux 内核是 Linux 的实际内核映像。它采用压缩格式，但在开始时有一段解压代码，用于解压 Linux 内核映像并将其放入高端内存中。

1.13.2 Linux 启动程序

Linux 内核可以由几个不同的启动加载程序启动。最受欢迎的 Linux 启动加载程序是 GRUB 和 LILO。另外，HD 启动程序（Wang 2015）也可用来启动 Linux。

1.13.3 Linux 启动

在启动期间，Linux 启动加载程序首先会定位 Linux 内核映像（文件）。然后：

- 加载 BOOT+SETUP 至实模式内存的 0x90000 处。
- 加载 Linux 内核至高端内存的 1MB 处。

对于常见的 Linux 内核映像，该加载程序还会将初始 ramdisk 映像 initrd 加载到高端内存中。然后转移控制权，运行 0x902000 处的 SETUP 代码，启动 Linux 内核。首次启动时，Linux 内核会将 initrd 作为临时根文件系统，在 initrd 上运行。Linux 内核将执行一个 sh 脚本，该脚本指示内核加载实际根设备所需的模块。当实际根设备被激活并准备就绪时，内核将放弃初始 ramdisk，将实际根设备挂载为根文件系统，从而完成 Linux 内核的两阶段启动。

1.13.4 Linux 运行级别

Linux 内核以单用户模式启动。它可以模仿 System V Unix 的运行级别，以多用户模式运行。然后，创建并运行 INIT 进程 P1，后者将创建各种守护进程和供用户登录的终端进程。之后，INIT 进程将等待任何子进程终止。

1.13.5 登录进程

各登录进程将在其终端上打开三个文件流：stdin（用于输入），stdout（用于输出），stderr（用于错误输出）。然后等待用户登录。在使用 X-window 作为用户界面的 Linux 系统上，X-window 服务器通常会充当用户登录界面。用户登录后，（伪）终端默认属于用户。

1.13.6 命令执行

登录后，用户进程通常会执行命令解释程序 sh，后者将提示用户执行命令。sh 将直接执行一些特殊命令，如 cd（更改目录）、exit（退出）、logout（注销）、&。非特殊命令通常是可执行文件。对于非特殊命令，sh 会复刻子进程并等待该子进程终止。子进程会将其执行映像更改为文件，并执行新映像。子进程在终止时会唤醒父进程 sh，后者将提示执行另一个命令等。除简单的命令之外，sh 还支持 I/O 重定向和通过管道连接的复合命令。除内置命令外，用户还可开发程序，将其编译成二进制的可执行文件，并按命令运行程序。

1.14 使用 Ubuntu Linux

1.14.1 Ubuntu 版本

在各种 Linux 版本中，推荐使用 Ubuntu Linux 15.10 或更高版本，原因如下。

（1）易于安装。若用户连接了互联网，可在线安装。

（2）易于通过“sudo apt-get install”包安装额外的软件包。

（3）通过发布新版本来定期更新和改进。

（4）拥有庞大的用户群，很多问题和解决方案都发布在网上论坛中。

（5）方便连接无线网络和接入互联网。

1.14.2 Ubuntu Linux 的特性

以下提供了一些有关如何使用 Ubuntu Linux 的帮助信息。

（1）在台式机或笔记本电脑上安装 Ubuntu 时，需要输入用户名和密码来创建一个默认主目录为“/home/username”的用户账户。当 Ubuntu 启动时，它会立即在用户环境中运行，因为其已自动登录默认用户。按下“Ctrl+ALT+T”组合键打开一个伪终端。右键点击“Term”（终端）图标，选择“lock to launcher”（锁定到启动器），锁定菜单栏中的“Term”（终端）图标。随后，在菜单栏中选择 terminal->new terminal（终端 -> 新终端）来启动新终端。每个新终端运行一个 sh，提示用户执行命令。

（2）出于安全原因，用户应为**普通用户**，而不是根用户或**超级用户**。要运行任何特权命令，用户必须输入

```
sudo command
```

首先会验证用户的密码。

（3）用户的“PATH”（路径）环境变量设置通常不包括用户的当前目录。为在当前目录下运行程序，用户每次都必须输入 ./a.out。为方便起见，用户应更改路径设置，以包含当前目录。在用户的主目录中，创建一个包含以下代码的 .bashrc 文件：

```
PATH=$PATH:./
```

用户每次打开伪终端时，sh 都会先执行 .bashrc 文件来设置路径，以包含当前工作目录。

（4）很多用户可能都安装了 64 位的 Ubuntu Linux。本书中的一些编程练习和作业是针对 32 位机器的。在 64 位 Linux 下，使用

```
gcc -m32 t.c # compile t.c into 32-bit code
```

生成 32 位代码。若 64 位 Linux 不采用 -m32 选项，则用户必须安装适用于 gcc 的附加支持插件，才能生成 32 位代码。

（5）Ubuntu 具有友好的 GUI 用户界面。许多用户都习惯于使用 GUI，以至于产生了过度依赖，这往往需要反复拖动和点击指向设备，从而浪费了大量时间。在系统编程中，用户还需要学习如何使用命令行和 sh 脚本，它们比 GUI 要通用和强大得多。

（6）现今，大多数用户都可以连接到计算机网络。Ubuntu 可同时支持有线和无线网络连接。Ubuntu 在带有无线硬件的个人计算机 / 笔记本电脑上运行时，顶部会显示一个无线图

标，只需要通过一个简单的用户界面即可实现无线连接。打开无线图标。它将显示附近可用无线网络的列表。选择其中一个网络，并打开“Edit Connections”（编辑连接）子菜单，输入所需的登录名和密码来编辑连接文件。关闭“Edit”（编辑）子菜单。Ubuntu 将尝试自动登录到所选的无线网络。

1.15　Unix/Linux 文件系统组织

Unix/Linux 文件系统采用树形组织结构，如图 1.6 所示。

Unix/Linux 将所有能够存储或提供信息的事物都视为文件。从一般意义上来说，文件系统树的每个节点都是一个“FILE”(文件)。在 Unix/Linux 中，文件有以下类型。

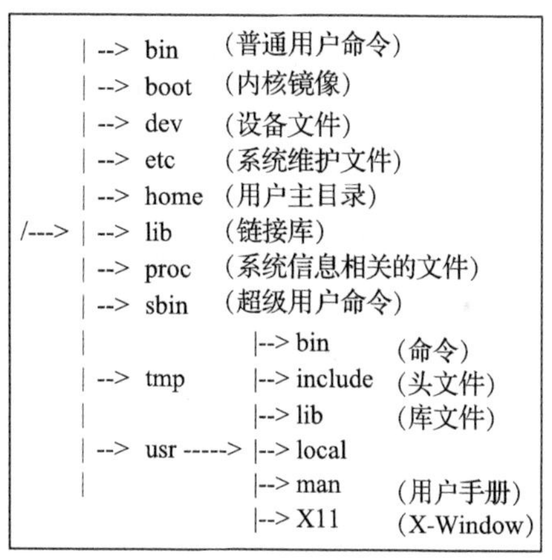

图 1.6　Unix/Linux 文件系统树

1.15.1　文件类型

（1）**目录文件**：一个目录可能包含其他目录和（非目录）文件。

（2）**非目录文件**：非目录文件要么是“REGULAR”（常规）文件，要么是“SPECIAL”（特殊）文件，但只能是文件系统树中的叶节点。非目录文件可以进一步分为：

- **常规文件**：常规文件也称为“ORDINARY”（普通）文件。这些文件要么包含普通文本，要么包含可执行的二进制代码。
- **特殊文件**：特殊文件是 /dev 目录中的条目。它们表示 I/O 设备，可进一步分为：
  - **字符特殊文件：字符 I/O**，如 /dev/tty0、/dev/pts/1 等。
  - **块特殊文件：块 I/O**，如 /dev/had、/dev/sda 等。
  - 其他类型，如网络（套接字）特殊文件、命名管道等。

（3）**符号链接文件**：属于常规文件，其内容为其他文件的路径名。因此，这些文件是指向其他文件的指针。例如，Linux 命令

```
ln -s aVeryLongFileName myLink
```

可创建一个符号链接文件“mylink”，其指向“aVeryLongFileName”。对“mylink”的访问将被重定向到实际文件“aVeryLongFileName”上。

1.15.2　文件路径名

Unix/Linux 文件系统树的根节点（用“/”符号表示）称为**根目录**，或简称为根。文件系统树的每个节点都由以下表单的路径名指定：

```
/a/b/c/d    OR    a/b/c/d
```

以“/”开头的路径名为**绝对**路径名，反之则为相对于进程**当前工作目录**（CWD）的**相对**路径名。当用户登录到 Unix/Linux 时，CWD 即被设为用户的主目录。CWD 可通过 cd（更改目录）命令更改。pwd 命令可打印 CWD 的绝对路径名。

1.15.3　Unix/Linux 命令

在使用操作系统时，用户必须学习如何使用系统命令。下面列出了 Unix/Linux 中最常用的命令。

- ls: ls dirname：列出 CWD 或目录的内容。
- cd dirname：更改目录。
- pwd：打印 CWD 的绝对路径名。
- touch filename：更改文件名时间戳（如果文件不存在，则创建文件）。
- cat filename：显示文件内容。
- cp src dest：复制文件。
- mv src dest：移动或重命名文件。
- mkdir dirname：创建目录。
- rmdir dirname：移除（空）目录。
- rm filename：移除或删除文件。
- ln oldfile newfile：在文件之间创建链接。
- find：搜索文件。
- grep：搜索文件中包含模式的行。
- ssh：登录到远程主机。
- gzip filename：将文件压缩为 .gz 文件。
- gunzip file.gz：解压 .gz 文件。
- tar -zcvf file.tgz .：从当前目录创建压缩 tar 文件。
- tar -zxvf file.tgz .：从 .tgz 文件中解压文件。
- man：显示在线手册页。
- zip file.zip filenames：将文件压缩为 .zip 文件。
- unzip file.zip：解压 .zip 文件。

1.15.4　Linux 手册页

Linux 将在线手册页保存在标准 /usr/man/ 目录下。在 Ubuntu Linux 中，手册页保存在 /usr/share/man 目录下。手册页分为不同类别，用 man1、man2 等表示。

```
/usr/man/
        |-- man1: 常用命令: ls、cat、mkdir 等
        |-- man2: 系统调用
        |-- man3: 库函数: strtok、strcat、basename、dirname 等
```

所有手册页均为压缩 .gz 文件。其中包含描述性文本，说明了如何使用附有输入参数和选项的命令。man 是一个程序，可读取手册页文件并以用户友好的格式显示其内容。下面是使用手册页的一些例子。

- man ls：　　显示 man1 中 ls 命令的手册页。
- man 2 open：　显示 man2 中 open 函数的手册页。
- man strtok：　显示 man 3 中 strtok 函数的手册页等。
- man 3 dirname：显示 man3（而非 man1）中 dirname 函数。

如有需要，应查看这些手册页，了解如何使用特定的 Linux 命令。很多所谓的 Unix/Linux 系统编程书籍实际上都是 Unix/Linux 手册页的浓缩版。

1.16 Ubuntu Linux 系统管理

1.16.1 用户账户

与在 Linux 中一样，用户账户信息保存在 /etc/passwd 文件中，该文件归超级用户所有，但任何人都可以读取。在表单的 /etc/passwd 文件中，每个用户都有一个对应的记录行：

```
loginName:x:gid:uid:usefInfo:homeDir:initialProgram
```

其中第二个字段“x”表示检查用户密码。加密的用户密码保存在 /etc/shadow 文件中。shadow 文件的每一行都包含加密的用户密码，后面是可选的过期限制信息，如过期日期和时间等。当用户尝试使用登录名和密码登录时，Linux 将检查 /etc/passwd 文件和 /etc/shadow 文件，以验证用户的身份。用户成功登录后，登录进程将通过获取用户的 gid 和 uid 来转换成用户进程，并将目录更改为用户的 homeDir，然后执行列出的 initialProgram，该程序通常为命令解释程序 sh。

1.16.2 添加新用户

对于大多数在个人计算机或笔记本电脑上运行 Ubuntu Linux 的用户来说，只有在个别情况下才会添加新用户。但我们可以假设：读者想以不同的用户身份添加一个家庭成员，以便使用同一台计算机。与在 Linux 中一样，Ubuntu 支持“adduser”（添加用户）命令，运行方式如下：

```
sudo adduer  username
```

它通过为新用户创建账户和默认的主目录 /home/username 来添加新用户。此后，Ubuntu 将在“About The Computer”（关于本电脑）菜单中显示用户名列表。新用户可通过选择新用户名来登录系统。

1.16.3 sudo 命令

出于安全原因，Ubuntu 禁用了**根**或**超级用户**账户，这可防止任何人以根用户身份登录（其实也不完全是；有一个方法可杜绝这种情况，但不便透露）。sudo（“超级用户执行”）允许用户以另一个用户（通常是超级用户）的身份执行命令。它在执行命令时临时将用户进程提升到超级用户特权级别。完成命令执行后，用户进程将恢复到原来的特权级别。为确保能够使用 sudo，用户名必须保存在 /etc/sudoers 文件中。为确保用户能够发出 sudo，只需在 sudoers 文件中添加一行，如下所示：

```
username ALL(ALL)   ALL
```

但 /etc/sudoers 文件的格式非常严格。文件中的任何语法错误都可能破坏系统安全性。Linux 建议只使用特殊命令 visudo 来编辑该文件，该命令可调用 vi 编辑器，但需要检查和验证。

参考文献

Bach M., "The Design of the UNIX Operating System", Prentice-Hall, 1986

Curry, David A., Unix Systems Programming for SRV4, O'Reilly, 1996

Haviland, Keith, Gray, Dian, Salama, Ben, Unix System Programming, Second Edition, Addison-Wesley, 1998

HP-UX, http://www.operating-system.org/betriebssystem/_english/bs-hpux.htm, 2017

IBM AIX, https://www.ibm.com/power/operating-systems/aix, 2017

Kernighan, B.W., Ritchie D.M, "The C Programming Language" 2nd Edition, 1988

Kerrisk, Michael, The Linux Programming Interface: A Linux and UNIX System Programming Handbook, 1st Edition, No Starch Press Inc, 2010

Leffler, S.J., McKusick, M. K, Karels, M. J. Quarterman, J., "The Design and Implementation of the 4.3BSD UNIX Operating System", Addison Wesley, 1989

Leffler, S.J., McKusick, M. K, Karels, M. J. Quarterman, J., "The Design and Implementation of the 4.4BSD UNIX Operating System", Addison Wesley, 1996

Linux, https://en.wikipedia.org/wiki/Linux , 2017

Love, Robert, Linux System Programming: Talking Directly to the Kernel and C Library, 2nd Edition, O'Reilly, 2013

McKusick, M. K, Karels, Neville-Neil, G.V., "The Design and Implementation of the FreeBSD Operating System", Addison Wesley, 2004

Robbins Kay, Robbins, Steve, UNIX Systems Programming: Communication, Concurrency and Threads: Communication, Concurrency and Threads, 2nd Edition, Prentice Hall, 2003

Pthreads: https://computing.llnl.gov/tutorials/pthreads, 2017

Rochkind, Marc J., Advanced UNIX Programming, 2nd Edition, Addison-Wesley, 2008

Stallman, R., Linux and the GNU System, 2017

Stevens, W. Richard, Rago, Stephan A., Advanced Programming in the UNIX Environment, 3rd Edition, Addison-Wesley, 2013

Solaris, http://www.operating-system.org/betriebssystem/_english/bs-solaris.htm, 2017

Sun OS, https://en.wikipedia.org/wiki/SunOS, 2017

Thompson, K., Ritchie, D. M., "The UNIX Time-Sharing System", CACM., Vol. 17, No.7, pp. 365-375, 1974

Thompson, K., Ritchie, D.M.; "The UNIX Time-Sharing System", Bell System Tech. J. Vol. 57, No.6, pp. 1905–1929, 1978

Unix System V, https://en.wikipedia.org/wiki/UNIX_System V, 2017

VirtualBox, https://www.virtualbox.org, 2017

Wang, K. C., Design and Implementation of the MTX Operating System, Springer International Publishing AG, 2015

第2章

Systems Programming in Unix/Linux

编 程 背 景

摘要

本章讲述了系统编程所需的背景信息；介绍了几种基于GUI的文本编辑器，比如vim、gedit和EMACS，可供读者编辑文件使用；展示了如何在命令和GUI模式下使用EMACS编辑器来编辑、编译和执行C语言程序；阐述了程序开发的步骤，这些编译链接步骤包括GCC、静态和动态链接、二进制可执行文件的格式和内容、程序执行和终止等；详细阐释了函数调用惯例和运行时堆栈的使用，包括参数传递、局部变量和栈帧；展示了C语言程序与汇编代码的链接；讨论了GNU make工具，并举例说明如何编写makefile；演示了如何使用GDB调试工具调试C语言程序；指出了C语言程序中常见的错误，并提出了在程序开发过程中防止此类错误的方法；提到了高级编程技术；描述了C语言中的结构和指针；通过详细的例子讲解了链表及其处理；最后还讲解了二叉树和树的遍历算法。本章以一个编程项目为例，向读者展示了如何用二叉树模拟Unix/Linux文件系统树中的操作。该项目从单根目录节点开始。它支持mkdir、rmdir、creat、rm、cd、pwd、ls操作，将文件系统树保存为文件，并从保存的文件中恢复文件系统树。在该项目中，读者可应用标记化字符串、解析用户指令及使用函数指针来调用指令处理函数等编程技巧。

2.1 Linux中的文本编辑器

2.1.1 vim

vim（Linux Vi和Vim Editor 2017）是Linux的标准内置编辑器。它是Unix原始默认vi编辑器的改进版本。与其他大多数编辑器不同，vim有3种不同的操作模式，分别是

- **命令模式**：用于输入命令。
- **插入模式**：用于输入和编辑文本。
- **末行模式**：用于保存文件并退出。

vim启动时，处于默认的**命令模式**，在该模式下，大多数键表示特殊命令。移动光标的命令键示例如下：

h：将光标向左移动一个字符	l：将光标向右移动一个字符
j：将光标向下移动一行	k：将光标向上移动一行

在X-window中使用vim时，也可以通过箭头键来完成光标的移动。要输入文本进行编辑，用户必须输入i（插入）或a（追加）命令将vim切换到**插入模式**：

- i：切换到插入模式，插入文本。
- a：切换到插入模式，追加文本。

要退出插入模式，请按ESC键一次或多次。在命令模式下，输入“：”进入**末行模式**，将文本保存为文件或退出vim：

- :w：写入（保存）文件。

- :q：退出 vim。
- :wq：保存并退出。
- :q!：不保存更改，强制退出。

虽然许多 Unix 用户已经习惯了 vim 不同的操作模式，但是其他用户可能认为与其他基于图形用户界面（GUI）的编辑器相比，vim 使用起来既不自然也不方便。以下类型的编辑器属于通常所说的所见即所得（WYSIWYG）编辑器。在 WYSIWYG 编辑器中，用户可以输入文本，用箭头键移动光标，和普通的文本输入一样。通常，通过输入一个特殊的 meta 键，接着输入一个字母键即可创建命令。例如：

- Ctrl+C：中止或退出。
- Ctrl+K：删除行到缓冲区。
- Ctrl+Y：从缓冲区内容中复制或粘贴。
- Ctrl+S：保存已编辑文本等。

由于不需要模式切换，大多数用户（尤其是初学者）更喜欢 WYSIWYG 类型的编辑器，而不是 vim。

2.1.2　gedit

gedit 是 GNOME 桌面环境默认的文本编辑器。它是 Ubuntu 及其他使用 GNOME GUI 用户界面的 Linux 的默认编辑器，包含用于编辑源代码和结构化文本（如标记语言）的工具。

2.1.3　emacs

emacs（GNU Emacs 2015）是一款强大的文本编辑器，可在多个不同的平台上运行。最受欢迎的 emacs 版本是 GNU Emacs，可在大多数 Linux 发行版中使用。

以上所有编辑器都支持直接输入和全屏模式下的文本编辑。它们还支持关键词搜索，可用新文本替换字符串。要使用这些编辑器，用户只需要学习一些基础知识即可，比如如何启动编辑器，输入文本进行编辑，将编辑后的文本保存为文件，然后退出编辑器。

根据 Unix/Linux 发行版的不同，在默认情况下，某些编辑器可能未安装。例如，Ubuntu Linux 通常安装有 gedit、nano 和 vim，但未安装 emacs。Ubuntu 的一个优点是，当用户试图运行一个未安装的命令时，它会提醒用户先安装。例如，如果用户输入

```
emacs filename
```

Ubuntu 将会显示一条消息，提示“尚未安装 emacs 程序。可输入 apt-get install emacs 进行安装”。同样，用户可通过 apt-get 命令安装其他缺失的软件包。

2.2　使用文本编辑器

所有文本编辑器都被设计成执行相同的任务，即允许用户输入文本、编辑文本并将编辑后的文本保存为文件。上文提到，有许多不同类型的文本编辑器。使用哪个文本编辑器是个人喜好问题。大多数 Linux 用户似乎更喜欢 gedit 或 emacs，因为它们采用 GUI，而且易于使用。就这两款文本编辑器而言，我们强烈建议用户使用 emacs。下面展示了一些使用 emacs 创建文本文件的简单示例。

2.2.1　使用 emacs

首先，从 X-window 的伪终端输入命令行

```
emacs [FILENAME]                # [ ] means optonal
```

使用一个可选文件名（如 t. c）调用 emacs 编辑器。emacs 将会在一个单独窗口中启动，如图 2.1 所示。emacs 窗口的顶部是一个菜单栏，每个菜单都可以打开，查看其他命令，用户可点击菜单图标调用命令。开始时，不要使用菜单栏，重点是创建 C 语言程序源文件的简单任务。emacs 启动时，如果已经有 t.c 文件，它会打开文件并将文件内容加载到一个缓冲区中进行编辑。否则，它会显示一个空的缓冲区，用户可在这里输入内容。图 2.1 所示为用户输入行。emacs 将任意 .c 文件识别为 C 语言程序的源文件。它将按照 C 代码行的惯例缩进，例如，它将自动匹配每个适当的左“{”右“}”缩进。实际上，它甚至可以检测出不完整的 C 语句，并显示不适当的缩进，以警告用户 C 源代码行中可能出现的语法错误。

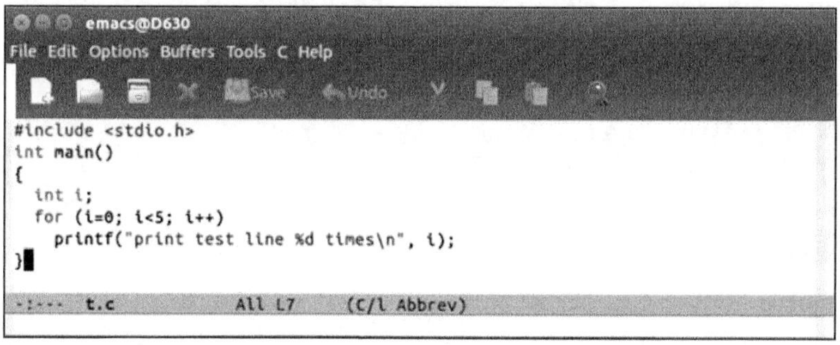

图 2.1　emacs 的使用方法 1

创建源文件后，按下 meta 键序列“Ctrl+X+C”，以保存文件并退出。如果缓冲区中有修改后未保存的文本，它会提示用户保存文件，如图 2.2 最后一行所示。输入 y 以保存文件并退出 emacs。或者，用户也可以点击菜单栏上的 Save 图标来保存并退出。

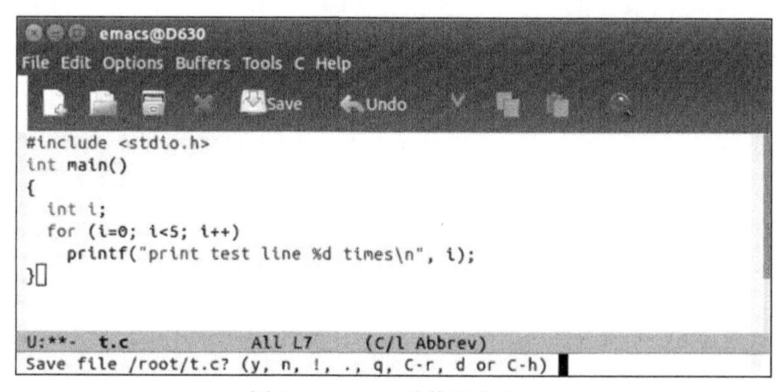

图 2.2　emacs 的使用方法 2

2.2.2　emacs 菜单

emacs 窗口的顶部是一个菜单栏，带有图标

```
File  Edit Options Buffers Tools  C  Help
```

- File 菜单支持打开文件、插入文件和保存文件的操作。它还支持打印正在编辑的缓冲

区、打开新窗口和新帧。

- Edit 菜单支持查找和替换操作。
- Options 菜单支持配置 emacs 操作的功能。
- Buffers 菜单支持缓冲区选择和显示。
- Tools 菜单支持编译源代码、执行二进制可执行文件和调试。
- C 菜单支持自定义编辑 C 源代码。
- Help 菜单为 emacs 的使用提供支持，如简单的 emacs 教程。

通常，点击一个菜单项将显示一个子菜单列表，允许用户选择单独操作。除了命令，用户还可以使用 emacs 菜单栏进行文本编辑，如撤销上次操作、剪切和粘贴、保存文件和退出等。

2.2.3 emacs 的集成开发环境

emacs 不仅是一款文本编辑器，它还为软件开发提供了一个集成开发环境（IDE），其中包括编译 C 语言程序、运行可执行的映像以及用 GDB 执行调试程序。我们将在 2.3 节中阐述 emacs 的 IDE 功能。

2.3 程序开发

2.3.1 程序开发步骤

执行程序的开发步骤如下。

（1）创建源文件：使用文本编辑器（如 gedit 或 emacs）创建一个或多个程序源文件。在系统编程中，最重要的编程语言是 C 语言和汇编语言。我们先讨论 C 语言程序。

C 语言中的标准注释行包含匹配的 “/*” 符号和 “*/” 符号对。为方便起见，除了标准注释外，我们还将使用 “//” 来表示 C 代码中的注释行。假设 t1.c 和 t2.c 是 C 语言程序的源文件。

```
/********************** t1.c file ****************************/
int g = 100;                  // initialized global variable
int h;                        // uninitialized global variable
static int s;                 // static global variable

main(int argc, char *argv[ ]) // main function
{
  int a = 1; int b;          // automatic local variables
  static int c = 3;          // static local variable
  b = 2;
  c = mysum(a,b);             // call mysum(), passing a, b
  printf("sum=%d\n", c);      // call printf()
}
/********************** t2.c file ***************************/
extern int g;                 // extern global variable
int mysum(int x, int y)       // function heading
{
   return x + y + g;
}
```

C 语言程序中的变量可分为**全局变量**、**局部变量**、**静态变量**、**自动变量**和**寄存器变量**等，如图 2.3 所示。

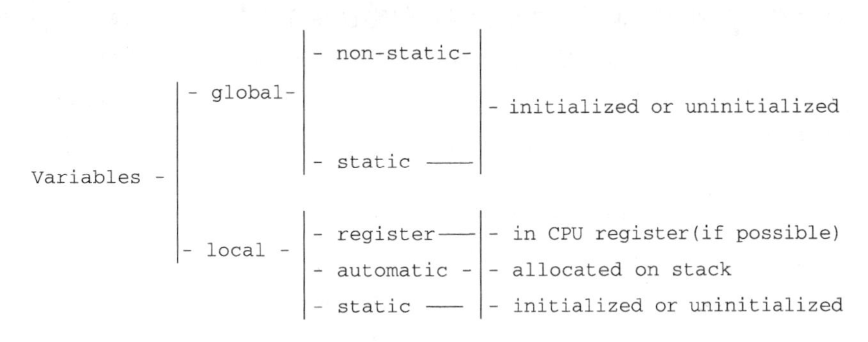

图 2.3　C 语言中的变量

全局变量在函数外定义。**局部变量**在函数内定义。全局变量具有唯一性，并且只有一个副本。**静态全局变量**仅对定义它们的文件可见。**非静态全局变量**则对同一程序的所有文件都可见。全局变量可以初始化，也可以不初始化。初始化的全局变量在编译时赋值。未初始化的全局变量在程序执行开始时清零。局部变量只对定义它们的函数可见。默认情况下，局部变量是**自动变量**，它们在函数调用时出现，按逻辑在函数退出时消失。对于**寄存器变量**，编译器试图把它们分配在 CPU 寄存器中。由于自动局部变量在函数调用前没有分配内存空间，因此在编译时不能初始化。**静态局部变量**具有永久性和唯一性，可以初始化。此外，C 语言还支持**易失性变量**，这些变量用作内存映射 I/O 的地址，或者通过中断处理程序或多个执行线程来访问的全局变量。易失性关键字可阻止 C 编译器优化用这些变量进行操作的代码。

在前面的 t1.c 文件中，g 是一个初始化的全局变量，h 是一个未初始化的全局变量，s 是一个静态全局变量。g 和 h 对整个程序都可见，但是 s 只在 t1.c 文件中可见。所以 t2.c 可通过声明 g 是外部变量引用，但是不能引用 s，因为 s 只在 t1.c 中可见。在 main() 函数中，局部变量 a、b 是自动变量，c 是静态变量。虽然局部变量 a 被定义为 int a = 1，但这不是初始化，因为 a 在编译时还不存在。在实际调用 main() 时，生成的代码将把 1 赋值给 a 的当前副本。

（2）用 gcc 把源文件转换成二进制可执行文件，如：

```
gcc  t1.c  t2.c
```

生成一个二进制可执行文件，文件名为 a.out。在 Linux 中，cc 链接到 gcc，所以它们是一样的。

（3）gcc 是什么？gcc 是一个程序，它包含三个主要步骤，如图 2.4 所示。

```
cc :      第1步              第2步         *      第3步
------   --------         -----------         --------
t1.c ->| 编译器 |->t1.s ->| 汇编器 | ->t1.o ->|        |
                                              | 链接器 | -> |a.out|
t2.c ->| 编译器 |->t2.s ->| 汇编器 | ->t2.o ->|        |
                                                  |
                                             C_library
```

图 2.4　程序开发步骤

第 1 步：将 C 源文件转换为汇编代码文件。cc 的第 1 步是调用 C 编译器，它将 .c 文件转换为包含目标机器**汇编代码**的 .s 文件。C 编译器处理过程本身有几个阶段，如预处理、词法分析、解析和代码生成等，但是在这里读者可忽略这些细节。

第 2 步：把汇编代码转换成**目标代码**。每台计算机都有自己的一套机器指令。用户可用汇编语言为具体的机器编写程序。汇编器是将汇编代码转换为二进制形式机器代码的程序。生成的 .o 文件称为目标代码。cc 的第 2 步是调用汇编器将 .s 文件转换成 .o 文件。每个 .o 文件都包含

- 一个文件头，包含代码段、数据段和 BSS 段的大小。
- 一个代码段，包含机器指令。
- 一个数据段，包含初始化全局变量和初始化静态局部变量。
- 一个 BSS 段，包含未初始化全局变量和未初始化静态局部变量。
- 代码中的指针以及数据和 BSS 中的偏移量的重定位信息。
- 一个符号表，包含非静态全局变量、函数名称及其属性。

第 3 步：链接。一个程序可能包含多个 .o 文件，这些文件相互依赖。此外，.o 文件可调用 C 库函数，比如 printf() 函数，该函数在源文件中不存在。cc 的最后一步是调用链接器，链接器将所有 .o 文件和必要的库函数组合成单一的二进制可执行文件。更具体地说，链接器执行以下操作：

- 将 .o 文件的所有代码段组合成单一代码段。对于 C 语言程序，组合代码段从默认的 C 启动代码 crt0.o 开始，该代码调用 main() 函数。这就是为什么每个 C 语言程序必定有一个唯一的 main() 函数。
- 将所有数据段组合成单一数据段。组合的数据段仅包含初始化全局变量和初始化静态局部变量。
- 将所有 BSS 段组合成单一 bss 段。
- 用 .o 文件中的重定位信息调整组合代码段中的指针以及组合数据段、bss 段中的偏移量。
- 用符号表来解析各个 .o 文件之间的交叉引用。比如，当编译器在 t1.c 中发现 c = mysum（a, b）时，它并不知道 mysum 在何处。所以，它在 t1.o 中留下一个空白（0）作为 mysum 的入口地址，但是在符号表的记录中，必须用 mysum 的入口地址替换空白。链接器将 t1.o 与 t2.o 合并时，它知道 mysum 在组合代码段中的位置。链接器只是用 mysum 的入口地址替换掉 t1.o 中的空白。对于其他交叉引用符号也是如此。由于静态全局变量不在符号表中，因此不能提供给链接器。从不同文件中引用静态全局变量的任何尝试都会产生交叉引用错误。同样，如果 .o 文件涉及任何未定义的符号或函数名称，链接器也会产生交叉引用错误。如果所有交叉引用都能够成功分辨，链接器会将产生的组合文件写为 a.out，这是二进制可执行文件。

2.3.2 静态与动态链接

创建二进制可执行文件的方式有两种，分别是**静态链接**和**动态链接**。在使用**静态库**的静态链接中，链接器将所有必要的库函数代码和数据纳入 a.out 文件中。这使得 a.out 文件完整、独立，但通常非常大。在使用**共享库**的动态链接中，库函数未包含在 a.out 文件中，但是对此类函数的调用以指令形式记录在 a.out 文件中。在执行动态链接的 a.out 文件时，操作

系统将 a.out 文件和共享库均加载到内存中，使加载的库代码在执行期间可供 a.out 文件访问。动态链接的主要优点是：

- 可减小每个 a.out 文件的大小。
- 许多执行程序可在内存中共享相同的库函数。
- 修改库函数不需要重新编译源文件。

动态链接所用的库称为**动态链接库**（DLL）。它们在 Linux 中称为**共享库**（.so 文件）。动态加载（DL）库是指仅按需加载的共享库。动态加载库可用作插件和动态加载模块。

2.3.3 可执行文件格式

虽然默认的二进制可执行文件名为 a.out，但是实际文件格式是可变的。大部分 C 编译器和链接器可以生成多种不同格式的可执行文件，其中包括：

（1）**二进制可执行平面文件**：二进制可执行平面文件仅包含可执行代码和初始化数据。该文件将作为一个整体加载到内存中，便于直接执行。比如，可启动操作系统映像通常是二进制可执行平面文件，该文件简化了引导装载程序。

（2）**a.out 可执行文件**：传统的 a.out 文件包含文件头，然后是代码段、数据段和 bss 段。有关 a.out 文件格式的详细信息将在下一节中给出。

（3）**ELF 可执行文件**：可执行的链接格式（ELF）（Youngdale 1995）文件包含一个或多个程序段。每个程序段均可加载至特定的内存地址。在 Linux 中，默认的二进制可执行文件为 ELF 文件，这种格式的文件更适合动态链接。

2.3.4 a.out 文件的内容

为了简便起见，我们首先讨论传统的 a.out 文件。ELF 文件将在后面的章节中讨论。a.out 文件包括以下部分：

（1）**文件头**：文件头包含 a.out 文件的加载信息和大小，其中

- tsize = 代码段大小
- dsize = 包含初始化全局变量和初始化静态局部变量的数据段的大小
- bsize = 包含未初始化全局变量和未初始化静态局部变量的 bss 段的大小
- total_size = 加载的 a.out 文件的总大小

（2）**代码段**：也称为**正文段**，其包含程序的可执行代码。代码段从标准 C 启动代码 crt0.o 开始，该代码调用 main() 函数。

（3）**数据段**：数据段包含初始化全局变量和初始化静态数据。

（4）**符号表**：可选，仅为运行调试所需。

注意，包含未初始化全局变量和未初始化静态局部变量的 **bss 段**不在 a.out 文件中，只有 bss 段的大小记录在 a.out 文件头中，自动局部变量也不在 a.out 文件中。图 2.5 显示了 a.out 文件的布局。

在图 2.5 中，符号“_brk”表示 bss 段结束。a.out 文件的总加载大小通常等于 _brk，即等于 tsize+ dsize+bsize。如果需要，_brk 可设置为更高值，以表示更大的加载量。bss 段上额外的内存空间为**堆**区，用于执行期间的动态内存分配。

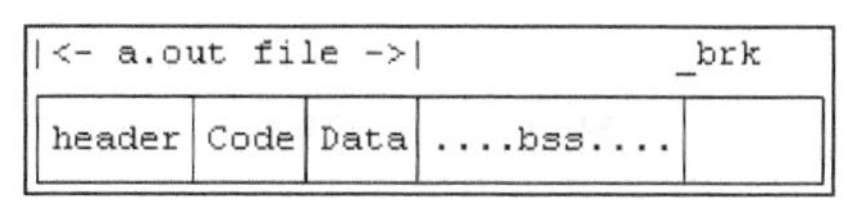

图 2.5 a.out 文件内容

2.3.5 程序执行过程

在类 Unix 操作系统中，在 sh 命令行

```
a.out one two three
```

执行 a.out 文件，以标记字符串作为**命令行参数**。为执行命令，sh 创建一个子进程并等待该子进程终止。子进程运行时，sh 使用 a.out 文件，按照以下步骤创建新的执行映像。

（1）读取 a.out 文件头，以确定所需的总内存大小，包括堆栈空间大小：

```
TotalSize = _brk + stackSize
```

其中，堆栈大小通常是操作系统内核为待启动程序选择的默认值。无法知道一个程序究竟需要多大的堆栈空间。比如，普通的 C 语言程序

```
main(){ main(); }
```

将因为任一计算机上的**堆栈溢出**而产生分段错误。因此，操作系统内核通常使用待启动程序的默认初始堆栈大小，并试着处理随后在运行期间可能出现的堆栈溢出问题。

（2）sh 从总大小中分配一个内存区给执行映像。从概念上讲，我们可假设分配的内存区是一个单独的连续内存。sh 将 a.out 文件的代码段和数据段加载到内存区中，堆栈区位于高位地址端。sh 将 bss 段清除为 0，使得所有未初始化全局变量和未初始化静态局部变量以初始值 0 开始。执行期间，堆栈向下朝低位地址延伸。

（3）然后，sh 放弃旧映像，开始执行新映像，如图 2.6 所示。

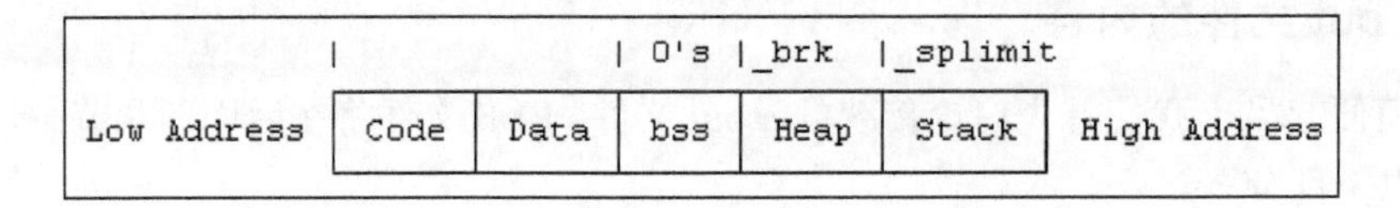

图 2.6　执行映像

在图 2.6 中，bss 段结束处的 _brk 为程序的初始“break”标记，_splimit 为堆栈大小限度。C 库函数 malloc()/free() 使用 bss 和堆栈（Stack）之间的堆（Heap）区在执行映像中进行动态内存分配。首次加载 a.out 文件时，_brk 和 _splimit 可能重合，因此初始堆大小为零。执行期间，进程可使用 brk（地址）或 sbrk（大小）系统调用将 _brk 更改到更高的地址，从而增加堆的大小。或者，malloc() 可隐式调用 brk() 或 sbrk() 来扩展堆大小。在执行期间，如果程序试图扩展 _splimit 下面的堆栈指针，就会发生堆栈溢出。在有内存保护的机器上，内存管理硬件将检测到这是一个错误，从而将进程捕获到操作系统内核。由于有最大值限制，操作系统内核可能会通过在进程地址空间中分配额外的内存来增大堆栈，从而允许继续执行。如果堆栈不能继续增大，则堆栈溢出将是毁灭性的。在没有适当硬件支持的机器上，必须通过软件实现堆栈溢出检测和处理。

（4）执行从 crt0.o 开始，调用 main()，将 argc 和 argv 作为参数传递给 main()，可以写成

```
int main( int argc, char *argv[ ] ) {  ....  }
```

其中 argc 为命令行参数的数量，每个 argv[] 指向对应的命令行参数字符串。

2.3.6 程序终止

可通过两种可能的方法终止正在执行 a.out 的进程。

（1）**正常终止**：如果程序执行成功，main() 最终会返回到 crt0.o，调用库函数 exit(0) 来终止进程。首先，exit(value) 函数会执行一些清理工作，如刷新 stdout、关闭 I/O 流等。然后，它发出一个 _exit(value) **系统调用**，使进入操作系统内核的进程终止。退出值 0 通常表示正常终止。如果需要，进程可直接调用 exit(value)，不必返回到 crt0.o。再直接一点，进程可能会发出 _exit(value) 系统调用以立即终止，不必先进行清理工作。当内核中的某个进程终止时，它会在进程结构体中将 _exit(value) 系统调用值记录为退出状态，通知它的父进程并使该进程成为僵尸进程。父进程可通过系统调用

```
pid = wait(int *status);
```

找到僵尸子进程，获得其 pid 和退出状态，它还会清空僵尸子进程的结构体，使该结构体可被另一个进程重新使用。

（2）**异常终止**：在执行 a.out 时，进程可能会遇到错误，如无效地址、非法指令、越权等，这些错误会被 CPU 识别为**异常**。当某进程遇到异常时，它会陷入操作系统内核。内核的陷入处理程序将陷入错误类型转换为一个幻数，称为**信号**，将信号传递给进程，使进程终止。在这种情况下，僵尸进程的退出状态就是信号数值，我们可以说该进程异常终止了。除了错误导致的陷入，信号也可能来自硬件或其他进程。例如，按下“Ctrl+C”组合键会产生硬件中断，它会向该终端上的所有进程发送数字 2 的信号 SIGINT，使进程终止。或者，用户可以使用命令

```
kill  -s signal_number pid    # signal_number = 1 to 31
```

向通过 pid 识别的目标进程发送信号。对于大多数信号数值，进程的默认操作是终止。信号和信号处理将在后面第 6 章讨论。

2.4 C 语言中的函数调用

接下来，我们来思考一下 a.out 在执行期间的运行时行为。程序的运行时行为主要来自函数调用。下面的讨论适用于在 32 位 Intel x86 处理器上运行 C 语言程序。在这些机器上，C 编译器生成的代码能够传递函数调用的堆栈参数。在执行期间，它用一个特殊 CPU 寄存器（ebp）来指向当前执行函数的栈帧。

2.4.1 32 位 GCC 中的运行时堆栈使用情况

思考下面的 C 语言程序，它由左侧显示的 main() 函数组成，调用右侧显示的 sub() 函数。

```
------------------------------------------------------------
 main()                          |    int sub(int x, int y)
 {                               |    {
   int a, b, c;                  |        int u, v;
   a = 1; b = 2; c = 3;          |        u = 4; v = 5;
   c = sub(a, b);                |        return x+y+u+v;
   printf("c=%d\n", c);          |    }
 }                               |
------------------------------------------------------------
```

（1）在执行 a.out 时，在内存中创建一个进程映像，它看起来（逻辑上）类似于图 2.7 所示的关系图，其中数据段包含初始化数据和 bss。

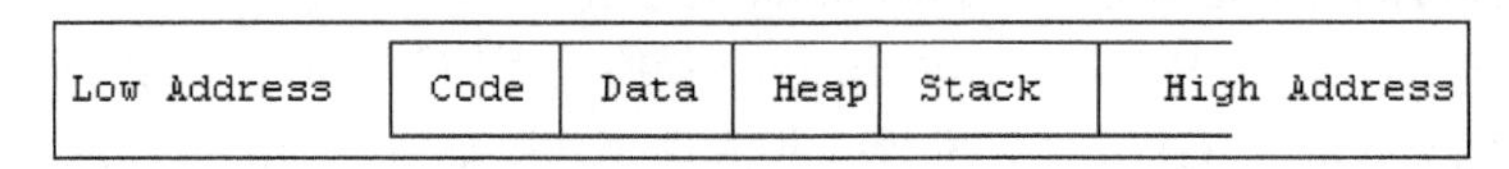

图 2.7　进程执行映像

（2）每个 CPU 都有以下寄存器或同等寄存器，括号中的条目表示 x86 CPU 的寄存器：

- PC（IP）：指向 CPU 要执行的下一条指令。
- SP（SP）：指向栈顶。
- FP（BP）：指向当前激活函数的栈帧。
- 返回值寄存器（AX）：函数返回值的寄存器。

（3）在每个 C 语言程序中，main() 函数均由 C 启动代码 crt0.o 调用。当 crt0.o 调用 main() 时，它将返回地址（当前 PC 寄存器）压栈，用 main() 的入口地址替换 PC，使 CPU 进入 main()。为了方便，我们按从左到右的顺序显示堆栈内容。当控制权转移到 main() 函数时，栈顶包含保存的返回 PC，如图 2.8 所示，其中 XXXX 表示 crt0.o 调用 main() 函数之前的堆栈内容，SP 指向保存的返回 PC，也就是 crt0.o 调用 main() 函数的位置。

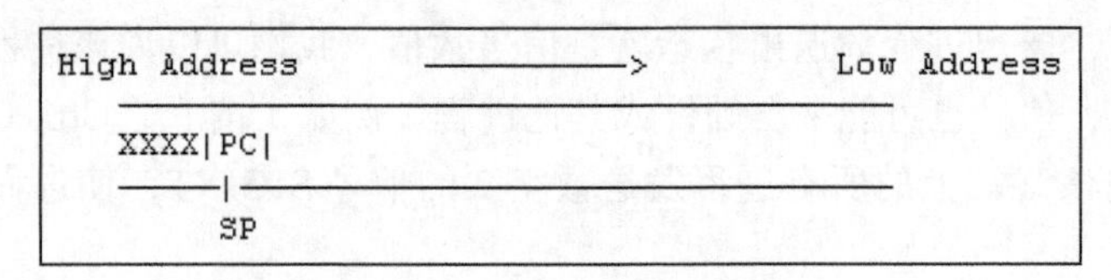

图 2.8　函数调用中的堆栈内容

（4）在每个 C 函数的入口处，编译后的代码都会完成如下功能：

- 将 FP 压栈 #，这将在堆栈上保存 CPU 的 FP 寄存器。
- 让 FP 指向保存的 FP# 建立栈帧。
- 向下移动 SP 为堆栈上的自动局部变量分配空间。
- 编译后的代码可能会继续向下移动 SP，在堆栈上分配一些临时工作空间，用 temps 表示。

在本示例中，有 3 个自动局部变量“int a, b, c”，每个 sizeof(int) = 4 字节。输入 main() 后，堆栈内容如图 2.9 所示，其中分配了 a、b、c 的空间，但尚未定义内容。

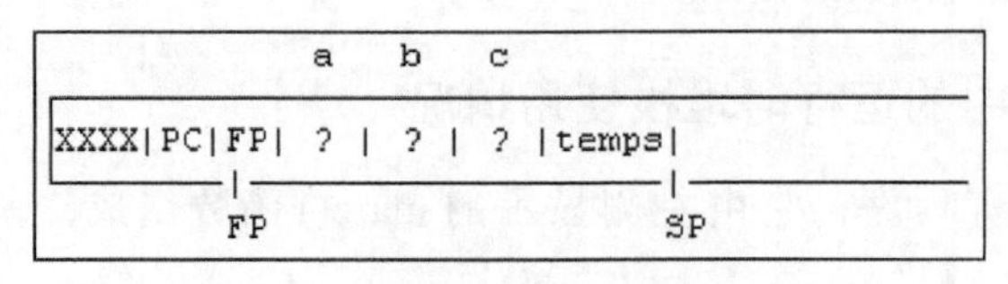

图 2.9　堆栈内容：分配局部变量

（5）然后 CPU 开始执行代码“a=1；b = 2；c = 3；”，分别将值 1、2、3 放入 a、b、c 的内存位置。假设 sizeof(int) 是 4 字节。a、b、c 的位置在 −4、−8、−12 字节处，这些位置是 FP 指向的起点。可用汇编代码表示为 −4（FP）、−8（FP）、−12（FP），其中 FP 是栈帧指针。例如，在 32 位 Linux 中，C 语言中 b=2 的汇编代码是

```
movl   $2, -8(%ebp)       # b=2 in C
```

其中 $2 表示 2 的值，%ebp 表示 ebp 寄存器。

（6）main() 通过“c = sub(a, b);”调用 sub()，编译后的函数调用代码包括

- 按倒序推送参数，即将值 b=2 和 a=1 推送到堆栈中。
- 调用 sub，将当前的 PC 推送到堆栈中，并用 sub 的入口地址替换 PC，使 CPU 进入 sub()。

当第一次进入 sub() 时，栈顶包含一个返回地址，前面是调用者的参数 a、b，如图 2.10 所示。

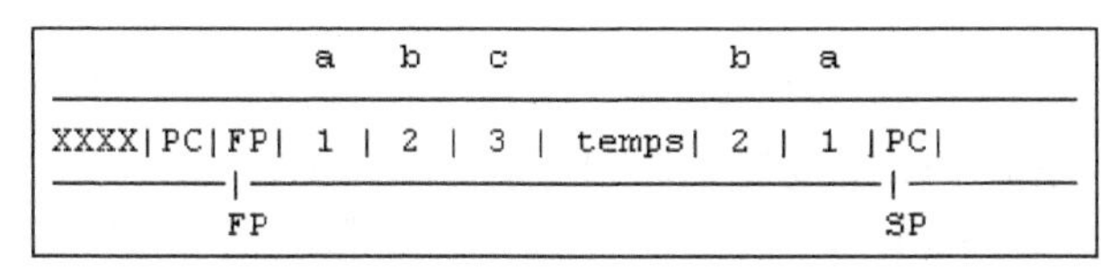

图 2.10　堆栈内容：传递参数

（7）因为 sub() 函数是用 C 语言编写的，所以它的动作和 main() 函数的动作完全一样：

- 推送 FP，并让 FP 指向保存的 FP。
- 向下移动 SP 为局部变量 u、v 分配空间。
- 编译后的代码可能会继续向下移动 SP，在堆栈上提供一些临时空间。

堆栈内容如图 2.11 所示。

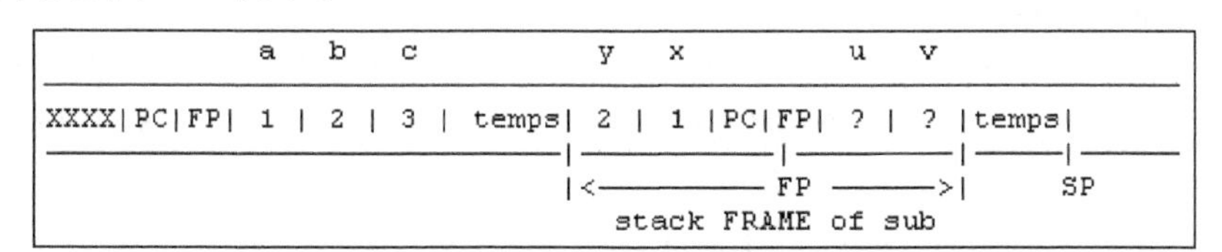

图 2.11　被调用函数的堆栈内容

虽然是在某函数（如 sub()）中执行，但它只能访问全局变量、调用者传入的参数和局部变量，不能访问其他内容。全局变量和静态局部变量位于组合数据段中，可通过固定基址寄存器引用。参数和自动局部变量在每次调用函数时都有不同的副本。所以问题是：如何引用参数和自动局部变量？在本示例中，参数 a、b 对应于参数 x、y，位置分别是 8（FP）和 12（FP）。类似地，自动局部变量 u、v 的位置分别是 −4（FP）和 −8（FP）。对某函数可见的栈区（即参数和自动局部变量）叫作函数的**栈帧**，就像是某个人可以看见的电影画面。因此，FP 称为**栈帧指针**。某个函数的栈帧看起来如图 2.12 所示。

```
|<————————— Stack Frame of a Function ————————>|
| parameters | retPC | savedFP | local variables |
                         |
                   CPU.FP register
```

图 2.12　函数栈帧

从上面的讨论中，读者应该能够推断出一系列函数调用的结果，例如

```
crt0.o --> main() --> A(par_a) --> B(par_b) --> C(par_c)
```

对于每个函数调用，堆栈都会（向低位地址）增加一个被调函数的帧。栈顶的帧是 CPU 的帧指针指向的当前执行函数的栈帧。保存的 FP（向后）指向其调用者的帧，后者保存的 FP

向后指向调用者的调用者，等等。因此，函数调用序列在堆栈中保持为一个链表，如图 2.13 所示。

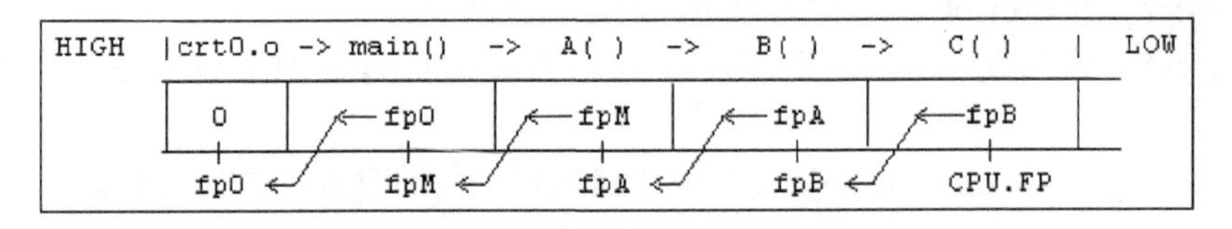

图 2.13 函数调用序列

按照惯例，当 crt0.o 从操作系统内核输入时，CPU 的 FP = 0。因此，栈帧链表的末尾为 0。当某函数返回时，它的栈帧被释放，堆栈向后收缩。

当 sub() 执行 C 语句 return x+y+u+v 时，它对表达式求值，并将结果值放入返回寄存器（AX）中。然后，它会通过以下操作分配局部变量：

- 将 FP 复制到 SP 中；　　　# SP 现在指向堆栈中保存的 FP
- 弹出堆栈到 FP；　　　　　# 该操作恢复了 FP，它现在指向调用者的栈帧，
　　　　　　　　　　　　　# 将返回的 PC 留在栈顶

（在 x86 CPU 上，上述操作相当于 leave 指令。）

- 然后，执行 RET 指令，该指令将栈顶弹出到 PC 寄存器中，使 CPU 从保存的调用者返回地址执行指令。

（8）返回后，调用者函数在返回寄存器（AX）中捕获返回值。然后，清除堆栈中的参数 a、b（将 SP 加 8）。该操作将堆栈恢复到函数调用之前的原始状态。然后，继续执行下一条指令。

注意，一些编译器，如 GCC 4，按地址递增顺序分配自动局部变量，例如，“int a, b ;”意味着（a 的地址）<（b 的地址）。使用这种分配方案，堆栈内容可能如图 2.14 所示。

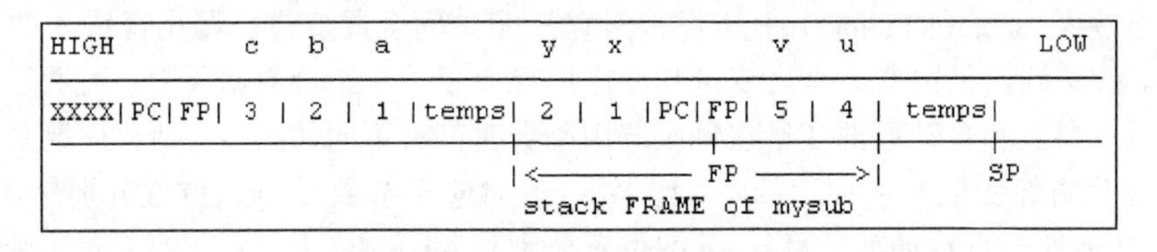

图 2.14 堆栈内容与反向分配方案

在这种情况下，也会以“反向顺序”分配自动局部变量，使它们与参数顺序一致，但是栈帧的概念和用法保持不变。

2.4.2 long jump

在一系列函数调用中，例如

```
main() --> A() --> B()-->C();
```

当一个被调用的函数结束时，通常会返回到正在调用的函数，例如 C() 返回到 B()，B() 返回到 A() 等。也可以通过 long jump 直接返回到调用序列中较早的某个函数。下面的程序演示了 Unix/Linux 中的 long jump。

```
/** longjump.c file: demonstrate long jump in Linux **/
#include <stdio.h>
#include <setjmp.h>
jmp_buf env;        // for saving longjmp environment

int main()
{
  int r, a=100;
  printf("call setjmp to save environment\n");
  if ((r=setjmp(env)) == 0){
     A();
     printf("normal return\n");
  }
  else
     printf("back to main() via long jump, r=%d a=%d\n", r, a);
}

int A()
{   printf("enter A()\n");
    B();
    printf("exit A()\n");
}

int B()
{
  printf("enter B()\n");
  printf("long jump? (y|n) ");
  if (getchar()=='y')
     longjmp(env, 1234);
  printf("exit B()\n");
}
```

在 longjump 程序中，setjmp() 将当前执行环境保存在 jmp_buf 结构体中，并返回 0。该程序继续调用 A()，然后调用 B()。而在函数 B() 中，如果用户选择不通过 long jump 返回，则函数将显示正常的返回顺序。如果用户选择通过 longjmp(env, value) 返回，则执行将以非零值返回到最后保存的环境。在这种情况下，B() 会绕过 A() 直接返回到 main()。long jump 的原理很简单。当一个函数结束时，它通过当前栈帧中的（callerPC, callerFP）返回，如图 2.15 所示。

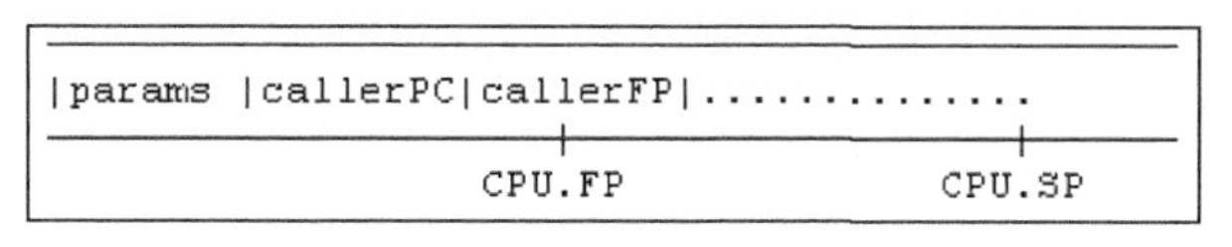

图 2.15　函数返回帧

如果我们用调用序列中较早函数的（savedPC, savedFP）替换（callerPC, callerFP），执行将直接返回到较早函数。除了（savedPC, savedFP），setjmp() 还可以保存 CPU 的通用寄存器和原始 SP，这样 longjmp() 就可以恢复返回函数的完整环境。long jump 可用来中止调用序列中的某个函数，从而使得从先前保存的已知环境中恢复执行。虽然它在用户模式程序中很少使用，但却是系统编程中的一种常见技术。例如，它可在信号捕捉器中使用，以绕过可导致异常或错误陷入的用户模式函数。我们将在第 6 章中论述这种技术。

2.4.3　64 位 GCC 中的运行时堆栈使用情况

在 64 位模式下，CPU 寄存器扩展到 rax、rbx、rcx、rdx、rbp、rsp、rsi、rdi、r8 到 r15，位宽都是 64 位。函数调用惯例与 32 位模式略有不同。当调用一个函数时，前 6 个参数依次进入 rdi、rsi、rdx、rcx、r8、r9。其他任何参数都和 32 位模式下一样通过堆栈。进入后，被调函数首先像往常一样建立栈帧（使用 rbp）。然后，它可能向下移动堆栈指针（rsp），在堆栈上提供局部变量和工作空间。GCC 编译器生成的代码可保持堆栈指针固定不动，默认保留**红色区域**栈区为 128 字节，当在某个函数中执行时，可使用 rsp 作为基址寄存器来访问堆栈内容。但是，GCC 编译器生成的代码仍使用栈帧指针 rbp 来访问参数和局部变量。下面我们通过一个例子来说明 64 位模式下的函数调用惯例。

示例 2.1：64 位模式下的函数调用惯例。

（1）下面的 t.c 文件包含 C 中的一个 main() 函数，它定义了 9 个局部 int（32 位）变量，从 a 到 i。它调用了一个有 8 个 int 参数的 sub() 函数。

```
/*********  t.c file ********/
#include <stdio.h>
int sub(int a, int b, int c, int d, int e, int f, int g, int h)
{
  int u, v, w;
  u = 9;
  v = 10;
  w = 11;
  return a+g+u+v; // use first and extra parameter, locals
}

int main()
{
  int a, b, c, d, e, f, g, h, i;
  a = 1;
  b = 2;
  c = 3;
  d = 4;
  e = 5;
  f = 6;
  g = 7;
  h = 8;
  i = sub(a,b,c,d,e,f,g,h);
}
```

（2）在 64 位 Linux 中，编译 t.c 生成 64 位汇编的 t.s 文件：

```
gcc -S t.c  # generate t.s file
```

然后，编辑 t.s 文件，删除编译器生成的非必要行，并添加注释来解释代码操作。下面给出了带有添加的注释行的简化 t.s 文件。

```
#------------ t.s file generated by 64-bit GCC compiler -------------
       .globl  sub
sub:                # int sub(int a,b,c,d,e,f,g,h)

# first 6 parameters a,  b,  c,  d,  e,  f are in registers
```

```
#                   rdi,rsi,rdx,rcx,r8d,r9d
# 2 extra parameters g,h are on stack.

# Upon entry, stack top contains g, h
#   ----------------------------------
#    ......| h | g | PC |    LOW address
#   ---------------|------------------
#                  rsp

# establish stack frame
        pushq  %rbp
        movq   %rsp, %rbp
# no need to shift rsp down because each function has a 128 bytes
# reserved stack area.
# rsp will be shifted down if function define more locals

# save first 6 parameters in registers on stack
        movl   %edi, -20(%rbp) # a
        movl   %esi, -24(%rbp) # b
        movl   %edx, -28(%rbp) # C
        movl   %ecx, -32(%rbp) # d
        movl   %r8d, -36(%rbp) # e
        movl   %r9d, -40(%rbp) # f

# access locals u, v, w at rbp -4 to -12
        movl   $9,   -4(%rbp)
        movl   $10,  -8(%rbp)
        movl   $11, -12(%rbp)

# compute x + g + u + v:
        movl   -20(%rbp), %edx   # saved a on stack
        movl   16(%rbp), %eax    # g at 16(rbp)
        addl   %eax, %edx
        movl   -4(%rbp), %eax    # u at -4(rbp)
        addl   %eax, %edx
        movl   -8(%rbp), %eax    # v at -8(rbp)
        addl   %edx, %eax

# did not shift rsp down, so just popQ to restore rbp
        popq   %rbp
        ret

#====== main function code in assembly ======
        .globl      main
main:
# establish stack frame
        pushq  %rbp
        movq   %rsp, %rbp

# shit rsp down 48 bytes for locals
        subq   $48, %rsp

# locals are at rbp -4 to -32
        movl   $1,  -4(%rbp)    # a=1
```

```
        movl   $2,   -8(%rbp)       # b=2
        movl   $3,  -12(%rbp)       # c=3
        movl   $4,  -16(%rbp)       # d=4
        movl   $5,  -20(%rbp)       # e=5
        movl   $6,  -24(%rbp)       # f=6
        movl   $7,  -28(%rbp)       # g=7
        movl   $8,  -32(%rbp)       # h=8

# call sub(a,b,c,d,e,f,g,h): first 6 parameters in registers
        movl   -24(%rbp), %r9d      # f in r9
        movl   -20(%rbp), %r8d      # e in r8
        movl   -16(%rbp), %ecx      # d in ecx
        movl   -12(%rbp), %edx      # c in edx
        movl    -8(%rbp), %esi      # b in esi
        movl    -4(%rbp), %eax      # a in eax but will be in edi
# push 2 extra parameters h,g on stack
        movl   -32(%rbp), %edi      # int h in edi
        pushq  %rdi                 # pushQ rdi ; only low 32-bits = h
        movl   -28(%rbp), %edi      # int g in edi
        pushq  %rdi                 # pushQ rdi ; low 32-bits = g

        movl   %eax, %edi           # parameter a in edi
        call   sub                  # call sub(a,b,c,d,e,f,g,h)

        addq   $16, %rsp            # pop stack: h,g, 16 bytes
        movl   %eax, -36(%rbp)      # i = sub return value in eax

        movl   $0, %eax             # return 0 to crt0.o
        leave
        ret

# GCC compiler version 5.3.0
        .ident    "GCC: (GNU) 5.3.0"
```

2.5 C 语言程序与汇编代码的链接

在系统编程中，经常需要访问和控制硬件，如 CPU 寄存器和 I/O 端口位置等。在这些情况下，需要汇编代码。因此，知道如何将 C 语言程序与汇编代码链接起来非常重要。

2.5.1 用汇编代码编程

1. 将 C 代码编译成汇编代码

```
/************* a.c file ********************/
#include <stdio.h>

extern int B();

int A(int x, int y)
{
  int d, e, f;
  d = 4; e = 5; f = 6;
  f = B(d,e);
}
```

```
======= compile a.c file into 32-bit assembly code ======
cc -m32 -S a.c ===> a.s file

====================================================
        .text
        .globl A
A:
        pushl    %ebp
        movl     %esp, %ebp

        subl     $24, %esp
        movl     $4, -12(%ebp)      # d=4
        movl     $5,  -8(%ebp)      # e=5
        movl     $6,  -4(%ebp)      # f=6

        subl     $8, %esp
        pushl     -8(%ebp)          # push e
        pushl    -12(%ebp)          # push d
        call     B

        addl     $16, %esp          # clean stack
        movl     %eax, -4(%ebp)     # f=return value in AX

        leave
        ret
======================================================
```

2. 汇编代码说明

GCC 生成的汇编代码由三部分组成：

（1）入口代码：又叫作 prolog，它建立栈帧，在堆栈上分配局部变量和工作空间。

（2）函数体代码：在 AX 寄存器中执行带有返回值的函数任务。

（3）退出代码：又叫作 epilog，它释放堆栈空间并返回到调用者。

下面解释了 GCC 生成的汇编代码以及堆栈内容：

```
A:                                    # A() start code location
```

（1）入口代码：

```
pushl    %ebp
movl     %esp, %ebp         # establish stack frame
```

入口代码首先将 FP（%bp）保存在堆栈上，并让 FP 指向调用者保存的 FP。堆栈内容变成

```
         SP
---------|--------------------------- LOW address
 xxx  |PC|FP|
---------|---------------------------
         FP

         subl     $24, %esp
```

然后将 SP 向下移动 24 字节，为局部变量和工作区域分配空间。

（2）函数体代码：

```
movl     $4, -20(%ebp)   // d=4
movl     $5, -16(%ebp)   // e=5
movl     $6, -12(%ebp)   // f=6
```

在一个函数内部，FP 指向一个固定位置，用作访问局部变量和参数的基址寄存器。可以看到，3 个局部变量 d、e、f 每个位长为 4 字节，与 FP 的偏移量分别为 -20、-16、-12 字节。当值分配给局部变量后，堆栈内容变为

```
          SP  ---   -24  ----> SP
          |                    |
---------- -4  -8 -12 -16 -20 -24|------------------ LOW address
 xxx |PC|FP|? |? | 6 | 5 | 4 |?  |
---------|---------f---e---d----|------------------
          FP


# call B(d,e): push parameters d, e in reverse order:

          subl     $8, %esp          # create 8 bytes TEMP slots on stack
          pushl    -16(%ebp)         # push e
          pushl    -20(%ebp)         # push d

                                                         SP
 -------------|-4 -8 -12 -16 -20 -24|- TEMP -|-------|--- LOW address
 xxx |retPC|FP|? |? | 6 | 5 | 4 |?  | ??| ?? | e | d |
 ----------|---------f---e---d----|--------------------
             FP

          call     B                 # B() will grow stack to the RIGHT

 # when B() returns:
          addl     $16, %esp         # clean stack
          movl     %eax, -4(%ebp)    # f = return value in AX
```

（3）退出代码：

```
#       leave
          movl   %ebp, %esp          # SAME as leave
          popl   %ebp

          ret                        # pop retPC on stack top into PC
```

2.5.2　用汇编语言实现函数

示例 2.2：获取 CPU 寄存器。由于这些函数很简单，因此不需要建立和释放栈帧。

```
#============= s.s file ==============
          .global get_esp, get_ebp
get_esp:
          movl    %esp, %eax
```

```
        ret
get_ebp:
        movl    %ebp, %eax
        ret
#====================================

int main()
{
   int ebp, esp;
   ebp = get_ebp();
   esp = get_esp();
   printf("ebp=%8x   esp=%8x\n", ebp, esp);
}
```

示例 2.3：假设 int mysum(int x, int y) 返回 x 和 y 的和。用汇编语言编写 mysum() 函数。由于函数必须使用其参数来计算和，所以我们列出了函数代码的入口代码、函数体代码和退出代码。

```
# ============ mysum.s file ==============================
        .text                    # Code section
        .global mysum, printf    # globals: export mysum, import printf
mysum:

# (1) Entry:(establish stack frame)

        pushl %ebp
        movl  %esp, %ebp

# Caller has pushed y, x on stack, which looks like the following
#            12   8    4     0
#-------------------------------------------------------
#          |  y | x |retPC| ebp|
#-----------------------------|----------------------
#                             ebp

# (2): Function Body Code of mysum: compute x+y in AX register
        movl  8(%ebp), %eax    # AX = x
        addl 12(%ebp), %eax    # AX += y

# (3) Exit Code: (deallocate stack space and return)
        movl  %ebp, %esp
        pop   %ebp
        ret
# =========== end of mysum.s file ==========================

int main()  # driver program to test mysum() function
{
  int a,b,c;
  a = 123; b = 456;
  c = mysum(a, b);
  printf("c=%d\n", c);   // c should be 579
}
```

2.5.3 从汇编中调用 C 函数

示例 2.4：访问全局变量并调用 printf()。

```
int a, b;
int main()
{
   a = 100; b = 200;
   sub();
}

#========== Assembly Code file ================
        .text
        .global sub, a, b, printf
sub:
        pushl   %ebp
        movl    %esp, %ebp

        pushl   b
        pushl   a
        pushl   $fmt          # push VALUE (address) of fmt
        call    printf        # printf(fmt, a, b);
        addl    $12, %esp

        movl    %ebp, %esp
        popl    %ebp
        ret

        .data
fmt:    .asciz   "a=%d  b=%d\n"
#=====================================
```

2.6 链接库

链接库中包含预编译的目标代码。在链接期间，链接器使用链接库完成链接过程。在 Linux 中，有两种链接库：用于静态链接的**静态链接库**和用于动态链接的**动态链接库**。在本节中，我们将介绍如何在 Linux 中创建和使用链接库。

假设我们有一个函数

```
// musum.c file
int mysum(int x, int y){ return x + y; }
```

想要创建一个包含 mysum() 函数的目标代码的链接库，可从不同的 C 语言程序中调用该函数，例如

```
// t.c file
int main()
{
   int sum = mysum(123,456);
}
```

2.6.1 静态链接库

下面的步骤说明如何创建和使用静态链接库。

```
(1). gcc  -c  mysum.c                # compile mysum.c into mysum.o
(2). ar  rcs  libmylib.a  mysum.o    # create static link library with member
                                       mysum.o
(3). gcc  -static  t.c  -L. -lmylib  # static compile-link t.c with libmylib.a
                                       as link library
(4). a.out                           # run a.out as usual
```

在编译链接过程中，-L. 指定链接库路径（当前目录），-l 指定链接库。注意，链接库（mylib）未指定前缀 lib 和后缀 .a。

2.6.2 动态链接库

下面的步骤说明如何创建和使用动态链接库。

```
(1). gcc  -c  -fPIC  mysum.c              # compile to Position Independent
                                            Code mysum.o
(2). gcc -shared  -o libmylib.so  mysum.o # create shared libmylib.so with
                                            mysum.o
(3). gcc  t.c  -L.  -lmylib               # generate a.out using shared library
                                            libmylib.so
(4). export LD_LIBRARY_PATH=./            # to run a.out, must export
                                            LD_LIBRARY=./
(5). a.out                                #  run a.out. ld will load libmylib.so
```

在上述两种情况下，如果链接库不在当前目录中，只需要更改 -L. 选项，设置 LD_LIBRARY_PATH，以指向包含链接库的目录。或者，用户也可将链接库放入标准 lib 目录中，如 /lib 或 /usr/lib，然后运行 ldconfig 来配置动态链接库路径。读者可参考 Linux ldconfig（man 8），以了解详细信息。

2.7 makefile

至此，我们已经可以用单个 gcc 命令来编译链接 C 语言程序的源文件了。为了方便，我们还可以使用包含所有命令的 **sh 脚本**。这些方案都有一个很大的缺点。如果只更改几个源文件，sh 命令或脚本仍会编译所有的源文件，包括未修改的文件，这不但没有必要，而且浪费时间。一个更好的方法是使用 Unix/Linux make 工具（GNU make 2008）。make 是一个程序，它按顺序读取 makefile 或 Makefile，以自动有选择地执行编译链接。本节将介绍 makefile 的基础知识，举例说明它们的用法。

2.7.1 makefile 格式

一个 make 文件由一系列**目标项**、**依赖项**和**规则**组成。目标项通常是要创建或更新的文件，但它也可能是 make 程序要引用的指令或标签。目标项依赖于一系列源文件、目标文件甚至其他目标项，具体描述见**依赖项列表**。规则是使用依赖项列表构建目标项所需的命令。图 2.16 显示了 makefile 的格式。

目标项	依赖项列表
target:	file1 file2 ... fileN
	规则
<tab>	command1
<tab>	command2
<tab>	other command

图 2.16　makefile 格式

2.7.2 make 程序

当 make 程序读取 makefile 时，它通过比较依赖项列表中源文件的时间戳来确定要构建哪些目标项。如果任何依赖项在上次构建后有较新的时间戳，make 将执行与目标项有关的规则。假设我们有一个 C 语言程序包含三个源文件：

```
1). type.h file:             // 头文件
    int mysum(int x, int y)  // types, constants, etc

2). mysum.c file:            // C 语言中的函数
    #include <stdio.h>
    #incldue "type.h"
    int mysum(int x, int y)
    {
        return x+y;
    }

3). t.c file:                // C 语言中的 main() 函数
    #include <stdio.h>
    #include "type.h"
    int main()
    {
        int sum = mysum(123,456);
        printf("sum = %d\n", sum);
    }
```

通常，我们会使用 sh 命令

```
gcc -o myt main.c mysum.c
```

生成一个名为 myt 的二进制可执行文件。下面我们将使用 makefile 演示 C 语言程序的编译链接。

2.7.3 makefile 示例

示例 2.5：makefile。

（1）创建一个名为 mk1 的 makefile，包括：

```
myt: type.h t.c mysum.c        # target: dependency list
        gcc -o myt t.c mysum.c # rule: line MUST begin with a TAB
```

在本示例中，生成的可执行文件名 myt 通常与目标项名称匹配。这允许 make 通过将目标项时间戳与依赖项列表中的时间戳进行比较，来决定稍后是否再次构建目标项。

（2）使用 mk1 作为 makefile 运行 make：make 通常使用默认的 makefile 或 Makefile，即当前目录中出现的 makefile。它可以通过 -f 标志直接使用另一个 makefile，如：

```
make -f mk1
```

make 将构建目标文件 myt，并将命令执行显示为：

```
gcc -o myt  t.c  mysum.c
```

（3）再次运行 make 命令，将会显示消息：

```
make: 'myt' is up to date
```

在这种情况下，make 不会再次构建目标，因为在上次构建后没有任何文件更改。

（4）相反，如果依赖项列表中的任何文件有更改，make 将再次执行 rule 命令。一种简单的文件修改方法是使用 touch 命令，修改文件的时间戳。那么，如果我们输入 sh 命令：

```
touch type.h      // or touch *.h, touch *.c, etc.
make -f mk1
```

make 将重新编译链接源文件，以生成新的 myt 文件。

（5）如果我们从依赖项列表中删除一些文件名，即使这些文件有更改，make 也不会执行 rule 命令。读者可以尝试自行验证。

可以看出，mk1 是一个非常简单的 makefile，它与 sh 命令的差别不大。但是，我们可以改进 makefile，使之更加灵活和通用。

示例 2.6：makefile 中的宏。

（1）创建一个名为 mk2 的 makefile，包括：

```
CC = gcc                    # define CC as gcc
CFLAGS = -Wall              # define CLAGS as flags to gcc
OBJS = t.o mysum.o          # define Object code files
INCLUDE = -Ipath            # define path as an INCLUDE directory

myt: type.h $(OBJS)         # target: dependency: type.h and .o files
        $(CC) $(CFLAGS) -o t $(OBJS) $(INCLUDE)
```

在 makefile 中，宏定义的符号——$（符号）被替换为它们的值，如 $(CC) 被替换为 gcc，$(CFLAGS) 被替换为 -Wall 等。对于依赖项列表中的每个 .o 文件，make 首先会将相应的 .c 文件编译成 .o 文件。但是，这只适用于 .c 文件。由于所有 .c 文件都依赖于 .h 文件，所以我们必须在依赖项列表中显式地包含 type.h（或任何其他 .h 文件）。或者，我们可以定义其他目标项来指定 .o 文件对 .h 文件的依赖关系，如：

```
t.o:      t.c type.h        # t.o depend on t.c and type.h
       gcc -c t.c
mysum.o:  mysum.c type.h    # mysum.o depend type.h
       gcc -c mysum.c
```

如果我们将上述目标项添加到 makefile 中，.c 文件或 type.h 中的任何更改都将触发 make 重新编译 .c 文件。如果 .c 文件的数量很小，则会很有效。如果 .c 文件的数量很大，则会很烦琐。因此，有更好的方法将 .h 文件包含在依赖项列表中，稍后将进行展示。

（2）以 mk2 作为 makefile 运行 make。

```
make -f mk2
```

（3）按前面一样运行生成的二进制可执行文件 myt。

示例 2.5 和示例 2.6 的简单 makefile 足以编译链接大多数小型 C 语言程序。以下显示了 makefile 的一些附加功能。

示例 2.7：按名称编译目标。

当 make 在 makefile 上运行时，通常会尝试在 makefile 中构建第一个目标。通过指定一个目标名称可以更改 make 的行为，从而 make 将设置特定的命名目标。以名为 mk3 的

makefile 为例，其中新功能以粗体字母突出显示。

```
# ------------------ mk3 file --------------------
CC = gcc                  # define CC as gcc
CFLAGS = -Wall            # define CLAGS as flags to gcc
OBJS = t.o mysum.o        # define Object code files
INCLUDE = -Ipath          # define path as an INCLUDE directory

all: myt install          # build all listed targets: myt, install

myt: t.o mysum.o          # target: dependency list of .o files
        $(CC) $(CFLAGS) -o myt $(OBJS) $(INCLUDE)

t.o:    t.c  type.h       #  t.o depend on t.c and type.h
       gcc -c t.c
mysum.o: mysum.c type.h #  mysum.o depend mysum.c and type.h
      gcc -c mysum.c

install: myt              # depend on myt: make will build myt first
       echo install myt to /usr/local/bin
       sudo mv myt /usr/local/bin/  # install myt to /usr/local/bin/

run:    install             # depend on install, which depend on myt
       echo run executable image myt
       myt || /bin/true  # no make error 10 if main() return non-zero

clean:
      rm -f *.o 2> /dev/null              # rm all *.o files
      sudo rm -f /usr/local/bin/myt       # rm myt
```

读者可以通过输入以下 make 命令测试 mk3 文件：

```
(1). make [all] -f mk3          # build all targets: myt and install
(2). make install -f mk3        # build target myt and install myt
(3). make run -f mk3            # run /usr/local/bin/myt
(4). make clean -f mk3          # remove all listed files
```

makefile 变量：makefile 支持变量。在 makefile 中，% 是一个与 sh 中的 * 类似的通配符变量。makefile 还可以包含**自动变量**，这些变量在匹配规则后由 make 设置。自动变量规定了对目标和依赖项列表中元素的访问，从而用户不必显式指定任何文件名。自动变量对于定义一般模式规则非常有用。以下列出了 make 的一些自动变量。

- $@：当前目标名
- $<：第一个依赖项名
- $^：所有依赖项名
- $*：不包含扩展名的当前依赖项名
- $?：比当前目标更新的依赖项列表

另外，make 还支持**后缀规则**，后缀规则并非目标，而是 make 程序的指令。我们通过一个例子来说明 make 变量和后缀规则。

在 C 语言程序中，.c 文件通常依赖于所有 .h 文件。如果任何 .h 文件发生更改，则必须重新编译所有 .c 文件。为了确保这一点，我们可以定义一个包含所有 .h 文件的依赖项列表，并在 makefile 中指定一个目标：

```
DEPS = type.h              # list ALL needed .h files
%.o: %.c $(DEPS)           # for all .o files: if its .c or .h file changed
        $(CC) -c -o $@     # compile corresponding .c file again
```

在上面的目标中，%.o 代表所有 .o 文件，$@ 设置为当前目标名称，即当前 .o 文件名。这样可以避免为单个 .o 文件定义单独的目标。

示例 2.8：使用 make 变量和后缀规则。

```
# ---------- mk4 file -------------
CC = gcc
CFLAGS = -I.
OBJS = t.o mysum.o
AS = as      # assume we have .s files in assembly also
DEPS = type.h                # list all .h files in DEPS

.s.o: # for each fname.o, assemble fname.s into fname.o
        $(AS) -o $< -o $@  # -o $@ REQUIRED for .s files

.c.o: # for each fname.o, compile fname.c into fname.o
        $(CC) -c $< -o $@  # -o $@ optional for .c files

%.o:  %.c $(DEPS) # for all .o files: if its .c or .h file changed
        $(CC) -c -o $@ $<  # compile corresponding .c file again

myt: $(OBJS)
        $(CC) $(CFLAGS) -o $@ $^
```

在 makefile mk4 中，.s.o: 和 .c.o: 行并非目标，而是**后缀规则**对 make 程序的指令。这些规则指定，对于每个 .o 文件，如果它们的时间戳不同，即 .s 或 .c 文件已更改，则应创建一个与之对应的 .s 或 .c 文件。在所有目标规则中，$@ 表示当前目标，$< 表示依赖项列表中的第一个文件，$^ 表示依赖项列表中的所有文件。例如，在 myt 目标的规则中，-o $@ 指定输出文件名为当前目标，即 myt。$^ 表示它包含依赖项列表中的所有文件，即 t.o 和 mysum.o。如果我们将 $^ 更改为 $< 并点触所有 .c 文件，make 将生成“未定义的 mysum 引用”错误。这是因为 $< 仅指定依赖项列表中的第一个文件（t.o），make 只会重新编译 t.c 而不会编译 mysum.c，由于缺少 mysum.o 文件而生成一个链接错误。从示例中可以看出，我们可以使用 make 变量编写通用压缩 makefile。其缺点是这样的 makefile 很难理解，特别是对于初级程序员。

子目录中的 makefile

大型 C 语言编程项目通常由数十到数百个源文件组成。为了便于维护，源文件通常被放到不同级别的目录中，每个目录都有自己的 makefile。make 很容易进入子目录以通过命令执行该目录中的本地 makefile。

```
(cd DIR; $(MAKE))    OR    cd DIR && $(MAKE)
```

执行子目录中的本地 makefile 后，控制返回到当前目录形式，make 从这里继续。我们通过一个真实的例子说明 make 的这一高级功能。

示例 2.9：PMTX 系统 makefile。

PMTX（Wang 2015）是一个类 Unix 操作系统，专为 32 位保护模式 Intel x86 架构而设计。它使用 32 位 GCC 汇编器、编译器和链接器生成 PMTX 内核映像。PMTX 的源文件放在三个子目录中：

- Kernel：PMTX 内核文件；一些 GCC 汇编文件，主要用 C 语言编写。
- Fs：文件系统源文件；均用 C 语言编写。
- Driver：设备驱动源文件；均用 C 语言编写。

编译链接步骤由 makefile 在不同目录中指定。PMTX 源目录中的顶层 makefile 非常简单。顶层 makefile 首先清理目录。然后进入 Kernel 子目录以在 Kernel 目录中执行 makefile。Kernel makefile 首先为 Kernel 中的 .s 和 .c 文件生成 .o 文件。然后，它指示 make 进入 Driver 和 Fs 子目录，通过执行它们的本地 makefile 生成 .o 文件。最后，它将所有 .o 文件链接到内核映像文件。以下显示了 PMTX 系统的各种 makefile。

```
#------------- PMTX Top level Makefile -------------
all: pmtx_kernel

pmtx_kernel:
        make clean
        cd Kernel && $(MAKE)

clean: # rm mtx_kerenl, *.o file in all directories

#------------- PMTX Kernel Makefile ----------------
AS = as -Iinclude
CC = gcc
LD = ld
CPP = gcc -E -nostdinc
CFLAGS = -W -nostdlib -Wno-long-long -I include -fomit-frame-pointer

KERNEL_OBJS = entry.o init.o t.o ts.o traps.o trapc.o queue.o \
fork.o exec.o wait.o io.o syscall.o loader.o pipe.o mes.o signal.o \
threads.o sbrk.o mtxlib.o

K_ADDR=0x80100000       # kernel start virtual address

all: kernel

.s.o:   # build each .o if its .s file has changed
        ${AS} -a $< -o $*.o > $*.map

pmtx_kernel: $(KERNEL_OBJS)       # kernel target: depend on all OBJs
        cd ../Driver && $(MAKE) # cd to Driver, run local Makefile
        cd ../Fs && $(MAKE)     # cd to Fs/,     run local Makefile

# link all .o files with entry=pm_entry, start VA=0x80100000
        ${LD} --oformat binary -Map k.map -N -e pm_entry \
            -Ttext ${K_ADDR} -o $@ \
             ${KERNEL_OBJS} ../DRIVER/*.o ../FS/*.o
clean:
        rm -f *.map *.o
        rm -f ../DRIVER.*.map ../DRIVER/*.o
        rm -f ../FS/*.map ../FS/*.o
```

PMTX 内核 makefile 首先从所有 .s（汇编）文件生成 .o 文件。然后顶层 makefile 通过 KERNEL_OBJ 中的依赖项列表从 .c 文件生成其他 .o 文件，然后进入 Driver 和 Fs 目录以执行本地 makefile，在这些目录中生成 .o 文件。最后，它链接所有 .o 文件以生成 pmtx_kernel 映像文件，该文件为 PMTX OS 内核。相反，由于 Fs 和 Driver 中的所有文件均用 C 语言编写，因此它们的 makefile 仅将 .c 文件编译为 .o 文件，因此没有 .s 或 ld 相应目标和规则。

```
#------------- PMTX Driver Makefile ----------------
CC=gcc
CPP=gcc -E -nostdinc
CFLAGS=-W -nostdlib -Wno-long-long -I include -fomit-frame-pointer

DRIVER_OBJS = timer.o pv.o vid.o kbd.o fd.o hd.o serial.o pr.o atapi.o
driverobj: ${DRIVER_OBJS}

#------------- PMTX Fs Makefile -------------------
CC=gcc
CPP=gcc -E -nostdinc
CFLAGS=-W -nostdlib -Wno-long-long -I include -fomit-frame-pointer

FS_OBJS = fs.o buffer.o util.o mount_root.o alloc_dealloc.o \
          mkdir_creat.o cd_pwd.o rmdir.o link_unlink.o stat.o touch.o \
          open_close.o read.o write.o dev.o mount_umount.o
fsobj:    ${FS_OBJS}
#--------------- End of Makefile -------------------
```

2.8 GDB 调试工具

GNU 调试工具（GDB）（GDB 调试 2002；GDB 2017）是一个交互式调试工具，可以调试用 C、C ++ 和其他几种语言编写的程序。在 Linux 中，man gdb 命令显示 gdb 的手册页，其中提供了如何使用 GDB 的简要说明。读者可以在列出的参考文献中找到有关 GDB 的更多详细信息。在本节中，我们将介绍 GDB 的基础知识，并说明如何使用 GDB 在 X-window 下的 EMACS 集成开发环境（IDE）中调试 C 语言程序，这在所有 Linux 系统中都可用。由于不同 Linux 发行版的图形用户界面（GUI）可能不同，因此以下讨论针对 Ubuntu Linux 15.10 或更高版本，但也应该适用于其他 Linux 发行版，例如 Slackware Linux 14.2 等。

2.8.1 在 emacs IDE 中使用 GDB

（1）**源代码**：在 X-window 下，打开一个伪终端。使用 EMACS 创建 makefile，如下所示。

makefile：

```
t: t.c
       gcc -g -o t t.c
```

然后使用 EMACS 编辑 C 语言源文件。由于这里旨在说明 GDB 的用法，我们将使用一个非常简单的 C 语言程序。

```
/******** Source file: t.c Code ********/
#include <stdio.h>
int sub();
int g, h;    // globals
```

```
int main()
{
  int a, b, c;
  printf("enter main\n");
  a = 1;
  b = 2;
  c = 3;
  g = 123;
  h = 456;
  c = sub(a, b);
  printf("c = %d\n", c);
  printf("main exit\n");
}

int sub(int x, int y)
{
  int u,v;
  printf("enter sub\n");
  u = 4;
  v = 5;
  printf("sub return\n");
  return x+y+u+v+g+h;
}
```

（2）**编译源代码**：当EMACS运行时，它会在编辑窗口顶部显示一个菜单和一个工具栏（见图2.17）。

图2.17　EMACS菜单和工具栏

可以打开每个菜单以显示子菜单。打开EMACS的Tools菜单，然后选择Compile。EMACS将在编辑窗口底部显示一个提示行

```
make -k
```

并等待用户响应。EMACS通常通过makefile编译链接源代码。如果在上面显示的同一目录中已有一个makefile，请按Enter键让EMACS继续。读者也可以手动输入命令行，而不是在makefile中。

```
gcc -g  -o  t  t.c
```

为了生成一个二进制可执行文件以便GDB进行调试，需要使用-g标记。使用-g标记，GCC编译链接器将在二进制可执行文件中构建一个符号表，以便GDB在执行过程中访问变量和函数。如果没有-g标记，GDB将无法调试生成的可执行文件。编译完成后，EMACS将在源代码窗口下方的一个单独窗口中显示编译结果，包括警告或错误消息（如有）。

（3）**启动GDB**：打开EMACS的Tools菜单并选择Debugger（调试工具）。

EMACS将在编辑窗口底部显示一个提示行，并等待用户响应。

```
gdb -i=mi  t
```

按 Enter 键启动 GDB 调试工具。GDB 将在上部窗口中运行，并在 EMACS 编辑窗口顶部显示一个菜单和一个工具栏，如图 2.18 所示。

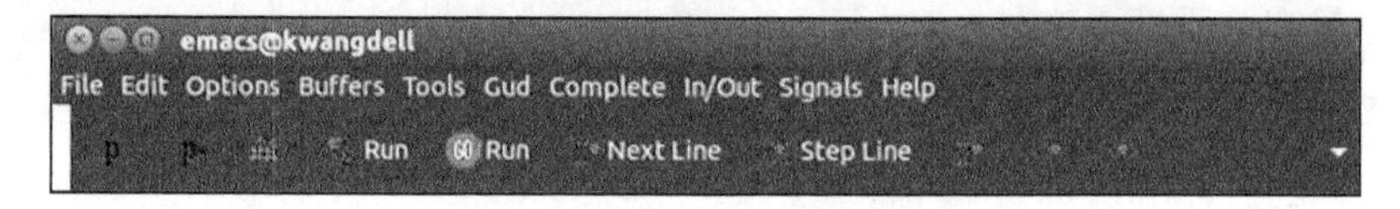

图 2.18 GDB 菜单和工具栏

用户现在可以输入 GDB 命令来调试程序。例如，要设置断点，请输入 GDB 命令：

```
b main        # set break point at main
b sub         # set break point at sub
b 10          # set break point at line 10 in program
```

当用户输入 Run（r）命令（或选择工具栏中的 Run）时，GDB 将在同一 GDB 窗口中显示程序代码。其他多窗口页面 / 窗口可以通过子菜单 GDB-Frames 或 GDB-Windows 激活。以下步骤演示了在 GDB 多窗口布局中同时使用命令和工具栏进行调试的过程。

（4）**多窗口 GDB**：从 GDB 菜单中，选择 Gud -> GDB-MI -> Display Other Windows，其中 -> 表示按子菜单操作。GDB 将在不同的窗口中显示 GDB 缓冲区，如图 2.19 所示。

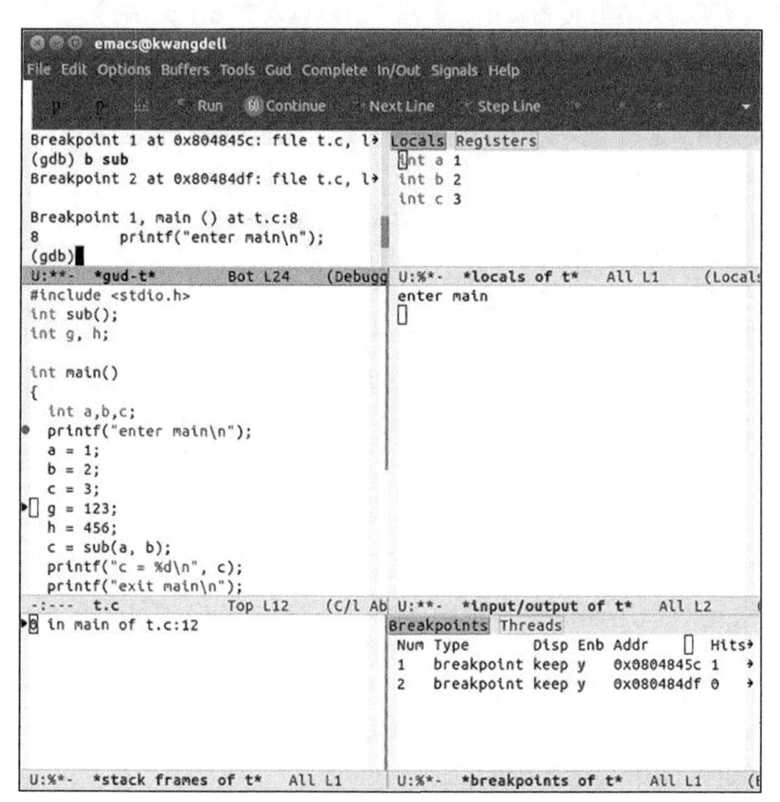

图 2.19 GDB 的多窗口模式

图 2.19 显示了 6 个 GDB 窗口，每个窗口显示一个特定的 GDB 缓冲区。

- Gud-t：用户命令和 GDB 消息的 GDB 缓冲区。
- t.c：显示执行进度的程序源代码。

- 栈帧：显示函数调用序列的栈帧。
- 本地寄存器：显示当前执行函数中的局部变量。
- 输入 / 输出：程序 I/O。
- 断点：显示当前断点设置。

它还会在工具栏中显示一些常用的 GDB 命令，例如 Run、Continue、Next line、Step line，允许用户选择一个操作而不是输入命令。

在程序执行时，GDB 以一个**黑色三角标志**显示执行进度，该标志指向要执行的下一行程序代码。执行将在每个断点处停止，中断用户与 GDB 的交互。由于我们将 main 设置为断点，因此当 GDB 开始运行时，执行将在 main 处停止。当执行在断点处停止时，用户可以在 GDB 窗口中输入命令与 GDB 交互，例如设置 / 清除断点、显示 / 更改变量等。然后，输入 Continue（c）命令或在工具栏中选择 Continue 以继续执行程序。输入 Next（n）命令或在 GDB 工具栏中选择 Next line 或 Step line 以单行模式执行。

图 2.19 显示 main() 函数中已经完成了几行程序代码的执行，下一行标记位于语句

```
g = 123;
```

此时，局部变量 a、b、c 已经赋值，这些值在 Locals registers 窗口中显示为 a = 1，b = 2，c = 3。全局变量不会显示在任何窗口中，但用户可以输入 Print（p）命令

```
p g
p h
```

打印全局变量 g 和 h，这两个变量仍为 0，因为它们尚未被赋予任何值。

图 2.19 还显示了以下 GDB 窗口：

- input/output 窗口显示 main() 函数 printf 语句的输出。
- stack frames 窗口显示目前执行仍在 main() 函数中。
- breakpoints 窗口显示当前断点设置等。

这种多窗口显示设置为用户提供了有关执行程序状态的完整信息。

用户可以输入 Continue 或在工具栏中选择 Continue 以继续执行。当执行到 sub() 函数时，将在断点处再次停止。图 2.20 显示程序执行目前在 sub() 中，并且执行已经传递到

```
printf("return from sub\n");
```

之前的语句。此时，Locals Registers 窗口显示 sub() 的局部变量 u=4 和 v=5。input/output 窗口显示 main() 和 sub() 的打印结果。Stack frames 窗口显示 sub() 为顶层帧，main() 为下一帧，与函数调用序列一致。

（5）**附加 GDB 命令**：在每个断点处或以单行模式执行时，用户可以通过 GDB 工具栏或选择 Gud 菜单（其中包括 GDB 工具栏中的所有命令）中的子菜单项手动输入 GDB 命令。以下列出了一些附加 GDB 命令及其含义。

- 清除断点：
  clear line#：清除 line# 的断点。
  clear name：清除函数（名字为 name）中的断点。
- 更改变量值：

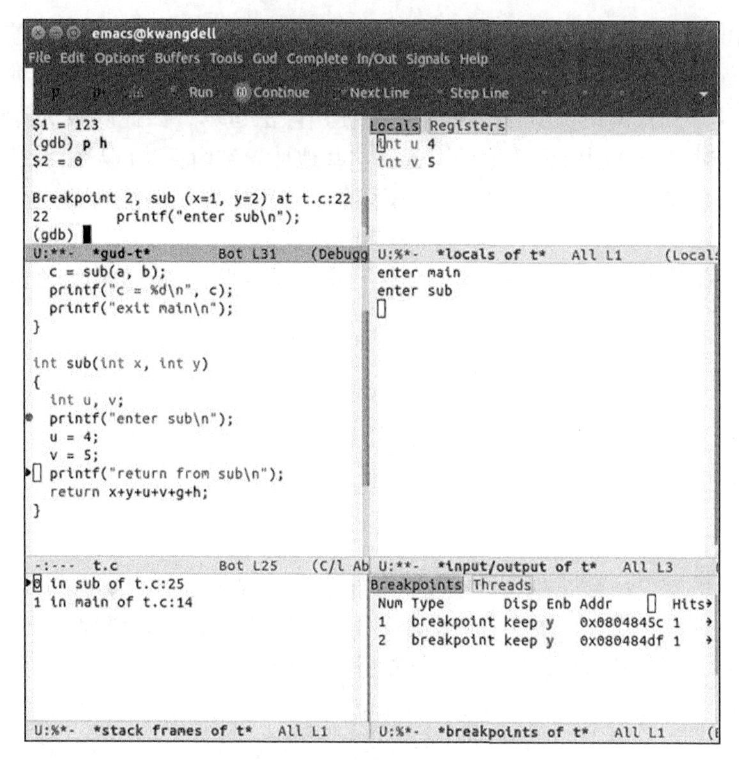

图 2.20 GDB 的多窗口模式

set var a = 100：设变量 a 为 100。

set var b = 200：设变量 b 为 200，以此类推。

- 监视变量值改变：

  watch c：监视变量 c 值改变；每当 c 的值变化时，将显示其旧值和新值。

- 回溯追踪（bt）：

  bt stackFrame#：回溯追踪栈帧。

2.8.2 有关使用调试工具的建议

GDB 是一个功能强大的调试工具，使用非常简单。但是，读者应该注意的是，所有调试工具只能提供有限的帮助。在某些情况下，即使像 GDB 这样的强大调试工具用处也不大。程序开发的最佳方法是仔细设计程序的算法，然后根据算法编写程序代码。许多编程初学者往往不做任何规划便编写程序代码，只是希望他们的程序能够运行，而这往往事与愿违。当他们的程序无法运行或无法产生正确的结果时，他们会立即使用调试工具，尝试追踪程序执行以找出问题所在。过度依赖调试工具通常会适得其反，因为这可能浪费更多的时间。在下文中，我们将指出一些常见的编程错误，并说明如何在 C 语言程序中避免这些错误。

2.8.3 C 语言程序中的常见错误

正在执行中的程序可能会遇到许多类型的运行时错误，例如非法指令、越权错误、除数为零、无效地址等。CPU 将这些错误识别为**异常**，从而将导致进程陷入操作系统内核中。

如果用户未做出任何措施处理这些错误，则进程将以一个信号编号结束，该信号编号指示异常原因。如果程序用C语言编写，并在用户模式下由一个进程执行，则绝对不应发生非法指令和越权等异常。在系统编程中，程序很少使用除法运算，因此除数为零的异常也很少见。主要的运行时错误类型由无效地址引起，这些地址导致内存访问异常，从而导致糟糕且常见的分段错误消息。在下文中，我们列出了C语言程序中最可能导致运行时内存访问异常的原因。

（1）未初始化的指针或含有错误值的指针：考虑以下带有行号以便参考的代码段。

```
1. int *p;          // global, initial value = 0
   int main()
   {
2.     int *q;      // q is local on stack, can be any value
3.     *p = 1;      // dereference a NULL pointer
4.     *q = 2;      // dereference a pointer with unknown value
   }
```

第1行定义了一个全局整数指针p，该指针位于运行时映像的BSS段中，初始值为0。因此，它是一个NULL指针。第3行试图取消引用NULL指针，这将导致分段错误。

第2行定义了一个局部整数指针q，该指针位于栈上，因此它可以是任何值。第4行试图取消引用指向未知内存位置的指针q。如果位置不在程序的可写入存储区内，则会由于内存违规访问而导致分段错误。如果位置在程序的可写入存储区内，则可能不会导致即时错误，但稍后可能会由于数据或栈内容损坏而导致其他错误。后一种运行时错误极难诊断，因为这些错误可能已经通过程序执行而传播。下面显示了这些指针的正确使用方法。修改上面的代码段，如下所示。

```
int  x, *p;              // or  int *p = &x;
int main()
{
    int *q;
    p = &x;              // let p point at x
    *p = 1;
    q = (int *)malloc(sizeof(int);  // q point at allocate memory
    *q = 2;
}
```

原理很简单。使用任何指针时，程序员必须确保指针不为NULL或被设置为指向有效的内存地址。

（2）数组下标越界：在C语言程序中，每个数组都定义为具有有限数量的N个元素。数组的下标必须在[0，N-1]范围内。如果数组下标在运行时超出范围，则可能导致无效的内存访问，这可能会破坏程序数据区或导致分段错误。我们通过一个例子加以说明。以下列代码段为例。

```
       #define N 10
1.     int a[N], i; //  An array of N elements, followed by int i
       int main()
       {
2.         for (i=0; i<N; i++)       // index i in range
               a[i] = i+1;           // set a[ ] values = 1 to N
```

```
3.      a[N] = 123456789;        // set a[N] to a LARGE value
4.      printf("i = %d\n", i);   // print current i value
5.      printf("%d\n", a[i]));   // segmentation fault !
    }
```

第 1 行定义了一个由 N 个元素组成的数组，后跟下标变量 i。第 2 行表示数组下标的正确使用，数组下标在 [0，N–1] 范围内。第 3 行将 a[N] 设置为一个较大的值，这实际上会更改变量 i，因为 a[N] 和 i 处于同一存储位置。第 4 行输出 i 的当前值，该值不再是 N，而是较大的值。第 5 行最有可能导致分段错误，因为 a [123456789] 试图访问程序数据区之外的存储位置。

（3）字符串指针和 char 数组使用不当：C 语言库中的许多字符串运算函数都使用 char * 参数来定义。具体以 strcpy() 函数为例，该函数定义为：

```
char * strcpy(char *dest, char *src)
```

它将字符串从 src 复制到 dest。关于 strcpy() 的 Linux 手册页明确指定 dest 字符串必须足够大才能接收复制。许多程序员（包括一些"有经验"的研究生）经常忽略规范，而试图按以下方式使用 strcpy()。

```
char *s;  // s is a char pointer
strcpy(s, " this is a string");
```

这将会导致代码段错误，因为 s 没有指向任何具有足够空间来接收 src 字符串的内存位置。如果 s 是全局变量，它就是一个 NULL 指针。在这种情况下，strcpy() 将立即导致分段错误。如果 s 是局部变量，则它可能指向任意存储位置。在这种情况下，strcpy() 可能不会导致即时错误，但稍后可能会由于内存内容损坏而导致其他错误。即使使用调试工具（例如 GDB），这些错误也非常微妙且难以诊断。strcpy() 的正确使用方法是确保 dest 不仅是一个字符串指针，而且是一个具有足够空间接收复制的字符串的实际内存区，例如：

```
char s[128];  // s is a char array
strcpy(s, " this is a string");
```

尽管可以在 char *s 和 char s[128] 中使用相同的 s 变量作为地址，但读者必须注意它们之间存在根本区别。

（4）assert 宏：大多数类 Unix 系统（包括 Linux）都支持 assert(condition) 宏，可以在 C 语言程序中用以检查是否满足指定条件。如果条件表达式的计算结果为 FALSE（0），则程序将中止并显示一条错误消息。例如，考虑以下代码段，其中 mysum() 函数旨在返回由 n≤128 个元素组成的整数数组（由 int *ptr 指向）的总和。由于该函数从程序的其他代码中调用，可能会传入无效参数，因此必须确保指针 ptr 不为 NULL，且数组大小 n 不超过极限。可以通过在函数的入口点输入 assert() 语句来完成这些操作，如下所示。

```
#define LIMIT 128
int mysum(int *ptr, int n)
{
    int i = 0, sum = 0;
    assert(ptr != NULL);       // assert ptr not NULL
    assert(n <= LIMIT);        // assert n <= LIMIT
    while(i++ < n)
       sum += *ptr++
}
```

当执行到 mysum() 函数时，如果任一 assert(condition) 失败，该函数将中止并显示一条错误消息。

（5）在程序代码中使用 fprintf() 和 getchar()：编写 C 语言程序时，在程序代码的关键位置输入 fprintf(stderr, message) 语句以显示预期结果通常很有用。由于 stderr 的 fprintf() 无缓冲，因此打印的消息或结果将在执行下一个 C 语句之前立即显示。如果需要，程序员还可以使用 getchar() 停止程序流程，从而允许用户在继续执行操作之前检查执行结果。我们引用一个简单的程序任务对此进行说明。优先级队列是按优先级排序的单向链表，高优先级条目位于最前面。具有相同优先级的条目按先进先出（FIFO）的顺序排序。编写一个 enqueue() 函数，该函数按优先级将项目插入优先级队列。

```
typedef struct entry{
        struct entry *next;
        char name[64];       // entry name
        int priority;        // entry priority
}ENTRY;

void printQ(ENTRY *queue)   // print queue contents
{
   while(queue){
      printf("[%s %d]-> ", queue->name, queue->priority);
      queue = queue->next;
   }
   printf("\n");
}

void enqueue(ENTRY **queue, ENTRY *p)
{
   ENTRY *q = *queue;
   printQ(q);          // show queue before insertion
   if (q==0 || p->priority > q->priority){ // first in queue
      *queue = p;
      p->next = q;
   }
   else{ // not first in queue; insert to the right spot
     while (q->next && p->priority <= q->priority)
           q = q->next;
     p->next = q->next;
     q->next = p;
   }
   printQ(q);               // show queue after insertion
}
```

在上面的示例代码中，如果 enqueue() 函数代码不正确，则可能会在队列中的错误位置插入一个条目，如果其他程序代码依赖于队列内容的正确性，则会导致其他程序代码失败。在这种情况下，使用 assert() 语句或依赖调试工具可能没有多少帮助，因为所有指针都有效，并且使用调试工具追踪较长的链表将过于烦琐。

相反，我们在插入新条目之前打印队列，然后在插入后再次打印。这时，可允许用户直接查看队列是否正确维护。验证代码有效后，用户可以注释掉 printQ() 语句。

2.9 C 语言结构体

结构体是包含变量或数据对象集合的复合数据类型。C 语言结构体类型由 struct 关键字定义。假设我们需要一个包含以下字段的节点结构体。

- next：指向下一个节点结构体的指针。
- key：一个整数。
- name：一个由 64 个字符组成的数组。

此类结构体可以定义为：

```
struct node{
   struct node *next;
   int  key;
   char name[64];
};
```

然后，“struct node”可以用作派生类型来定义该类型的变量，如：

```
struct node x, *nodePtr;
```

这将 x 定义为节点结构体，将 nodePtr 定义为节点指针。或者，可以使用 C 语言的 typedef 语句将“struct node”定义为派生类型。

```
typedef struct node{
   struct node *next;
   int  key;
   char name[64];
}NODE;
```

于是，NODE 就是一个派生类型，可用于定义该类型的变量，如：

```
NODE x, *nodePtr;
```

以下总结了 C 语言结构体的属性。

（1）定义 C 语言结构体时，该结构体的每个字段都必须具有一个编译器已知的类型，但自引用指针除外。这是因为指针的大小始终相同，例如 32 位架构中的 4 个字节。例如，在上面的 NODE 类型中，字段 next 是一个

```
struct node *next;
```

这是正确的，因为编译器知道 struct node 是一个类型（尽管还不完整）以及要为下一个指针分配多少字节。相反，以下语句

```
typedef struct node{
   NODE *next;      // error
   int  key;
   char name[64];
}NODE;
```

会导致编译时错误，因为尽管下一个是指针，但编译器仍不知道 NODE 类型是什么。

（2）每个 C 语言结构体数据对象都分配了一个连续内存块。C 语言结构体的单个字段通过使用 .operator（. 运算符）访问，该运算符标识一个特定的字段，如：

```
NODE x;   // x is a structure of NODE type
```

然后 x 的单个字段如下访问：

- x.next：这是指向另一个 NODE 类型对象的指针。
- x.key：这是一个整数。
- x.name：这是由 64 个字符组成的数组。

在运行时，每个字段都作为相对于结构体起始地址的**偏移量**进行访问。

（3）一个结构体的大小可以由 sizeof(struct type) 确定。C 编译器将计算该结构体的总字节数大小。由于内存排列限制，C 编译器可能会用额外字节填充结构体的某些字段。如果需要，用户可以用 PACKED 属性定义 C 语言结构体，这可以防止 C 编译器用额外字节填充字段，如：

```
typedef struct node{
    struct node *next;
    int  key;
    char name[2];
 }__attribute__((packed, aligned(1))) NODE;
```

在这种情况下，NODE 结构体的大小将为 10 字节。如果没有 packed 属性，则将为 12 字节，因为 C 编译器将用两个额外字节填充 name 字段，从而为进行内存排列将每个 NODE 对象变为 4 字节的倍数。

（4）假设"NODE x，y；"为两个相同类型的结构体。除了复制结构体的单个字段，我们还可以通过 C 语句 y = x 将 x 分配给 y。编译器生成的代码使用库函数 memncpy(&y, &x, sizeof(NODE)) 复制整个结构体。

（5）C 语言联合体与结构体类似。要定义一个联合体，只需将关键字 struct 替换成关键字 union，如：

```
union node{
    int  *ptr;       // pointer to integer
    int  ID;         // 4-byte integer
    char name[32];   // 32 chars
 }x;                 // x is a union of 3 fields
```

联合体成员的访问方式与结构体成员的访问方式完全相同。结构体和联合体之间的主要区别在于，结构体中的每个成员都具有唯一的存储区，而联合体的所有成员共享相同的存储区，可通过单个成员的属性对其进行访问。联合体的大小由最大的成员决定。例如，在联合体 x 中，成员 name 需要 32 字节。所有其他成员每个只需要 4 字节。因此，联合体 x 的大小为 32 字节。以下 C 语句演示了如何访问联合体的单个成员。

```
x.ptr = 0x12345678;             // use first 4 bytes of x
x.ID = 12345;                   // use first 4 bytes of x also
strcpy(x.name, "1234567890"); // uses first 11 bytes of x
```

2.9.1　结构体和指针

在 C 语言中，**指针**是指向其他数据对象的变量，即指针包含其他数据对象的地址。在 C 语言程序中，指针用 * attribute 定义，如：

```
TYPE *ptr;
```

它将 ptr 定义为指向 TYPE 数据对象的指针，其中 TYPE 在 C 语言中既可以是基本类型也可以是派生类型，例如 struct。在 C 语言编程中，结构体通常通过指向结构体的指针访问。例如，假设 NODE 是一种结构体类型。以下 C 语句

```
NODE x, *p;
p = &x;
```

将 x 定义为一个 NODE 类型数据对象，将 p 定义为指向 NODE 对象的指针。语句“p = &x;”将 x 的地址分配给 p，以便 p 指向数据对象 x。然后 * p 表示对象 x，x 的成员可以如下进行访问：

```
(*p).name, (*p).value, (*p).next
```

或者，C 语言允许我们使用“指向”运算符 -> 引用 x 的成员，如：

```
p->name, p->value, p->next;
```

这比“.”运算符更方便。实际上，使用 -> 运算符访问结构体的成员已成为 C 语言编程的标准做法。

2.9.2 C 语言类型转换

类型转换是一种使用转换运算符（TYPE）**变量**将变量从一种数据类型转换为另一种数据类型的方法。考虑以下代码段。

```
    char *cp, c = 'a';         // c is 1 byte
    int  *ip, i = 0x12345678; // i is 4 bytes

(1). i = c;              // i = 0x00000061; lowest byte = c
(2). c = i;              // c = 0x78 (c = lowest byte of i)
(3). cp = (char *)&i;    // typecast to suppress compiler warning
(4). ip = (int *)&c;     // typecast to suppress compiler warning
(5). c = *(char *)ip;    // use ip as a char *
(6). i = *(int *)cp;     // use cp as an int *
```

第（1）行和第（2）行不需要类型转换，即使赋值涉及不同的数据类型。赋值的结果值显示在注释中。第（4）行和第（5）行需要类型转换以禁止编译器警告。在给第（4）行和第（5）行赋值之后，*cp 仍为一个字节，即（4 字节）整数 i 的最低字节。*ip 是一个整数（0x00000061），最低字节为 'c' 或 0x61。第（6）行强制编译器将 ip 用作 char *，因此 *(char *)ip 取消引用单个字节。第（7）行强制编译器将 cp 用作 int *，因此 *(int *)cp 从 cp 所指向的位置开始间接引用一个 4 字节的值。

类型转换对于指针特别有用，允许同一指针指向不同大小的数据对象。下面显示了一个更实用的类型转换示例。在 Ext2/3 文件系统中，目录的内容为 dir_entries，定义为：

```
struct dir_entry{
   int  ino;           // inode number
   int  entry_len;     // entry length in bytes
   int  name_len;      // name_len
   char name[  ]       // name_len chars
};
```

目录内容由以下形式的dir_entries线性列表组成：

```
| ino elen nlen NAME | ino elen nlen NAME | . . .
```

由于条目的name_len不同，其中条目的长度可变。假设char buf[] 包含一个dir_entries列表。问题是如何遍历buf[] 中的dir_entries。为了按顺序遍历dir_entires，我们定义

```
struct dir_entry *dp = (struct dir_entry *)buf;  // typecasting
char *cp = buf;                               // no need for typecasting
// Use dp to access the current dir_entry;
// advance dp to next dir_entry:
cp = += dp->entry_len;        // advance cp by entry_len
dp = (struct dir_entry *)cp; // pull dp to where cp points at
```

适当的类型转换后，C语言代码的最后两行可简化为：

```
dp = (struct dir_entry *)((char *)dp + dp->rlen);
```

不再需要一个char *cp。

2.10　链表处理

结构体和指针通常用于构建和操作**动态数据结构**，例如链表、队列和树等。动态数据结构的最基本类型是链表。在本节中，我们将解释链表的概念、列表操作，并通过示例程序演示列表处理。

2.10.1　链表

假设节点为NODE类型结构体，如：

```
typedef struct node{
   struct node *next;    // next node pointer
   int  value;           // ID or key value
   char name[32];        // name field if needed
}NODE;
```

（单向）链表是一种由一系列节点组成的数据结构，这些节点通过节点的next指针链接在一起，即每个节点的next指针指向列表中的下一个节点。链表由NODE指针表示。例如，以下C语句定义两个链表，分别由list和head表示。

```
NODE *list, *head;          // define list and head as link lists
```

如果链表不包含任何节点，则该链表为空，即一个空链表只是一个NULL指针。在C语言编程中，符号NULL由以下宏（在stddef.h中）定义：

```
#define NULL (void *)0
```

是一个0值指针。为了方便起见，我们将使用NULL或0表示空指针。因此，可以通过为链表分配空指针将其初始化为空，如：

```
list = NULL;    // list = null pointer
head = 0;       // head = null pointer
```

2.10.2 链表操作

链表最常见的操作类型为

- **构建**：用一组节点初始化和构建列表。
- **遍历**：访问列表的全部元素。
- **搜索**：按键搜索元素。
- **插入**：在表中插入新元素。
- **删除**：从表中删除现有元素。
- **重新排序**：按（已更改的）元素键或优先级值对链表重新排序。

在下文中，我们应开发 C 语言程序，以演示链表处理。

2.10.3 构建链表

在许多 C 语言编程书籍中，动态数据结构通常由数据对象构成，数据对象由 C 语言库函数 malloc() 动态分配。在这种情况下，数据对象从程序的**堆区**分配。这没有问题，但某些书中似乎过于强调这样一种概念，即动态数据结构必须使用 malloc()，这并不正确。与数组等静态数据结构（元素数量固定）不同，动态数据结构包含的数据对象很容易修改，例如插入新的数据对象、删除现有数据对象、数据对象重新排序等。它与数据对象的位置无关，数据对象既可位于程序的数据区，也可位于堆区。我们将在下面的示例程序中阐明和演示这一点。

1. 数据区中的链表

示例程序 C2.1：该程序从节点结构体数组构建链表。该程序由一个 type.h 头文件和一个 C2.1.c 文件组成，包含 main() 函数。

（1）type.h 文件：首先，我们编写一个 type.h 文件，其中包含标准头文件，定义常量 N 和 NODE（节点）类型。这个 type.h 文件将被用作所有列表处理示例程序中的输入文件。

```
/********** type.h file ***********/
#include <stdio.h>
#include <stdlib.h>
#include <string.h>

#define N 10

typedef struct node{
  struct node *next;
  int  id;
  char name[64];

}NODE;
/******* end of type.h file *******/
```

（2）C2.1 程序代码：该程序会构建一个链表，并打印链表。

```
/*********************** C2.1.c file **********************/
#include "type.h"

NODE *mylist, node[N]; // in bss section of program run-time image
```

```
int printlist(NODE *p) // print list function
{
   while(p){
     printf("[%s %d]->", p->name, p->id);
     p = p->next;

   }
   printf("NULL\n");
}

int main()
{
  int i;
  NODE *p;
  for (i=0; i<N; i++){
     p = &node[i];
     sprintf(p->name, "%s%d", "node", i); // node0, node1, etc.
     p->id = i;              // used node index as ID
     p->next = p+1;          // node[i].next = &node[i+1];
  }
  node[N-1].next = 0;
  mylist = &node[0];         // mylist points to node[0]
  printlist(mylist);         // print mylist elements
}
```

（3）C2.1.c 代码说明：

- 变量的内存位置：下图显示了变量的内存位置，全部位于程序的数据区中。更准确地说，它们位于组合程序数据区的 bss 段。

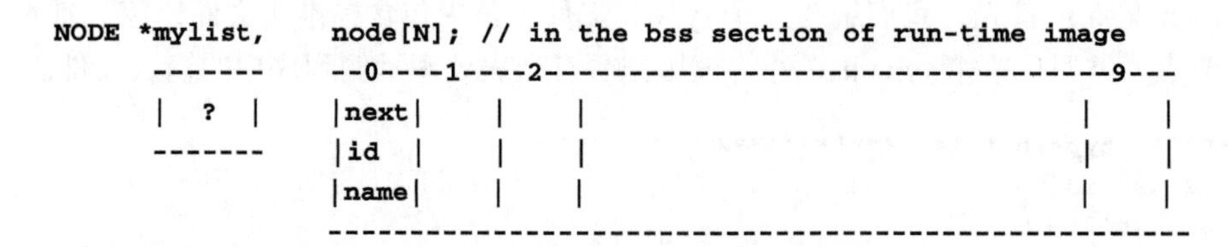

```
NODE *mylist,     node[N]; // in the bss section of run-time image
     -------      --0----1----2-----------------------------------9---
     |  ?  |      |next|    |    |                               |    |
     -------      |id  |    |    |                               |    |
                  |name|    |    |                               |    |
                  ----------------------------------------------------
```

- main() 函数：main() 函数构建一个链表，节点名为 “node0” ～ “node9”，节点 id 值为 0～9。链表以空指针结束。注意，链表指针和链表元素都在 node[N] 数组中，是程序的数据区。

（4）运行结果：读者可以编译并运行 C2.1 程序。链表打印结果为

```
[node0 0]->[ndoe1 1]-> ... [node9 9]->NULL
```

2. 堆区中的链表

示例程序 C2.2：本示例程序在程序的堆区构建链表。

```
/**************** C2.2.c file *****************/
#include "type.h"

NODE *mylist, *node;    // pointers, no NODE area yet

int printlist(NODE *p){ // same as in C2.1 }
```

```
int main()
{
  int i;
  NODE *p;
  node = (NODE *)malloc(N*sizeof(NODE)); // node->N*72 bytes in HEAP
  for (i=0; i < N; i++){
     p = &node[i];          // access each NODE area in HEAP
     sprintf(p->name, "%s%d", "node",i);
     p->id = i;
     p->next = p+1;         // node[i].next = &node[i+1];
  }
  node[N-1].next = 0;
  mylist = &node[0];
  printlist(mylist);
}
```

变量的内存位置：在本示例程序中，变量 mylist 和 node 都是 NODE 指针，它们是未初始化全局变量。因此，它们位于程序的 bss 段，如图 2.21 所示。该程序没有为实际节点结构体提供任何内存位置。

```
NODE *mylist, *node;    // pointers, no NODE area yet
```

当程序启动时，使用语句

```
node = (NODE *)malloc(N*sizeof(NODE));
```

在程序的堆区分配 N*72 字节，用作链表的节点。因为节点是一个 NODE 指针，指向一个包含 N 个相邻节点的存储区，所以我们可将存储区视为一个数组，并使用 node[i]（i=0 到 N−1）访问每个节点元素。这是一条普遍原则。当某个指针指向某个包含 N 个相邻数据对象的存储区时，我们可以用数组元素序列的形式访问该存储区。在本例中，假设

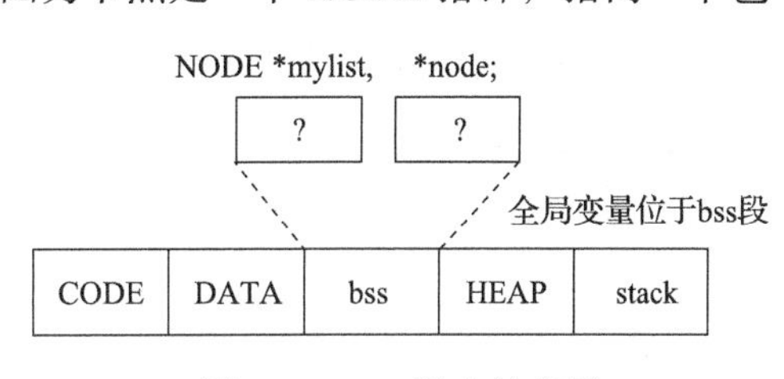

图 2.21 bss 段中的变量

```
int *p = (int *)malloc(N*sizeof(int));
```

那么，p 所指向的存储区可按 *(p + i) 或 p[i] 访问，i=0 到 N−1。

示例程序 C2.3：本示例程序在程序的堆区构建一个链表，但是使用单独分配的节点。

```
/**************** C2.3.c file *****************/
#include "type.h"

NODE *mylist, *node;    // pointers, no NODE area yet

int printlist(NODE *p){ // same as in L1.c }

int main()
{
  int i;
  NODE *p, *q;
  for (i=0; i < N; i++){
     p = (NODE *)malloc(sizeof(NODE));  // allocate a node in heap
```

```
        sprintf(p->name, "%s%d", "node",i);
        p->id = i;
        p->next = 0;        // node[i].next = 0;
        if (i==0){
           mylist = q = p; // mylist -> node0; q->current node
        }
        q->next = p;
        q = p;
    }
    printlist(mylist);
}
```

2.10.4　链表遍历

链表遍历是按顺序访问全部列表元素。最简单的链表遍历是打印链表元素。为此，我们可以实现一个通用 printlist() 函数，如下文所示。

```
int printlist(char *listname, NODE *list)
{
   printf("%s = ", name);
   while(list){
     printf("[%s %d]->", list->name, list->id);
     list = list->next;
   }
  printf("NULL\n");
}
```

另外，printlist() 函数可以递归实现。

```
void rplist(NODE *p)
{
   if (p == 0){
      printf("NULL\n");
      return;
   }
   printf("[%s %d]->", p->name, p->id);
   rplist(p->next);          // recursive call to rplist(p->next)
}

int printlist(char *listname, NODE *list)
{
   printf("%s = ", name);  // print list name
   rplist(list);           // recursively print each element
}
```

除了打印链表之外，还有许多其他列表操作实质上是列表遍历问题的变体。这也是计算机相关工作面试时最喜欢问的一个问题。

该示例可实现一个函数，该函数计算了链表中节点值的和。

```
int sum(NODE *list) // return SUM of node values in a link list
{
     int sum = 0;
     while(list){
```

```
        sum += list->value;  // add value of current node
        list = list->next;   // step to next node
    }
    return sum;
}
```

另外，sum() 函数可以递归实现，如：

```
int sum(NODE *list) // return SUM of node values in a link list
{
    if (list == 0)
        return 0;
    return list->value + sum(list->next);
}
```

使用C语言的“?”运算符，可进一步简化为

```
int sum(NODE *list)
{ return (list==0)? 0: list->value + sum(list->next); }
```

当使用单节点指针遍历单向链表时，只能访问当前节点，而不能访问前一个节点。解决这个问题的一个简单方法是在遍历链表时使用两个指针。我们通过一个例子来说明这种方法。

使用两个指针遍历（单）链表。

```
NODE *p, *q;
p = q = list;          // both p and q start form list
while(p){
   // access current node by p AND previous node by q
   q = p;              // let q point at current node
   p = p->next;        // advance p to next node
}
```

在上面的代码段中，p 指向当前节点，q 指向前一个节点（如果有）。这样我们就可以访问当前节点和前面的节点。该方法可用于删除当前节点、在当前节点之前插入等操作。

2.10.5 搜索链表

搜索操作在链表中搜索具有给定键值的元素，键值既可以是整数值也可以是字符串。如果元素存在，将会返回一个指向该元素的指针；如果元素不存在，则会返回 NULL。假设键值是一个整数值，我们可以实现一个 search() 函数，如下文所示。

```
NODE *search(NODE *list, int key)
{
   NODE *p = list;
   while(p){
     if (p->key == key)  // found element with key
        return p;        // return node pointer
     p = p->next;        // advance p to next node
   }
   return 0;             // not found, return 0
}
```

2.10.6 插入操作

插入操作按照指定的条件将新元素插入链表中。若是单向链表，最简单的插入是插入链表的末尾。这相当于遍历链表到最后一个元素，然后在最后添加一个新元素。我们通过一个例子加以说明。

示例程序 C2.4：本示例程序向链表末尾插入新节点来构建链表。

（1）插入函数：insert() 函数将一个新节点指针 p 输入链表中。由于插入操作可能会修改链表，所以必须通过引用来传递链表参数，即传递链表的地址。通常，链表中的所有元素都应该有唯一键值。为确保任何两个元素的键值都不相同，插入函数可以先进行检查。如果节点键值已经存在了，则会拒绝插入请求。我们把这个问题留给读者作为练习。

```
int insert(NODE **list, NODE *p) // insert p to end of *list
{
   NODE *q = *list;
   if (q == 0)            // if list is empty:
      *list = p;          // insert p as first element
   else{                  // otherwise, insert p to end of list
      while(q->next)      // step to LAST element
         q = q->next;
      q->next = p;        // let LAST element point to p
   }
   p->next = 0;           // p is the new last element
}
```

（2）C2.4.c 文件：

```
/***************** C2.4.c file **************/
#include "type.h"
NODE *mylist;
int main()
{
  char line[128], name[64];
  int id;
  NODE *p;
  mylist = 0;  // initialize mylist to empty list
  while(1){
     printf("enter node name and id value : ");
     fgets(line, 128, stdin);    // get an input line
     line[strlen(line)-1] = 0;   // kill \n at end
     if (line[0] == 0)           // break out on empty input line
        break;
     sscanf("%s %d", name, &id); // extract name string and id value
     p = (NODE *)malloc(sizeof(NODE));
     if (p==0) exit(-1);         // out of HEAP memory
     strcpy(p->name, name);
     p->id = id;
     insert(&mylist, p);         // insert p to list end
     printlist(mylist);
  }
}
```

2.10.7 优先级队列

优先级队列是一个按优先级排序的链表，优先级值高的在前面。在优先级队列中，优先级相同的节点按照先进先出（FIFO）的顺序进行排列。我们可以将 insert() 函数修改为 enqueue() 函数，按优先级将节点插入优先级队列中。假设每个节点都有一个 priority（优先级）字段。下文给出的是 enqueue() 函数。

```
int enqueue(NODE **queue, NODE *p) // insert p into queue by priority
{
   NODE *q = *queue;
   if (q==0 || p->priority > q->priority){
      *queue = p;
      p->next = q;
   }
   else{
      while(q->next && p->priority <= q->next->priority)
         q = q->next;
      p->next = q->next;
      q->next = p;
   }
}
```

2.10.8 删除操作

给定一个链表和一个链表元素键值，若链表中存在该元素，则将其从链表中删除。因为删除操作可能会修改链表，所以必须通过引用传递链表参数。这里，我们假设元素键值是整数。delete() 函数返回删除的节点指针，如果它之前是动态分配的，则可用于释放节点。

```
NODE *delete(NODE **list, int key)
{
  NODE *p, *q;
  if (*list == 0)            // empty list
      return 0;              // return 0 if deletion failed
  p = *list;
  if (p->key == key){        // found element at list beginning
    *list = p->next;         // modify *list
    return p;
  }
  // element to be deleted is not the first one; try to find it
  q = p->next;
  while(q){
    if (q->key == key){
       p->next = q->next;  // delete q from list
       return q;
    }
    p = q;
    q = q->next;
  }
  return 0;                  // failed to find element
}
```

2.10.9 循环链表

在循环链表中，最后一个元素指向第一个元素。通常，空的循环链表是一个 NULL（空）指针。图 2.22 显示的是一个循环链表。

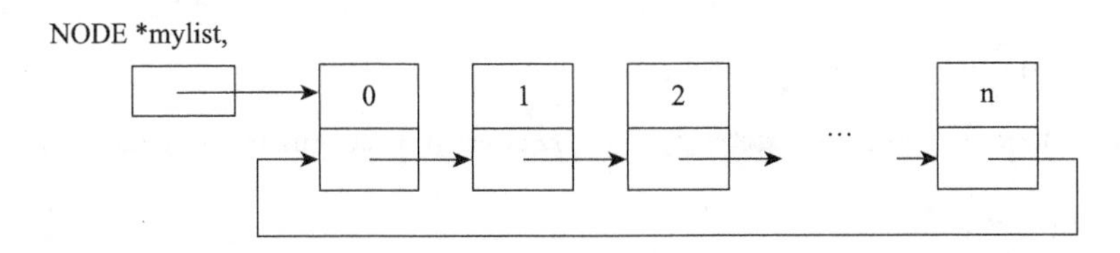

图 2.22　循环链表

在遍历循环列表时，必须正确地检测“链表结束”条件，以终止遍历。假设链表是一个非空循环链表。

```
NODE *list, *p = list;
while(p->next != list){    // stop after last element
   // access current node p;
   p = p->next;
}
```

2.10.10 可扩充 C 语言结构体

在原始的 C 语言中，必须完全指定结构体的所有成员，使得每个结构体的大小在编译时是固定的。实际上，在某些情况下，结构体的大小可能会发生变化。为了方便，将 C 语言扩展为支持包含不完全指定字段的**可扩充结构体**。例如，思考下面的结构体，其中的最后成员是一个未指定大小的数组。

```
struct node{
    struct node *next;
    int ID;
    char name[ ];    // unspecified array size
};
```

在上述可扩充结构体中，name[] 表示不完全指定字段，必须是最后一项条目。可扩充结构体的大小由指定字段决定。对于上面的示例，结构体大小为 8 字节。要使用该结构体，用户必须为实际结构体分配所需的内存，如：

```
struct node *sp = malloc(sizeof(struct node) + 32);
strcpy(sp->name, "this is a test string");
```

第一行分配了一个 8+32 字节的存储区，其中 32 字节用于字段 name。下一行将一个字符串复制到所分配存储区的 name 字段中。如果用户没有为可扩充结构体分配内存，而是直接使用它们，会怎么样？下面的代码行说明了可能出现的结果。

```
struct node x;              // define a variable x of size only 8 bytes
strcpy(x.name, "test");     // copy "test" into x.name field
```

在这种情况下，x 只有 8 字节，但是 name 字段无存储区。第二个语句在运行时将字符串“test”复制到紧随 x 之后的存储区，覆盖其中的内容。这可能会破坏程序的其他数据对象，稍后会导致崩溃。

2.10.11 双向链表

双向链表简称 dlist，在双向链表中，列表元素按向前和向后两个方向链接在一起。在 dlist 中，每个元素有两个指针，用 next 和 prev 表示，如下文所示。

```
typedef struct node{
   struct node *next;     // pointer to next node
   struct node *prev;     // pointer to previous node
   // other fields, e.g. key, name, etc.
}NODE;                    // list NODE type
```

在 dlist 节点结构体中，next 指针指向下一个节点，允许从左到右向前遍历 dlist。prev 指针指向上一个节点，允许从右到左向后遍历 dlist。与单向链表只有一个节点指针不同，双向链表可以有三个不同的方向，每种链表都有其优缺点。在本节中，我们将使用示例程序来显示双向链表操作，其中包括：

（1）在列表末尾插入新元素来构建 dlist。

（2）在列表前面插入新元素来构建 dlist。

（3）按向前和向后方向打印 dlist。

（4）在 dlist 中搜索具有给定键值的元素。

（5）从 dlist 中删除元素。

2.10.12 双向链表示例程序

示例程序 C2.5：双向链表第 1 版。

在第一个 dlist 示例程序中，我们假设一个 dlist 有一个 NODE 指针，该 NODE 指针指向第一个列表元素（如有）。图 2.23 显示了这类 dlist。

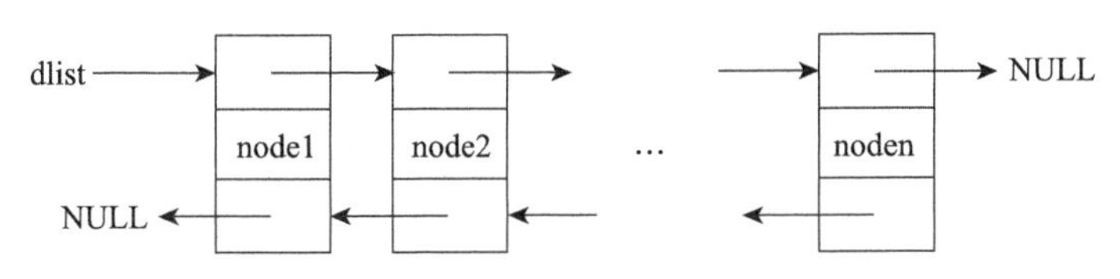

图 2.23　双向链表第 1 版

由于第一个元素没有前项，它的 prev 指针必定为空。同样，最后一个元素没有后项，它的 next 指针也必定为空。空的 dlist 就是一个 NULL 指针。下面列出了示例程序 C2.5 的 C 代码。

```
/********** dlist Program C2.5.c **********/
#include <stdio.h>
#include <stdlib.h>
#include <string.h>
typedef struct node{
  struct node *next;
  struct node *prev;
  int key;
}NODE;
```

```
NODE *dlist;          // dlist is a NODE pointer

int insert2end(NODE **list, int key) // insert to list END
{
  NODE *p, *q;
  //printf("insert2end: key=%d ", key);
  p = (NODE *)malloc(sizeof(NODE));
  p->key = key;
  p->next = 0;
  q = *list;
  if (q==0){            // list empty
    *list = p;
    p->next = p->prev = 0;
  }
  else{
    while(q->next)      // step to LAST element
       q = q->next;
    q->next = p;        // add p as last element
    p->prev = q;
  }
}

int insert2front(NODE **list, int key) // insert to list FRONT
{
  NODE *p, *q;
  //printf("insert2front: key=%d ", key);
  p = (NODE *)malloc(sizeof(NODE));
  p->key = key;
  p->prev = 0;          // no previous node
  q = *list;
  if (q==0){            // list empty
    *list = p;
     p->next = 0;
  }
  else{
    p->next = *list;
    q->prev = p;
    *list = p;
  }
}

void printForward(NODE *list) // print list forward
{
  printf("list forward =");
  while(list){
    printf("[%d]->", list->key);
    list = list->next;
  }
  printf("NULL\n");
}

void printBackward(NODE *list) // print list backward
{
  printf("list backward=");
```

```
  NODE *p = list;
  if (p){
     while (p->next)     // step to last element
       p = p->next;
     while(p){
       printf("[%d]->", p->key);
       p = p->prev;
     }
  }
  printf("NULL\n");
}

NODE *search(NODE *list, int key) // search for an element with a key
{
  NODE *p = list;
  while(p){
    if (p->key==key){
      printf("found %d at %x\n", key, (unsigned int)p);
      return p;
    }
    p = p->next;
  }
  return 0;
}

int delete(NODE **list, NODE *p) // delete an element pointed by p
{
  NODE *q = *list;
  if (p->next==0 && p->prev==0){         // p is the only node
    *list = 0;
  }
  else if (p->next==0 && p->prev != 0){ // last but NOT first
    p->prev->next = 0;
  }
  else if (p->prev==0 && p->next != 0){ // first but NOT last
    *list = p->next;
    p->next->prev = 0;

  }
  else{                                  // p is an interior node
    p->prev->next = p->next;
    p->next->prev = p->prev;
  }
  free(p);
}

int main()
{
  int i, key;
  NODE *p;
  printf("dlist program #1\n");

  printf("insert to END\n");
  dlist = 0;                      // initialize dlist to empty
  for (i=0; i<8; i++){
```

```
    insert2end(&dlist, i);
  }
  printForward(dlist);
  printBackward(dlist);

  printf("insert to FRONT\n");
  dlist = 0;                    // initialize dlist to empty
  for (i=0; i<8; i++){
    insert2front(&dlist, i);
  }
  printForward(dlist);
  printBackward(dlist);

  printf("test delete\n");
   while(1){
    printf("enter a key to delete: ");
    scanf("%d", &key);
    if (key < 0) exit(0);      // exit if negative key
    p = search(dlist, key);
    if (p==0){
       printf("key %d not found\n", key);
       continue;
    }
    delete(&dlist, p);
    printForward(dlist);
    printBackward(dlist);
  }
}
```

图 2.24 显示了运行示例程序 C2.5 的输出。首先，它在列表末尾插入新元素来构造 dlist，并按向前和向后两个方向打印得到的列表。接着，它在列表前面插入新元素来构造 dlist，并打印在两个方向得到的列表。然后，它展示了如何搜索具有给定键值的元素。最后，它演示了如何从列表中删除元素，并在每次删除操作之后打印列表来验证结果。

示例程序 C2.5 的讨论：第 1 版 dlist 的主要优点是只需要一个 NODE 指针来表示 dlist。这类 dlist 的缺点是：

（1）很难访问列表的最后一个元素。就像在列表末尾插入新元素并向后打印列表的情况一样，要想访问最后一个元素，必须使用 next 指针先跳到最后一个元素，如果列表有很多元素，这将会非常耗时。

（2）当插入新元素时，必须检测和处理列表是否为空或者新元素是否是第一个或最后一个元素等各种情况。这会让代码不统一，没有完全实现双向链表的功能。

（3）同样，在删除元素时，还必须检测和处理各种不同的情况。

可以看出，插入 / 删除代码的复杂性主要有两种情况：

- 链表为空，在这种情况下，我们必须插入第一个元素。
- 插入 / 删除的节点是第一个或最后一个节点，在这种情况下，我们必须修改链表指针。

或者，可以把一个双向链表看作是由两个单向链表合并而成。因此，我们可以用一个包含两个指针的 Listhead（链表头）结构体来表示双向链表，如：

```
typedef struct listhead{
      struct node *next;          // head node pointer
      struct node *prev;          // tail node pointer
  }DLIST;                         // doubly link list type
  DLIST dlist;                    // dlsit is a structure
  dlist.next = dlist.prev = 0; // initialize dlist as empty
```

```
dlist program #1
insert to END
list forward =[0]->[1]->[2]->[3]->[4]->[5]->[6]->[7]->NULL
list backward=[7]->[6]->[5]->[4]->[3]->[2]->[1]->[0]->NULL
insert to FRONT
list forward =[7]->[6]->[5]->[4]->[3]->[2]->[1]->[0]->NULL
list backward=[0]->[1]->[2]->[3]->[4]->[5]->[6]->[7]->NULL
test delete
enter a key to delete: 0
found 0 at 9567088
list forward =[7]->[6]->[5]->[4]->[3]->[2]->[1]->NULL
list backward=[1]->[2]->[3]->[4]->[5]->[6]->[7]->NULL
enter a key to delete: 7
found 7 at 95670f8
list forward =[6]->[5]->[4]->[3]->[2]->[1]->NULL
list backward=[1]->[2]->[3]->[4]->[5]->[6]->NULL
enter a key to delete: 4
found 4 at 95670c8
list forward =[6]->[5]->[3]->[2]->[1]->NULL
list backward=[1]->[2]->[3]->[5]->[6]->NULL
enter a key to delete: █
```

图 2.24　示例程序 C2.5 的输出

图 2.25 显示了第 2 版的双向链表。

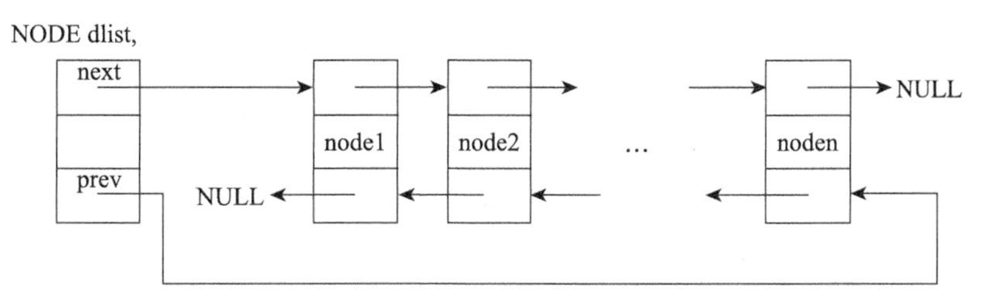

图 2.25　双向链表第 2 版

使用链表头结构体的唯一好处是，可通过链表的 prev 指针直接访问上一个元素。其缺点与 dlist 第 1 版完全相同。具体来说：

（1）由于 dlist 是一个结构体，所以必须通过引用（指针）来传递给所有的链表操作函数。

（2）在插入和删除函数时，我们必须检测和处理链表是否为空、元素是第一个元素还是最后一个元素等各种情况。在这些情况下，在处理时间方面，双向链表并不比单向链表有优势。

为了实现双向链表的功能，需要一种更好的方法来表示双向链表。在下一个示例中，我们将对双向链表进行如下定义。dlist 用链表头表示，是一个 NODE 结构体，但是 dlist 经过了初始化，两个指针均指向链表头，如：

```
NODE dlist;                         //  dlist is a NODE structure
dlist.next = dlist.prev = &dlist;   //  initialize both pointers to the NODE
                                        structure
```

该 dlist 可看作是由两个循环链表合并而成，链表头是初始虚拟元素。图 2.26 显示的是此类双向链表。

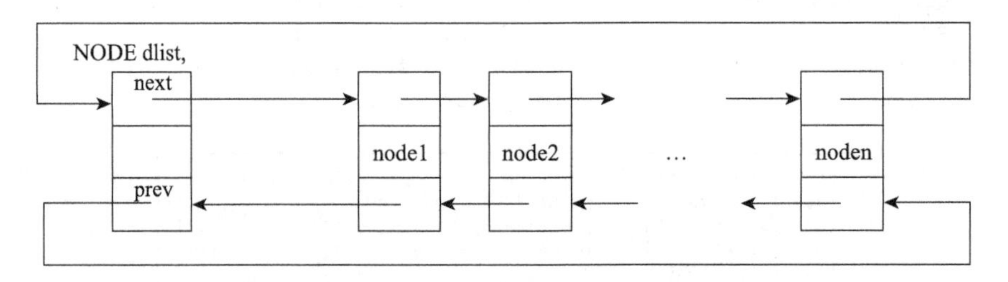

图 2.26　双向链表第 3 版

由于在第 3 版 dlist 中没有空指针，每个节点均可视为内部节点，这大大简化了链表操作代码。我们用一个例子来论述这类 dlist。

示例程序 C2.6：该示例程序假设 dlist 用一个 NODE 结构体表示，该结构体经过了初始化，两个指针均指向 NODE 结构体自身。

```
/************** C2.6.c Program Code *************/
#include <stdio.h>
#include <stdlib.h>
#include <string.h>
typedef struct node{
  struct node *next;
  struct node *prev;
  int key;
}NODE;

NODE dlist; // dlist is NODE struct using only next & prev pointers

int insert2end(NODE *list, int key)
{
  NODE *p, *q;
  //printf("insert2end: key=%d\n", key);
  p = (NODE *)malloc(sizeof(NODE));
  p->key = key;
  p->prev = 0;

  q = list->prev;      // to LAST element
  p->next = q->next;
  q->next->prev = p;
  q->next = p;
  p->prev = q;
}

int insert2front(NODE *list, int key)
{
  NODE *p, *q;
  //printf("insertFront key=%d\n", key);
  p = (NODE *)malloc(sizeof(NODE));
  p->key = key;
  p->prev = 0;

  q = list->next;     // to first element
```

```
  p->prev = q->prev;
  q->prev->next = p;
  q->prev = p;
  p->next = q;
}

void printForward(NODE *list)
{
  NODE *p = list->next;      // use dlist's next pointer
  printf("list forward =");
  while(p != list){          // detect end of list
    printf("[%d]->", p->key);
    p = p->next;
  }
  printf("NULL\n");
}

void printBackward(NODE *list)
{
  printf("list backward=");
  NODE *p = list->prev;      // use dlist's prev pointer
  while(p != list){          // detect end of list
    printf("[%d]->", p->key);
    p = p->prev;
  }
  printf("NULL\n");
}

NODE *search(NODE *list, int key)
{
  NODE *p = list->next;
  while(p != list){          // detect end of list
    if (p->key==key){
      printf("found %d at %x\n", key, (unsigned int)p);
      return p;
    }
    p = p->next;
  }
  return 0;
}

int delete(NODE *list, NODE *p)
{
  p->prev->next = p->next;
  p->next->prev = p->prev;
  free(p);
}

int main()
{
  int i, key;
  NODE *p;
  printf("dlist program #3\n");

  printf("insert to END\n");
```

```
    dlist.next = dlist.prev = &dlist; // empty dlist
    for (i=0; i<8; i++){
      insert2end(&dlist, i);
    }
    printForward(&dlist);
    printBackward(&dlist);

    printf("insert to front\n");
    dlist.next = dlist.prev = &dlist; // empty dlist to begin
    for (i=0; i<8; i++){
      insert2front(&dlist, i);
    }

    printForward(&dlist);
    printBackward(&dlist);

    printf("do deletion\n");
    while(1){
      printf("enter key to delete: ");
      scanf("%d", &key);
      if (key < 0) exit(0);     // exit if key negative
      p = search(&dlist, key);
      if (p==0){
         printf("key %d not found\n", key);
         continue;
      }
      delete(&dlist, p);
      printForward(&dlist);
      printBackward(&dlist);
    }
}
```

图 2.27 显示了运行示例程序 C2.6 的输出。虽然输出与图 2.24 相同，但它们的处理代码有很大的不同。

```
dlist program #3
insert to END
list forward =[0]->[1]->[2]->[3]->[4]->[5]->[6]->[7]->NULL
list backward=[7]->[6]->[5]->[4]->[3]->[2]->[1]->[0]->NULL
insert to front
list forward =[7]->[6]->[5]->[4]->[3]->[2]->[1]->[0]->NULL
list backward=[0]->[1]->[2]->[3]->[4]->[5]->[6]->[7]->NULL
do deletion
enter key to delete: 0
found 0 at 8a71088
list forward =[7]->[6]->[5]->[4]->[3]->[2]->[1]->NULL
list backward=[1]->[2]->[3]->[4]->[5]->[6]->[7]->NULL
enter key to delete: 7
found 7 at 8a710f8
list forward =[6]->[5]->[4]->[3]->[2]->[1]->NULL
list backward=[1]->[2]->[3]->[4]->[5]->[6]->NULL
enter key to delete: 4
found 4 at 8a710c8
list forward =[6]->[5]->[3]->[2]->[1]->NULL
list backward=[1]->[2]->[3]->[5]->[6]->NULL
enter key to delete:
```

图 2.27　示例程序 C2.6 的输出

示例程序 C2.6 主要有以下优点。由于每个节点均可视为内部节点，所以不再需要检测和处理空链表、第一元素、最后元素等各种情况。因此，插入和删除操作都大为简化。唯一需要修改的是在用代码段搜索或遍历链表期间要检测链表结束条件。

```
NODE *p = list.next;
while (p != &list){
  p = p->next;
}
```

2.11 树

树是由多级链表构成的动态数据结构。作为一个数据结构，树被定义为节点，由一个值和其他节点引用列表组成。象征性地，树的递归定义是

```
node: value [&node[1], ..., &node[k]]
```

其中每个节点 [i] 本身是一个树（可能为空）。树的第一个节点称为树的**根节点**。指向其他节点列表的每个节点称为**父节点**，其他节点称为父节点的**子节点**。在树中，每个节点都有唯一一个父节点，但每个节点的子节点数目可能不同，也可能为零。树可以用图来表示，图通常是上下颠倒的，根节点在上面。图 2.28 为一个普通树示例。

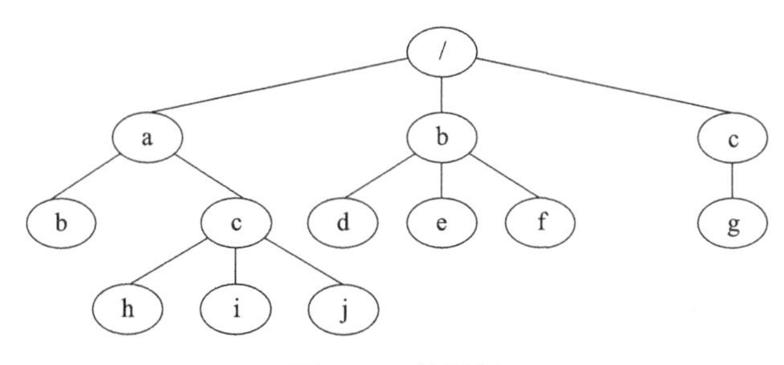

图 2.28 普通树

2.12 二叉树

最简单的树是**二叉树**，其中每个节点都有两个指针，分别是左指针和右指针。我们可以为二叉树定义包含单键值的节点类型，如：

```
typedef struct node{
        int   key;
        struct node *left;
        struct node *right;
}NODE;
```

每个普通树都可用二叉树的形式实现。我们将在 2.13 节中阐述。

2.12.1 二叉搜索树

二叉搜索树（BST）是具有以下性质的二叉树：

- 所有节点键值都不同，即该树不包含具有重复键值的节点。
- 节点的左子树只包含键值小于该节点键值的节点。
- 节点的右子树只包含键值大于该节点键值的节点。
- 左右子树也是二叉搜索树。

例如，图 2.29 显示了二叉搜索树。

二叉搜索树实现了节点键之间的排序，这样就可以快速完成查找最小键、最大键和搜索给定键值的操作。BST 的搜索深度取决于树的形状。如果是对称二叉树，即每个节点有两个子节点，那么对于一个有 n 个节点的二叉树，搜索深度为 log2（n）。对于非对称 BST，如果节点都在左边或者都在右边，搜索深度最差为 n。

图 2.29　二叉搜索树

2.12.2　构建二叉搜索树

示例程序 C2.7：构建二叉搜索树。

```
/************ C2.7.c file ************/
#include <stdio.h>
#include <stdlib.h>
typedef struct node{
    int key;
    struct node *left, *right;
}NODE;
#define N 7
int nodeValue[N] = {50,30,20,40,70,60,80};
// create a new node
NODE *new_node(int key)
{
    NODE *node = (NODE *)malloc(sizeof(NODE));
    node->key = key;
    node->left = node->right = NULL;
    return node;
}
// insert a new node with given key into BST
NODE *insert(NODE *node, int key)
{
    if (node == NULL)
       return new_node(key);
    if (key < node->key)
       node->left  = insert(node->left, key);
    else if (key > node->key)
       node->right = insert(node->right, key);
    return node;
}
int main()
{
  int i;
  NODE *root = NULL;
  root = insert(root, nodeValue[0]);
  for (i=1; i<N; i++){
      insert(root, nodeVlaue[i]);
  }
}
```

图 2.30 显示了示例程序 C2.7 生成的 BST。

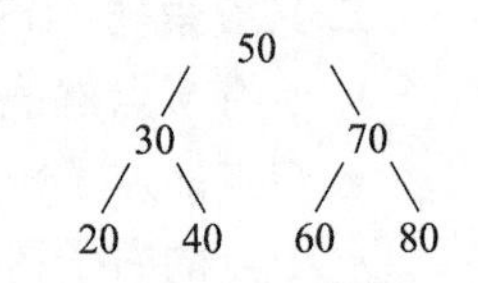

图 2.30　二叉搜索树

2.12.3 二叉树遍历算法

在数据结构课程中，读者应该已经学过了下面的二叉树遍历算法。

- **先序遍历**：node；node.left；node.right
- **中序遍历**：node.left; node; node.right
- **后序遍历**：node.left; node.right; node

2.12.4 深度优先遍历算法

以上算法均属于**深度优先**（DF）搜索/遍历算法，使用堆栈进行回溯。因此，它们可通过递归自然地实现。例如，可通过递归实现在BST中搜索给定键值，基本上是BST的中序遍历。

```
/******** Search BST for a give key ********/
NODE *search(NODE *t, int key)
{
    if (t == NULL || t->key == key)
       return t;
    if (key < t->key)                   // key is less than node key
       return search(t->left, key);
    else
       return search(t->right, key); // key is greater than node key
}
```

2.12.5 广度优先遍历算法

也可以采用**广度优先**（BF）遍历算法对二叉树进行遍历，用**队列**存储尚未遍历的二叉树部分。对二叉树应用广度优先遍历算法时，可逐层打印二叉树，如下示例程序C2.8所示。

示例程序C2.8：逐层打印二叉树。

```
/***** C2.8.c file: print binary tree by levels *****/
#include <stdio.h>
#include <stdlib.h>

typedef struct node{
    struct node *left;
    int key;
    struct node *right;
}NODE;

typedef struct qe{         // queue element structure
    struct qe    *next;    // queue pointer
    struct node  *node;    // queue contents
}QE;                       // queue element type

int enqueue(QE **queue, NODE *node)
{
  QE *q = *queue;
  QE *r = (QE *)malloc(sizeof(QE));
  r->node = node;
  if (q == 0)
     *queue = r;
```

```
  else{
    while (q->next)
       q = q->next;
    q->next = r;
  }
  r->next = 0;
}

NODE *dequeue(QE **queue)
{
  QE *q = *queue;
  if (q)
     *queue = q->next;
  return q->node;
}

int qlength(QE *queue)
{
  int n = 0;
  while (queue){
    n++;
    queue = queue->next;
  }
  return n;
}

// print a binary tree by levels, each level on a line
void printLevel(NODE *root)
{
  int nodeCount;
  if (root == NULL) return;
  QE queue = 0;                // create a FIFO queue
  enqueue(&queue, root);       // start with root
  while(1){
     nodeCount = qlength(queue);
     if (nodeCount == 0) break;
     // dequeue nodes of current level, enqueue nodes of next level
     while (nodeCount > 0){
        NODE *node = dequeue(&queue);
        printf("%d ", node->key);
        if (node->left != NULL)
           enqueue(&queue, node->left);
        if (node->right != NULL)
           enqueue(&queue, node->right);
        nodeCount--;
     }
     printf("\n");
  }
}
NODE *newNode(int key) // create a new node
{
  NODE *t = (NODE *)malloc(sizeof(NODE));
  t->key = key;
  t->left = NULL;
  t->right = NULL;
```

```
  return t;
}
int main()  // driver program to test printLevel()
{
  queue = 0;
  // create a simple binary tree
  NODE *root = newNode(1);
  root->left = newNode(2);
  root->right = newNode(3);
  root->left->left = newNode(4);
  root->left->right = newNode(5);
  root->right->leftt = newNode(6);
  root->right->right = 0; // right child = 0
  printLevel(root);
  return 0;
}
```

在示例程序 C2.8 中，每个队列元素都有一个 next 指针和一个 NODE 指针指向树中的节点。二叉树会经过遍历，但不会修改；只有它的节点（指针）将会在队列元素中使用，先进入队列再从队列中删除。printLevel() 函数从队列中的根节点开始。在每次迭代时，它会退出当前级别的节点，打印其键值，并在队列中输入左侧子节点（如有），然后输入右侧子节点（如有）。当队列变为空时，迭代结束。对于在 main() 函数中构造的简单二叉树，程序输出为：

```
1
2 3
4 5 6
```

将二叉树的每一层打印在单独的行中。

2.13　编程项目：Unix/Linux 文件系统树模拟器

在概述了 C 语言结构体、链表处理和二叉树的背景信息之后，我们准备将这些概念和技术整合到一个编程项目中来解决一个实际问题。该编程项目是设计和实现一个 Unix/Linux 文件系统树模拟器。

2.13.1　Unix/Linux 文件系统树

Unix 文件系统的逻辑结构体是一个普通树，如图 2.28 所示。Linux 文件系统的结构体相同，因此本节论述也适用于 Linux 文件系统。文件系统树通常是上下颠倒的，根节点在上面。为简便起见，假设文件系统只包含目录（DIR）和常规文件（FILE），即没有特殊文件，这些文件是 I/O 设备。DIR 节点拥有可变数量的**子节点**。同一父节点的子节点称为**兄弟节点**。在 Unix/Linux 文件系统中，每个节点都由表单 /a/b/c 或 a/b/c 的唯一**路径名**表示。如果路径名以 / 开头，则表示**绝对**路径，说明从根开始。否则，它是相对于**当前工作目录**（CWD）的相对路径。

2.13.2　用二叉树实现普通树

普通树可用二叉树的形式实现。对于每个节点，让 childPtr 指向最大的子节点，让 siblingPtr 指向最大的兄弟节点。方便起见，每个节点还有一个 parentPtr 指向其父节点。对于**根节点**，parentPtr 和 siblingPtr 都指向自己。例如，图 2.31 给出了与图 2.28 的普通树等同的二叉树。在图 2.31 中，细线表示 childPtr 指针，粗线表示 siblingPtr 指针。为清晰起见，未显示空指针。

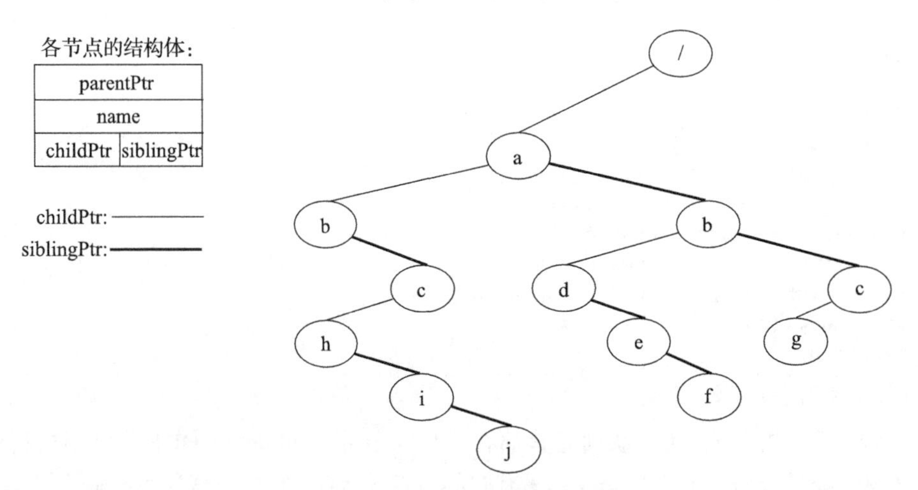

图 2.31 用二叉树实现普通树

2.13.3 项目规范及要求

本项目是设计并实现一个 C 语言程序来模拟 Unix/Linux 文件系统树。该程序的工作流程如下。

（1）从“/”节点开始，该节点也是当前工作目录（CWD）。

（2）提示用户输入命令。有效命令有：mkdir、rmdir、cd、ls、pwd、creat、rm、save、reload、menu、quit。

（3）执行命令，发出适当的跟踪消息。

（4）重复步骤（2）至“quit”命令，该命令会终止程序。

2.13.4 命令规范

- mkdir 路径名：为给定的路径名创建一个新目录。
- rmdir 路径名：如果目录为空，则删除该目录。
- cd [路径名]：将当前工作目录更改为路径名，如果没有路径名则更改为“/”。
- ls [路径名]：列出路径名或当前工作目录的目录内容。
- pwd：打印当前工作目录的（绝对）路径名。
- creat 路径名：创建一个 FILE 节点。
- rm 路径名：删除 FILE 节点。
- save 文件名：将当前文件系统树保存为文件。
- reload 文件名：从一个文件构造一个文件系统树。
- menu：显示有效命令菜单。
- quit：保存文件系统树，然后终止程序。

2.13.5 程序结构体

有许多设计和实现这类模拟器程序的方法。下文概述了该程序组织形式的建议。

（1）**节点类型**：定义一个 C 语言结构体节点类型，包含：

64 个字符：节点名称字符串。

字符：节点类型，目录节点类型为“D”，文件节点类型为“F”。

节点指针：*childPtr、*siblingPtr、*parentPtr。

（2）**全局变量**：

```
NODE *root, *cwd;                   // root and CWD pointers
char line[128];                     // user input command line
char command[16], pathname[64];     // command and pathname strings
char dname[64], bname[64];          // dirname and basename string holders
(Others as needed)
```

（3）**main() 函数**：该程序的 main 函数如下所示。

```
int main()
{
  initialize();    //initialize root node of the file system tree
  while(1){
    get user input line = [command pathname];
    identify the command;
    execute the command;
    break if command="quit";
  }
```

（4）**获取用户输入**：假设每个用户输入行都包含一个带有可选路径名的命令。读者可以使用 scanf() 从 stdin 读取用户输入。下面介绍一种更好的方法：

```
fgets(line, 128, stdin);  // get at most 128 chars from stdin
line[strlen(line)-1] = 0; // kill \n at end of line
sscanf(line, "%s %s", command, pathname);
```

sscanf() 函数按格式从 line[ ] 区提取项，可以是字符、字符串或整数。因此，它比 scanf() 更灵活。

（5）**识别命令**：由于每条命令都是一个字符串，大多数读者可能会在一系列 if-else-if 语句中使用 strcmp() 来识别该命令，如：

```
if (!strcmp(command, "mkdir)
    mkdir(pathname);
else if (!strcmp(command, "rmdir"
    rmdir(pathname);
else if . . .
```

这需要多行字符串比较。一种更好的方法是使用一个包含命令字符串（指针）的命令表，以一个 NULL 指针结束。

```
char *cmd[] = {"mkdir", "rmdir", "ls", "cd", "pwd", "creat", "rm",
               "reload", "save", "menu", "quit", NULL};
```

对于某个给定的命令，在命令表中搜索命令字符串并返回其索引，如下面的 findCmd() 函数所示。

```
int findCmd(char *command)
{
   int i = 0;
   while(cmd[i]){
     if (!strcmp(command, cmd[i]))
         return i;  // found command: return index i
     i++;
   }
   return -1;         // not found: return -1
}
```

例如，命令 = “creat”，

```
int index = findCmd("creat");
```

返回 index 5，可用于调用相应的 creat() 函数。

（6）**main() 函数**：假设我们为每个命令编写了一个相应的操作函数，如 mkdir()、rmdir()、ls()、cd() 等。main() 函数可简化如下。

```
int main()
{
  int index;
  char line[128], command[16], pathname[64];
  initialize();    //initialize root node of the file system tree
  while(1){
    printf("input a commad line : ");
    fgets(line,128,stdin);
    line[strlen(line)-1] = 0;
    sscanf(line, "%s %s", command, pathname);
    index = fidnCmd(command);
    switch(index){
        case 0 : mkdir(pathname);     break;
        case 1 : rmdir(pathname);     break;
        case 2 : ls(pathname);        break;
        etc.
      default: printf("invalid command %s\n", command);
    }
  }
}
```

该程序可使用命令索引作为切换表中的不同情况。如果命令的数量很小，这会很有效。如果命令数量很大，最好使用函数指针表。假设我们已经实现了命令函数：

```
int mkdir(char *pathname){..........}
int rmdir(char *pathname){..........}
......
```

定义一个**函数指针表**，其中函数名的顺序与其索引相同，如：

```
                                           0    1    2  3  4    5    6
int (*fptr[ ])(char *)={(int (*)())mkdir,rmdir,ls,cd,pwd,creat,rm,. .};
```

链接器将使用函数的入口地址填充表格。给定一个命令索引，我们可以直接调用对应的函数，将参数路径名传递给函数，如

```
int r = fptr[index](pathname);
```

2.13.6 命令算法

每个用户命令调用相应的作用函数来实现该命令。下面介绍作用函数的算法。

mkdir 路径名

（1）将路径名分解为 dirname 和 basename，例如：

- 绝对：路径名 =/a/b/c/d。则，dirname=/a/b/c，basename=d
- 相对：路径名 = a/b/c/d。则，dirname=a/b/c，basename=d

（2）搜索 dirname 节点：

- 绝对路径名：起始于“ / ”
- 相对路径名：起始于当前工作目录（CWD）
  - 如果不存在：错误消息和返回 FAIL
  - 如果存在，但不是 DIR：错误消息并返回 FAIL

（3）(dirname 存在而且是 DIR)：

- 在 dirname 节点中（下）搜索 basename：
- 如已存在：错误消息并返回 FAIL
- 在 dirname 下添加一个新的 DIR 节点
- 返回 SUCCESS

rmdir 路径名

（1）如果路径名是绝对值，起始于“/”

否则　起始于当前工作目录（CWD），指向当前工作目录节点

（2）搜索路径名节点：

- 将路径名标记为组件字符串
- 从头开始，搜索每个组件
- 若失败，则返回 ERROR

（3）路径名存在：

- 检查它的 DIR 类
- 检查 DIR 是否为空；如果不为空，则不能 rmdir

（4）从父节点的子节点列表中删除节点

creat 路径名

与 mkdir 相同，只是节点类型为“F”。

rm 路径名

与 rmdir 相同，只是检查它是否为文件，而不是是否为空。

cd 路径名

（1）找到路径节点

（2）检查它是否为 DIR

（3）将 CWD 更改为指向 DIR

ls 路径名

（1）找到路径节点

（2）按照 [TYPE NAME] [TYPE NAME]……的形式列出所有子节点

pwd

从 CWD 开始，通过递归实现 pwd：

（1）保存当前节点的名称（字符串）

（2）跟踪 parentPtr 到父节点，直到根节点

（3）在每个名称字符串中添加 / 来打印名称

将树保存到文件中　模拟器程序在内存中构建文件系统树。当程序退出时，所有内存内容将丢失。我们可以把当前的树保存为文件，以便以后用于恢复树，而不是每次都构建一个新树。要想把当前的树保存为文件，我们需要打开一个可供写入模式的文件。下面的代码段显示了如何打开一个文件流进行写入，然后向它写入（文本）行。

```
FILE *fp = fopen("myfile", "w+");   // fopen a FILE stream for WRITE
fprintf(fp, "%c %s", 'D', "string\n"); // print a line to file
fclose(fp);                          // close FILE stream when done
```

save(文件名)　该函数将树的绝对路径名保存为打开的可供写入的文件的（文本）行。假设文件系统树如图 2.32 所示。

```
类型   路径名
----  -------
 D      /
 D      /A
 F      /A/x
 F      /A/y
 D      /B
 F      /B/z
 D      /C
 D      /C/E
```

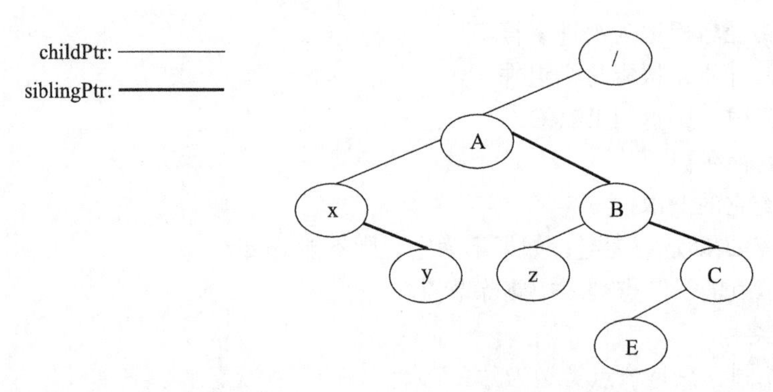

图 2.32　文件系统树，其中大写名称 A、B、C、E 是 DIR，小写名称 x、y、z 是 FILE。树可以用（文本）行表示

路径名由二叉树的**先序遍历**生成：

```
print node        name;    // current node
print node.left   name;    // left pointer = childPtr
print node.right  name;    // right pointer = siblingPtr
```

每个打印函数打印节点的绝对路径名，与 pwd() 基本相同。由于根节点始终存在，所以可从保存文件中省略。

reload(文件名) 该函数从文件中重新构建树。首先，将树初始化为空，即只有根节点。然后，读取文件的每一行。如果行包含“D 路径名”，则调用 mkdir(路径名) 来创建目录。如果行包含“F 路径名”，则调用 creat(路径名) 来创建文件。

它们将会重新构建先前保存的树。

退出命令 将当前树保存到一个文件中。然后终止程序执行。

在模拟器程序的后续运行中，用户可以使用 reload 命令来恢复先前保存的树。

额外编程帮助

（1）将路径名标记为组件：给定一个路径名，例如“/a/b/c/d”，下面的代码段展示了如何将路径名标记为组件字符串。

```
int tokenize(char *pathname)
{
  char *s;
  s = strtok(path, "/");  // first call to strtok()
  while(s){
     printf("%s ", s);
     s = strtok(0, "/");  // call strtok() until it returns NULL
  }
}
```

strtok() 函数通过指定的分隔符字符“/”将字符串划分为子字符串。子字符串留在原字符串中，从而破坏了原字符串。为保留原始路径名，用户必须将路径名的副本传给 strtok() 函数。为访问标记化子字符串，用户必须确保可通过以下一种方案访问它们。

- 复制的路径名是一个全局变量，例如 char path[128]，包含标记化的子字符串。
- 如果复制的路径名 path[] 是函数中的局部变量，则只访问函数中的子字符串。

（2）dir_nam 和 base_name: 对于模拟器程序，通常还需要将路径名分解为 dir_name 和 base_name，其中 dir_name 是路径名的目录部分，base_name 是路径名的最后一个组成部分。例如，如果 pathname=“/a/b/c”，那么 dir_name=“/a/b”而且 base_name=“c”。这可以通过库函数 dirname() 和 basename() 来完成，这两个函数也会破坏路径名。以下代码段展示了如何将路径名分解为 dir_name 和 base_name。

```
#include <libgen.h>
char dname[64], bname[64]; // for decomposed dir_name and base_name

int dbname(char *pathname)
{
  char temp[128];  // dirname(), basename() destroy original pathname
  strcpy(temp, pathname);
  strcpy(dname, dirname(temp));
  strcpy(temp, pathname);
  strcpy(bname, basename(temp));
}
```

2.13.7 示例解决方案

编程项目的示例解决方案可在本书的网站上下载。应作者要求，可以将项目解决方案的源代码提供给讲师。

2.14 习题

1. 参见 2.5.1 节中 GCC 生成的汇编代码。在函数 A(int x, int y) 中，显示如何访问参数 x 和 y。
2. 参见 2.5.2 节中的示例 2.2，编写汇编代码函数来获取 CPU 的 ebx、ecx、edx、esi 和 edi 寄存器。
3. 假设：Intel x86 CPU 寄存器 CX=N，BX 指向一个包含 N 个整数的存储区。以下代码段在汇编中实现一个简单的循环，它用整数数组的每个连续元素加载 AX。

```
loop:  movl  (%ebx), %eax    # AX = *BX (consider BX as int * in C)
       addl  $4, %ebx        # ++BX
       subl  $1, %ecx        # CX--;
       jne                   # jump to loop if CX NON-zero
```

在汇编中实现一个函数 int Asum(int *a, int N)，计算并返回数组 int a[N] 中 N 个整数的和。使用以下代码测试 Asum() 函数：

```
int a[100], N = 100;
int main()
{
    int i, sum;
    for (i=0; i<N; i++) // set a[] values to 1 to 100
        a[i] = i+1;
    sum = Asum(a, N);   // a[] is an int array, a is int *
    printf("sum = %d\n", sum);
}
```

和的值应为 5050。

4. (1) 每个 C 语言程序都必须有一个 main() 函数，为什么？
   (2) 以下程序由 C 语言中的 t.c 文件和 32 位汇编中的 ts.s 文件组成。程序的 main() 函数是用汇编语言编写的，调用 C 语言中的 mymain()。使用以下代码编译链接该程序：

```
gcc -m32 ts.s t.c
```

它以 a.out one two three 的形式运行。

```
# ******* ts.s file *******
        .global main, mymain    # int mymain() is in C
main:    pushl %ebp
         movl  %esp, %ebp
(2).1:   WRITE assembly CODE TO call mymain(argc, argv, env)
         call mymain
         leave
         ret

/*********** t.c file of a C program *************/
int mymain(int argc, char *argv[], char *env[ ])
{
    int i;
    printf("argc=%d\n", argc);
    i = 0;
    while(argv[i]){
      printf("argv[%d] = %s\n", i, argv[i]);
      i++;
    }
```

```
    (2).2: WRITE C code to print all env[ ] strings
    }
```

将标签（2）.1 和（2）.2 中缺失的代码补充完整，使该程序正常工作。

5. 在 2.7.3 节的示例 2.8 中，后缀规则

```
.s.o: # build each .o file if its .s file has changed
      ${AS} -a $< -o $*.o > $*.map
```

$(AS) -a 告诉汇编器生成一个汇编清单，该清单被重定向到一个映射文件。假设 PMTX 内核目录包含以下汇编代码文件：

```
entry.s, init.s, traps.s, ts.s
```

生成的 .o 和 .map 文件的文件名是什么？

6. 参见 2.10.6 节中的插入函数。重写 insert() 函数，以确保链表中没有重复键。
7. 参见 2.10.8 节中的 delete() 函数。将 delete() 函数重写为

```
NODE *delete(NODE **list, NODE *p)
```

从列表中删除 p 指向的给定节点。

8. 自行设计一个多级优先队列（MPQ）。
   （1）为 MPQ 设计一个数据结构。
   （2）设计并实现一种算法，将优先级在 1 到 n 之间的队列元素插入 MPQ。
9. 假设双向链表第 1 版用一个单 NODE 指针表示，如 2.10.12 节中的示例程序 C2.5 中所示。
   （1）实现 insertAfter(NODE **dlist, NODE *p, int key) 函数，将具有给定键值的新节点插入给定节点 p 之后的 dlist 中。如果 p 不存在，例如 p=0，则插入列表末尾。
   （2）实现 insertBefore(NODE **dlist, NODE *p, int key) 函数，将具有给定键值的新节点插入给定节点 p 之前。如果 p 不存在，例如 p=0，则插入列表开头。
   （3）编写一个完整的 C 语言程序，使用 insertAfter() 和 insertBefore() 函数构建一个 dlist（双向链表）。
10. 假设双向链表第 2 版用一个列表头结构体表示。
   （1）实现 insertAfter(NODE *dlist, NODE *p, int key) 函数，将具有给定键值的新节点插入给定节点 p 之后的 dlist 中。如果 p 不存在，则插入列表末尾。
   （2）实现 insertBefore(NODE *dlist, NODE *p, int key) 函数，将具有给定键值的新节点插入给定节点 p 之前。如果 p 不存在，则插入列表开头。
   （3）编写一个完整的 C 语言程序，使用 insertAfter() 和 insertBefore() 函数构建一个 dlist（双向链表）。
11. 假设双向链表用示例程序 C2.6 中的一个初始化表头表示。
   （1）实现 insertBefore（NODE *dlist, NODE *p, NODE *new）函数，将新节点插入 dlist 中的当前节点 p 前面。
   （2）实现 insertAfter(NODE *dlist, NODE *p, NODE *new) 函数，将新节点插入 dlist 中的当前节点 p 后面。
   （3）编写一个完整的 C 语言程序，使用 insertAfter() 和 insertBefore() 函数构建一个 dlist。
12. 参见 2.9.2 节中的二叉搜索树（BST）程序 C2.7。
   （1）证明 insert() 函数不会两次插入具有相同键值的节点。
   （2）重写 main() 函数，该函数用 rand()%100 生成的键值构建 BST，因此键值的范围为 [0-99]
13. 2.10.1 节中的“逐层打印树”程序有一个缺点，即它不显示任何子节点是否为空。如果任何子节点为 NULL（空），则要修改程序以打印“-”。

14. 对于编程项目，需要实现重命名命令，更改文件树中的节点名称。

```
rename(char *pathname, char *newname)
```

15. 对于编程项目，需要设计并实现一个 mv 命令，将 pathname（路径名）1 移动到 pathname2。

```
mv(char *pathname1, char *pathname2);
```

16. 对于编程项目，需要讨论是否可以使用中序遍历或后序遍历来保存和重构二叉树。

参考文献

Debugging with GDB, https://ftp.gnu.org/Manuals/gdb/html_node/gdb_toc.html, 2002
Linux vi and vim editor, http://www.yolinux.com/TUTORIALS/LinuxTutorialAdvanced_vi.html, 2017
GDB: The GND Project Debugger, www.gnu.org/software/gdb/, 2017
GNU Emacs, https://www.gnu.org/s/emacs, 2015
GNU make, https://www.gnu.org/software/make/manual/make.html, 2008
Wang, K. C., Design and Implementation of the MTX Operating system, Springer A.G, 2015
Youngdale, E. The ELF Object File Format: Introduction, Linux Journal, April, 1995

Unix/Linux进程管理

摘要

本章讨论了Unix/Linux中的进程管理；阐述了多任务处理原则；介绍了进程概念；并以一个编程示例来说明多任务处理、上下文切换和进程处理的各种原则和方法。多任务处理系统支持动态进程创建、进程终止，以及通过休眠与唤醒实现进程同步、进程关系，以及以二叉树的形式实现进程家族树，从而允许父进程等待子进程终止；提供了一个具体示例来阐释进程管理函数在操作系统内核中是如何工作的；然后，解释了Unix/Linux中各进程的来源，包括系统启动期间的初始进程、INIT进程、守护进程、登录进程以及可供用户执行命令的sh进程；接着，对进程的执行模式进行了讲解，以及如何通过中断、异常和系统调用从用户模式转换到内核模式；再接着，描述了用于进程管理的Unix/Linux系统调用，包括fork、wait、exec和exit；阐明了父进程与子进程之间的关系，包括进程终止和父进程等待操作之间关系的详细描述；解释了如何通过INIT进程处理孤儿进程，包括当前Linux中的subreaper进程，并通过示例演示了subreaper进程；接着，详细介绍了如何通过exec更改进程执行映像，包括execve系统调用、命令行参数和环境变量；解释了I/O重定向和管道的原则及方法，并通过示例展示了管道编程的方法；读者可借助本章的编程项目整合进程管理的各种概念和方法，实现用于执行命令的sh模拟器。sh模拟器的功能与标准sh完全相同。它支持简单命令、具有I/O重定向的命令和通过管道连接的多个命令的执行。

3.1 多任务处理

一般来说，多任务处理指的是同时进行几项独立活动的能力。比如，我们经常看到有人一边开车一边打电话。从某种意义上说，这些人正在进行多任务处理，尽管这样非常不好。在计算机技术中，多任务处理指的是同时执行几个独立的任务。在单处理器（单CPU）系统中，一次只能执行一个任务。多任务处理是通过在不同任务之间多路复用CPU的执行时间来实现的，即将CPU执行操作从一个任务切换到另一个任务。不同任务之间的执行切换机制称为上下文切换，将一个任务的执行环境更改为另一个任务的执行环境。如果切换速度足够快，就会给人一种同时执行所有任务的错觉。这种逻辑并行性称为“并发”。在有多个CPU或处理器内核的多处理器系统中，可在不同CPU上实时、并行执行多项任务。此外，每个处理器也可以通过同时执行不同的任务来实现多任务处理。多任务处理是所有操作系统的基础。总体上说，它也是并行编程的基础。

3.2 进程的概念

操作系统是一个多任务处理系统。在操作系统中，任务也称为进程。在实际应用中，任务和进程这两个术语可以互换使用。在第2章中，我们把执行映像定义为包含执行代码、数据和堆栈的存储区。进程的正式定义：

进程是对映像的执行。

操作系统内核将一系列执行视为使用系统资源的单一实体。系统资源包括内存空间、I/O设备以及最重要的CPU时间。在操作系统内核中，每个进程用一个独特的数据结构表示，叫作进程控制块（PCB）或任务控制块（TCB）等。在本书中，我们直接称它为PROC结构体。与包含某个人所有信息的个人记录一样，PROC结构体包含某个进程的所有信息。在实际操作系统中，PROC结构体可能包含许多字段，而且数量可能很庞大。首先，我们来定义一个非常简单的PROC结构体来表示进程。

```
typedef struct proc{
    struct proc *next;      // next proc pointer
    int  *ksp;              // saved sp: at byte offset 4
    int   pid;              // process ID
    int   ppid;             // parent process pid
    int   status;           // PROC status=FREE|READY, etc.
    int   priority;         // scheduling priority
    int   kstack[1024];     // process execution stack
 }PROC;
```

在PROC结构体中，next是指向下一个PROC结构体的指针，用于在各种动态数据结构（如链表和队列）中维护PROC结构体。ksp字段是保存的堆栈指针。当某进程放弃使用CPU时，它会将执行上下文保存在堆栈中，并将堆栈指针保存在PROC.ksp中，以便以后恢复。在PROC结构体的其他字段中，pid是标识一个进程的进程ID编号，ppid是父进程ID编号，status是进程的当前状态，priority是进程调度优先级，kstack是进程执行时的堆栈。操作系统内核通常会在其数据区中定义有限数量的PROC结构体，表示为：

```
PROC proc[NPROC];          // NPROC a constant, e.g. 64
```

用来表示系统中的进程。在一个单CPU系统中，一次只能执行一个进程。操作系统内核通常会使用正在运行的或当前的全局变量PROC指针，指向当前正在执行的PROC。在有多个CPU的多处理器操作系统中，可在不同CPU上实时、并行执行多个进程。因此，在一个多处理器系统中正在运行的[NCPU]可能是一个指针数组，每个指针指向一个正在特定CPU上运行的进程。为简便起见，我们只考虑单CPU系统。

3.3　多任务处理系统

为了让读者更好地理解多任务处理和进程，我们先来学习一个旨在说明多任务处理、上下文切换和进程处理原则的编程示例。该程序实现了一个模拟操作系统内核模式各项操作的多任务环境。多任务处理系统，简称MT，由以下几个部分组成。

3.3.1　type.h文件

type.h文件定义了系统常数和表示进程的简单PROC结构体。

```
/*********** type.h file ************/
#define NPROC    9          // number of PROCs
#define SSIZE 1024          // stack size = 4KB

// PROC status
#define FREE     0
#define READY    1
```

```
#define SLEEP    2
#define ZOMBIE   3

typedef struct proc{
    struct proc *next;       // next proc pointer
    int  *ksp;               // saved stack pointer
    int   pid;               // pid = 0 to NPROC-1
    int   ppid;              // parent pid
    int   status;            // PROC status
    int   priority;          // scheduling priority
    int   kstack[SSIZE];     // process stack
}PROC;
```

后面，我们在扩展 MT 系统时，应向 PROC 结构体中添加更多的字段。

3.3.2 ts.s 文件

ts.s 在 32 位 GCC 汇编代码中可实现进程上下文切换，后面小节中将会讲到。

```
#------------- ts.s file file -------------------
       .globl  running, scheduler, tswitch
tswitch:
SAVE:   pushl %eax
        pushl %ebx
        pushl %ecx
        pushl %edx
        pushl %ebp
        pushl %esi
        pushl %edi
        pushfl
        movl   running,%ebx    # ebx -> PROC
        movl   %esp,4(%ebx)    # PORC.save_sp = esp
FIND:   call   scheduler
RESUME: movl   running,%ebx    # ebx -> PROC
        movl   4(%ebx),%esp    # esp = PROC.saved_sp
        popfl
        popl  %edi
        popl  %esi
        popl  %ebp
        popl  %edx
        popl  %ecx
        popl  %ebx
        popl  %eax
        ret
# stack contents = |retPC|eax|ebx|ecx|edx|ebp|esi|edi|eflag|
#                    -1    -2  -3  -4  -5  -6  -7  -8   -9
```

3.3.3 queue.c 文件

queue.c 文件可实现队列和链表操作函数。enqueue() 函数按优先级将 PROC 输入队列中。在优先级队列中，具有相同优先级的进程按先进先出（FIFO）的顺序排序。dequeue() 函数可返回从队列或链表中删除的第一个元素。printList() 函数可打印链表元素。

```
/***************** queue.c file *****************/
int enqueue(PROC **queue, PROC *p)
{
  PROC *q = *queue;
  if (q == 0 || p->priority > q->priority){
     *queue = p;
     p->next = q;
  }
  else{
     while (q->next && p->priority <= q->next->priority)
            q = q->next;
     p->next = q->next;
     q->next = p;
  }
}
PROC *dequeue(PROC **queue)
{
     PROC *p = *queue;
     if (p)
        *queue = (*queue)->next;
     return p;
}
int printList(char *name, PROC *p)
{
  printf("%s = ", name);
  while(p){
     printf("[%d %d]->", p->pid, p->priority);
     p = p->next;
  }
  printf("NULL\n");
}
```

3.3.4 t.c 文件

t.c 文件定义 MT 系统数据结构、系统初始化代码和进程管理函数。

```
/*********** t.c file of A Multitasking System *********/
#include <stdio.h>
#include "type.h"

PROC proc[NPROC];        // NPROC PROCs
PROC *freeList;          // freeList of PROCs
PROC *readyQueue;        // priority queue of READY procs
PROC *running;           // current running proc pointer

#include "queue.c"       // include queue.c file

/*******************************************************
  kfork() creates a child process; returns child pid.
  When scheduled to run, child PROC resumes to body();
*******************************************************/
int kfork()
{
  int  i;
  PROC *p = dequeue(&freeList);
```

```
  if (!p){
     printf("no more proc\n");
     return(-1);
  }

  /* initialize the new proc and its stack */
  p->status = READY;
  p->priority = 1;        // ALL PROCs priority=1,except P0
  p->ppid = running->pid;
  /************ new task initial stack contents ************
   kstack contains: |retPC|eax|ebx|ecx|edx|ebp|esi|edi|eflag|
                      -1   -2  -3  -4  -5  -6  -7  -8   -9
  **********************************************************/
  for (i=1; i<10; i++)               // zero out kstack cells
      p->kstack[SSIZE - i] = 0;
  p->kstack[SSIZE-1] = (int)body;    // retPC -> body()
  p->ksp = &(p->kstack[SSIZE - 9]);  // PROC.ksp -> saved eflag
  enqueue(&readyQueue, p);           // enter p into readyQueue
  return p->pid;
}

int kexit()
{
   running->status = FREE;
   running->priority = 0;
   enqueue(&freeList, running);
   printList("freeList", freeList);
   tswitch();
}

int do_kfork()
{
   int child = kfork();
   if (child < 0)
      printf("kfork failed\n");
   else{
      printf("proc %d kforked a child = %d\n", running->pid, child);
      printList("readyQueue", readyQueue);
   }
   return child;
}
int do_switch()
{
   tswitch();
}
int do_exit
{
  kexit();
}

int body()   // process body function
{
  int c;
  printf("proc %d starts from body()\n", running->pid);
  while(1){
```

```
    printf("***************************************\n");
    printf("proc %d running: parent=%d\n", running->pid,running->ppid);
    printf("enter a key [f|s|q] : ");
    c = getchar(); getchar();   // kill the \r key
    switch(c){
      case 'f': do_kfork();     break;
      case 's': do_switch();    break;
      case 'q': do_exit();      break;
    }
  }
}
// initialize the MT system; create P0 as initial running process
int init()
{
  int i;
  PROC *p;
  for (i=0; i<NPROC; i++){ // initialize PROCs
    p = &proc[i];
    p->pid = i;            // PID = 0 to NPROC-1
    p->status = FREE;
    p->priority = 0;
    p->next = p+1;
  }
  proc[NPROC-1].next = 0;
  freeList = &proc[0];     // all PROCs in freeList
  readyQueue = 0;          // readyQueue = empty

  // create P0 as the initial running process
  p = running = dequeue(&freeList); // use proc[0]
  p->status = READY;
  p->ppid = 0;             // P0 is its own parent
  printList("freeList", freeList);
  printf("init complete: P0 running\n");
}

/*************** main() function ***************/
int main()
{
   printf("Welcome to the MT Multitasking System\n");
   init();    // initialize system; create and run P0
   kfork();   // kfork P1 into readyQueue
   while(1){
     printf("P0: switch process\n");
     if (readyQueue)
         tswitch();
   }
}

/*********** scheduler *************/
int scheduler()
{
  printf("proc %d in scheduler()\n", running->pid);
  if (running->status == READY)
     enqueue(&readyQueue, running);
  printList("readyQueue", readyQueue);
```

```
    running = dequeue(&readyQueue);
    printf("next running = %d\n", running->pid);
}
```

3.3.5 多任务处理系统代码介绍

我们通过以下步骤来介绍 MT 系统的基本代码。

（1）虚拟 CPU：MT 系统在 Linux 下编译链接为

```
gcc -m32 t.c ts.s
```

然后运行 a.out。整个 MT 系统在用户模式下作为 Linux 进程运行。在 Linux 进程中，我们创建了多个独立执行实体（叫作任务），并通过我们自己的调度算法将它们调度到 Linux 进程中运行。对于 MT 系统中的任务，Linux 进程就像一个虚拟 CPU。为避免混淆，我们将 MT 系统中的执行实体叫作任务或进程。

（2）init()：当 MT 系统启动时，main() 函数调用 init() 以初始化系统。init() 初始化 PROC 结构体，并将它们输入 freeList 中。它还将 readyQueue 初始化为空。然后使用 proc[0] 创建 P0，作为初始运行进程。P0 的优先级最低，为 0。所有其他任务的优先级都是 1，因此它们将轮流从 readyQueue 运行。

（3）P0 调用 kfork() 来创建优先级为 1 的子进程 P1，并将其输入就绪队列中。然后 P0 调用 tswitch()，将会切换任务以运行 P1。

（4）tswitch()：tswitch() 函数实现进程上下文切换。它就像一个进程交换箱，一个进程进入时通常另一个进程会出现。tswitch() 由 3 个独立的步骤组成，下面将详细解释这些步骤。

（4）.1 tswitch() 中的 SAVE 函数：当正在执行的某个任务调用 tswitch() 时，它会把返回地址保存在堆栈上，并在汇编代码中进入 tswitch()。在 tswitch() 中，SAVE 函数将 CPU 寄存器保存到调用任务的堆栈中，并将堆栈指针保存到 proc.ksp 中。32 位 Intel x86 CPU 有许多寄存器，但在用户模式下，只有 eax、ebx、ecx、edx、ebp、esi、edi 和 eflag 对 Linux 进程可见，它是 MT 系统的虚拟 CPU。因此，我们只需要保存和恢复虚拟 CPU 的这些寄存器。下图显示了在执行 tswitch() 的 SAVE 函数后，调用任务的堆栈内容和保存的堆栈指针，其中 xxx 表示调用 tswitch() 之前的堆栈内容。

```
                                            proc.ksp
                                                |
|xxx|retPC|eax|ebx|ecx|edx|ebp|esi|edi|eflag|
```

在基于 Intel x86 的 32 位 PC 中，每个 CPU 寄存器的宽度为 4 字节，堆栈操作始终以 4 字节为单位。因此，我们可以将 PROC 结构体中的每个 PROC 堆栈定义为一个整数数组。

（4）.2 scheduler()：在执行了 tswitch() 中的 SAVE 函数之后，任务调用 scheduler() 来选择下一个正在运行的任务。在 scheduler() 中，如果调用任务仍然可以运行，则会调用 enqueue() 将自己按优先级放入 readyQueue 中。否则，它不会在 readyQueue 中，因此也就无法运行。然后，它会调用 dequeue()，将从 readyQueue 中删除的第一个 PROC 作为新的运行任务返回。

（4）.3 tswitch() 中的 RESUME 函数：当执行从 scheduler() 返回时，“运行”可能已经转而指向另一个任务的 PROC。运行指向的那个 PROC，就是当前正在运行的任务。tswitch() 中的 RESUME 函数将 CPU 的堆栈指针设置为当前运行任务的已保存堆栈指针。然后弹出保存的寄存器，接着是弹出 RET，使当前运行的任务返回到之前调用 tswitch() 的位置。

（5）kfork()：kfork() 函数创建一个子任务并将其输入 readyQueue 中。每个新创建的任务都从同一个 body() 函数开始执行。虽然新任务以前从未存在过，但我们可以假装它不仅存在过，而且运行过。它现在不运行是因为它调用了 tswitch()，所以提前放弃了使用 CPU。如果是这样的话，它的堆栈必须包含 tswitch() 中的 SAVE 函数保存的一个帧，而且它保存的 ksp 必须指向栈顶。由于新任务之前从未真正运行过，所以我们可以假设它的堆栈为空，而且当它调用 tswitch() 时，所有 CPU 寄存器内容都是 0。因此，在 kfork() 中，我们按以下方法初始化新任务的堆栈。

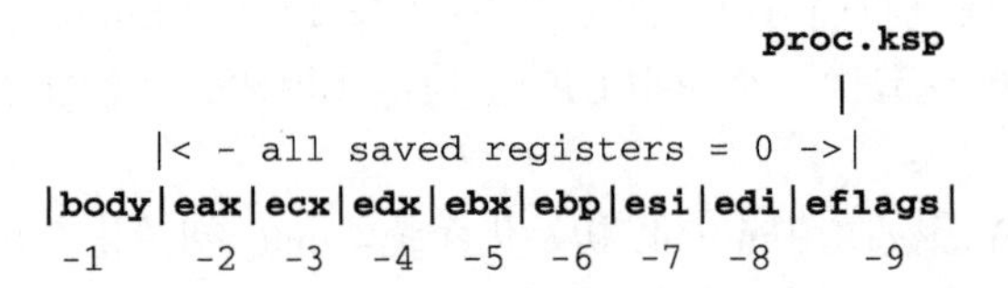

这里的索引 -i 意味着 SSIZE-i. 这是通过 kfork() 中的以下代码段来实现的。

```
/************** task initial stack contents **************
 kstack contains: |retPC|eax|ebx|ecx|edx|ebp|esi|edi|eflag|
                    -1    -2  -3  -4  -5  -6  -7  -8   -9
**********************************************************/
for (i=1; i<10; i++)              // zero out kstack cells
    p->kstack[SSIZE-i] = 0;
p->kstack[SSIZE-1] = (int)body;   // retPC -> body()
p->ksp = &(p->kstack[SSIZE-9]);   // PROC.ksp -> saved eflag
```

当新任务开始运行时，它首先执行 tswitch() 中的 RESUME 函数，使它返回 body() 函数的输入地址。逻辑上，当执行首次进入 body() 函数时，任务堆栈为空。一旦执行开始，任务的堆栈就会像第 2 章的堆栈使用和栈帧部分描述的那样增长和收缩。实际上，body() 函数从不返回，因此在堆栈中不需要有一个返回地址。对于初始堆栈内容的变化，读者可参阅习题 3.2。

（6）body()：为便于演示，所有创建的任务都执行同一个 body() 函数。这说明了进程和程序之间的区别。多个进程可执行同一个程序代码，但是每个进程都只能在自己的上下文中执行。例如，body() 中的所有（自动）局部变量都供进程专用，因为它们都是在每个进程堆栈中分配的。如果 body() 函数调用其他函数，则所有调用序列都会保存在每个进程堆栈中，等等。在 body() 中执行时，进程提示输入 char = [f|s|q] 命令，其中：

- f：kfork 一个新的子进程来执行 body()
- s：切换进程
- q：终止进程，并将进程以 freeList 中的 FREE 函数的形式返回

（7）空闲任务 P0：P0 的特殊之处在于它在所有任务中具有最低的优先级。在系统初始化之后，P0 创建 P1 并切换到运行 P1。当且仅当没有可运行任务时，P0 将会再次运行。在这种情况下，P0 会一直循环。当 readyQueue 变为非空时，它将切换到另一个任务。在基本 MT 系统中，如果所有其他进程都已终止，则 P0 将会再次运行。要想结束 MT 系统，用户可按下“Ctrl+C”组合键来终止 Linux 进程。

（8）运行多任务处理（MT）系统：在 Linux 下，输入：

```
gcc -m32 t.c s.s
```

编译链接 MT 系统并运行所得到的 a.out。图 3.1 为运行 MT 系统的样本输出示意图。图中显示了 MT 系统初始化，即初始进程 P0，它创建 P1 并将任务切换到运行 P1。P1 kfork 子进程 P2，并将任务切换到运行 P2。在进程运行时，读者可输入其他命令来测试系统。

```
Welcome to the MT Multitasking System
freeList = [1 0] -> [2 0] -> [3 0] -> [4 0] -> [5 0] -> [6 0] -> [7 0] -> [8 0] -> NULL
init complete: P0 running
P0: switch task
proc 0 in scheduler()
readyQueue = [1 1] -> [0 0] -> NULL
next running = 1
proc 1 starts from body()
***************************************
proc 1 running: Parent = 0
child = NULL
input a char [f|s|q] : f
proc 1 kforked a child = 2
readyQueue = [2 1] -> [0 0] -> NULL
***************************************
proc 1 running: Parent = 0
child = NULL
input a char [f|s|q] : s
proc 1 switching task
proc 1 in scheduler()
readyQueue = [2 1] -> [1 1] -> [0 0] -> NULL
next running = 2
proc 2 starts from body()
***************************************
proc 2 running: Parent = 1
child = NULL
input a char [f|s|q] : █
```

图 3.1 MT 系统的输出示例

3.4 进程同步

一个操作系统包含许多并发进程，这些进程可以彼此交互。进程同步是指控制和协调进程交互以确保其正确执行所需的各项规则和机制。最简单的进程同步工具是休眠和唤醒操作。

3.4.1 睡眠模式

当某进程需要某些当前没有的东西时，例如申请独占一个存储区域、等待用户通过标准输入来输入字符等，它就会在某个事件值上进入休眠状态，该事件值表示休眠的原因。为实现休眠操作，我们可在 PROC 结构体中添加一个 event 字段，并实现 ksleep(int event) 函数，使进程进入休眠状态。接下来，我们将假设对 PROC 结构体进行修改以包含加粗显示的添加字段。

```
typedef struct proc{
    struct proc *next;      // next proc pointer
    int   *ksp;             // saved sp: at byte offset 4
    int    pid;             // process ID
    int    ppid;            // parent process pid
    int    status;          // PROC status=FREE|READY, etc.
    int    priority;        // scheduling priority
    int    event;           // event value to sleep on
    int    exitCode;        // exit value
```

```
    struct proc *child;      // first child PROC pointer
    struct proc *sibling;    // sibling PROC pointer
    struct proc *parent;     // parent PROC pointer
    int   kstack[1024];      // process stack
}PROC;
```

ksleep() 的算法为：

```
/************ Algorithm of ksleep(int event) **************/

1. record event value in PROC.event: running->event = event;
2. change status to SLEEP:           running->status = SLEEP;
3. for ease of maintenance, enter caller into a PROC *sleepList
                                     enqueue(&sleepList, running);
4. give up CPU:                      tswitch();
```

由于休眠进程不在 readyQueue 中，所以它在被另一个进程唤醒之前不可运行。因此，在让自己进入休眠状态之后，进程调用 tswitch() 来放弃使用 CPU。

3.4.2 唤醒操作

多个进程可能会进入休眠状态等待同一个事件，这是很自然的，因为这些进程可能都需要同一个资源，例如一台当前正处于繁忙状态的打印机。在这种情况下，所有这些进程都将休眠等待同一个事件值。当某个等待时间发生时，另一个执行实体（可能是某个进程或中断处理程序）将会调用 kwakeup(event)，唤醒正处于休眠状态等待该事件值的所有程序。如果没有任何程序休眠等待该程序，kwakeup() 就不工作，即不执行任何操作。kwakeup() 的算法是：

```
/*********** Algorithm of kwakeup(int event) **********/
// Assume SLEEPing procs are in a global sleepList
   for each PROC *p in sleepList do{
         if (p->event == event){        // if p is sleeping for the event
            delete p from sleepList;
            p->status = READY;          // make p READY to run again
            enqueue(&readyQueue, p);    // enter p into readyQueue
         }
   }
```

注意，被唤醒的进程可能不会立即运行。它只是被放入 readyQueue 中，排队等待运行。当被唤醒的进程运行时，如果它在休眠之前正在试图获取资源，那么它必须尝试重新获取资源。这是因为该资源在它运行时可能不再可用。ksleep() 和 kwakeup() 函数一般用于进程同步，但在特殊情况下也用于同步父进程和子进程，这是我们接下来要论述的主题。

3.5 进程终止

在操作系统中，进程可能终止或死亡，这是进程终止的通俗说法。如第 2 章所述，进程能以两种方式终止：

- 正常终止：进程调用 exit(value)，发出 _exit(value) 系统调用来执行在操作系统内核中的 kexit(value)，这就是我们本节要讨论的情况。
- 异常终止：进程因某个信号而异常终止。信号和信号处理将在后面第 6 章讨论。

在这两种情况下，当进程终止时，最终都会在操作系统内核中调用 kexit()。kexit() 的一般算法见下文。

3.5.1 kexit() 的算法

```
/**************** Algorithm of kexit(int exitValue) *****************/
1. Erase process user-mode context, e.g. close file descriptors,
   release resources, deallocate user-mode image memory, etc.
2. Dispose of children processes, if any
3. Record exitValue in PROC.exitCode for parent to get
4. Become a ZOMBIE (but do not free the PROC)
5. Wakeup parent and, if needed, also the INIT process P1
```

MT 系统中的所有进程都以操作系统（OS）模拟内核模式运行。因此，它们没有任何用户模式上下文。因此，我们首先来讨论 kexit() 的步骤 2。在某些操作系统中，某个进程的执行环境可能依赖于其父进程的执行环境。例如，子进程的存储区可能在父进程的存储区内，因此除非父进程的所有子进程都已死亡，否则父进程不会死亡。在 Unix/Linux 中，进程只有非常松散的父子关系，它们的执行环境都是独立的。因此，在 Unix/Linux 中，进程可能随时死亡。如果有子进程的某个父进程首先死亡，那么所有子进程将不再有父进程，即成为孤儿进程。那么问题来了：如何处理这些孤儿进程呢？在人类社会，它们会被送到奶奶家里。但是，如果奶奶已经死了呢？根据这个推理，很明显，如果没有其他进程存在，那么肯定有一个进程不会死亡。否则，父进程与子进程的关系很快就会瓦解。在所有类 Unix 系统中，进程 P1（又叫作 INIT 进程）被选来扮演这个角色。当某个进程死亡时，它将其所有的孤儿子进程，不论死亡还是活跃，都送到 P1 中，即成为 P1 的子进程。同样，我们也要将 MT 系统中的 P1 指定为这类进程。因此，如果还有其他进程存在，P1 就不应该消失。剩下的问题是如何有效地实现 kexit() 的步骤 2。为了让一个濒死进程处理孤儿子进程，该进程必须能够确定它是否有子进程，如果有，则必须快速找到所有的子进程。如果进程的数量很少，例如像 MT 系统一样只有几个进程，通过搜索所有 PROC 结构体可以有效地回答这两个问题。例如，要确定某个进程是否有任何子进程，只需在 PROC 中搜索任何非空闲进程，并且搜索到的进程的 ppid 与前面进程的 pid 匹配即可。如果进程的数量很大，例如有数百甚至数千个进程，这种简单的搜索方案就会慢得让人难以忍受。因此，大多数大型操作系统内核通过维护进程家族树来跟踪进程关系。

3.5.2 进程家族树

通常，进程家族树通过个 PROC 结构中的一对子进程和兄弟进程指针以二叉树的形式实现，如：

```
PROC *child, *sibling, *parent;
```

其中，child 指向进程的第一个子进程，sibling 指向同一个父进程的其他子进程。为方便起见，每个 PROC 还使用一个 parent 指针指向其父进程。例如，图 3.2 左侧所示的进程树可以实现为右侧所示的二叉树，其中每个垂直链接都是 child 指针，每个水平链接都是 sibling 指针。清晰起见，图中没有显示 parent 指针和空指针。

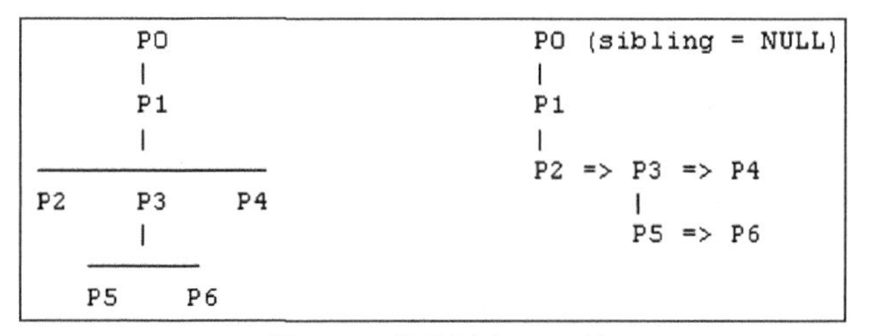

图 3.2 进程树和二叉树

使用进程树，更容易找到进程的子进程。首先，跟随child指针到第一个子进程。然后，跟随sibling指针遍历兄弟进程。要想把所有子进程都送到P1中，只需要把子进程链表分出来，然后把它附加到P1的子进程链表中（还要修改它们的ppid和parent指针）。

每个PROC都有一个退出代码（exitCode）字段，是进程终止时的进程退出值（exitValue）。在PROC.exitCode中记录exitValue之后，进程状态更改为ZOMBIE，但不释放PROC结构体。然后，进程调用kwakeup(event)来唤醒其父进程，其中事件必须是父进程和子进程使用的相同唯一值，例如父进程的PROC结构体地址或父进程的pid。如果它将任何孤儿进程送到P1中，也会唤醒P1。濒死进程的最后操作是进程最后一次调用tswitch()。在这之后，进程基本上死亡了，但还有一个空壳，以僵尸进程的形式存在，它通过等待操作被父进程埋葬（释放）。

3.5.3　等待子进程终止

在任何时候，进程都可以调用内核函数

```
pid = kwait(int *status)
```

等待僵尸子进程。如果成功，则返回的pid是僵尸子进程的pid，而status包含僵尸子进程的退出代码。此外，kwait()还会将僵尸子进程释放回freeList以便重用。kwait的算法是：

```
/******* Algorithm of kwait() *******/
  int kwait(int *status)
  {
     if (caller has no child) return -1 for error;
     while(1){    // caller has children
        search for a (any) ZOMBIE child;
        if (found a ZOMBIE child){
           get ZOMBIE child pid
           copy ZOMBIE child exitCode to *status;
           bury the ZOMBIE child (put its PROC back to freeList)
           return ZOMBIE child pid;
        }
        //**** has children but none dead yet ****
        ksleep(running);  // sleep on its PROC address
     }
  }
```

在kwait算法中，如果没有子进程，则进程会返回 -1，表示错误。否则，它将搜索僵尸子进程。如果它找到僵尸子进程，就会收集僵尸子进程的pid和退出代码，将僵尸进程释放到freeList并返回僵尸子进程的pid。否则，它将在自己的PROC地址上休眠，等待子进程终止。由于每个PROC地址都是一个唯一值，所有子进程也都知道这个值，所以等待的父进程可以在自己的PROC地址上休眠，等待子进程稍后唤醒它。相应地，当进程终止时，它必须发出：

```
kwakeup(running->parent);
```

以唤醒父进程。若不用父进程地址，读者也可使用父进程pid进行验证。在kwait()算法中，进程唤醒后，当它再次执行while循环时，将会找到死亡的子进程。注意，每个kwait()调用只处理一个僵尸子进程（如有）。如果某个进程有多个子进程，那么它可能需要多次调用kwait()来处理所有死亡的子进程。或者，某进程可以先终止，而不需要等待任何死亡子进程。当某进程死亡时，它所有的子进程都成了 P1 的子进程。在真实系统中，P1 在无限循环中执行，多次等待死亡的子进程，包括接收的孤儿进程。因此，在类 Unix 系统中，INIT 进程 P1 扮演着许多角色。

- 它是除 P0 之外所有进程的祖先。具体来说，它是所有用户进程的始祖，因为所有登录进程都是 P1 的子进程。
- 它就像孤儿院的院长，所有孤儿都会送到它这里，并叫它爸爸。
- 它又像是太平间管理员，因为它要不停地寻找僵尸进程，以埋葬它们死亡的空壳。

所以，在类 Unix 系统中，如果 INIT 进程 P1 死亡或被卡住，系统将停止工作，因为用户无法再次登录，系统内很快就会堆满腐烂的尸体。

3.6 MT 系统中的进程管理

本节为读者提供了一个编程练习，完善基础 MT 系统，实现 MT 系统的进程管理函数。具体来说，

（1）用二叉树的形式实现进程家族树。

（2）实现 ksleep() 和 kwakeup() 进程同步函数。

（3）实现 kexit() 和 kwait() 进程管理函数。

（4）添加“w”命令来测试和演示等待操作。

编程练习的示例解决方案可在线下载。图 3.3 为运行修改后 MT 系统的样本输出示意图。如图所示，P1 复刻出子进程 P2，P2 又复刻出子进程 P3 和 P4。然后，P2 执行“q”命令以终止。从图中可以看出，P2 成了僵尸进程，将所有的孤儿子进程都送给了 P1，现在，P1 的子进程列表包含 P2 的所有孤儿进程。当 P1 执行“w”命令时，它会找到僵尸子进程 P2，并将它返回到 freeList 中。或者，读者可以修改 child-exit 和 parent-wait 的顺序，以确认顺序不会造成影响。

```
proc 2 running: Parent=1  child = NULL
input a char [f|s|q|w] : f
readyQ = [1 1] -> [3 1] -> [0 0] -> NULL
proc 2 kforked a child = 3
readyQueue = [1 1] -> [3 1] -> [0 0] -> NULL
*************************************
proc 2 running: Parent=1  child = [3 READY ]->NULL
input a char [f|s|q|w] : f
readyQ = [1 1] -> [3 1] -> [4 1] -> [0 0] -> NULL
proc 2 kforked a child = 4
readyQueue = [1 1] -> [3 1] -> [4 1] -> [0 0] -> NULL
*************************************
proc 2 running: Parent=1  child = [3 READY ]->[4 READY ]->NULL
input a char [f|s|q|w] : q
proc 2 in kexit()
P2 child = [3 READY ]->[4 READY ]->NULL
P1 child = [2 ZOMBIE]->[3 READY ]->[4 READY ]->null
proc 2 in scheduler()
readyQueue = [1 1] -> [3 1] -> [4 1] -> [0 0] -> NULL
next running = 1
proc 1 resuming
*************************************
proc 1 running: Parent=0  child = [2 ZOMBIE]->[3 READY ]->[4 READY ]->NULL
input a char [f|s|q|w] : w
freeList = [5 0] -> [6 0] -> [7 0] -> [8 0] -> [2 0] -> NULL
proc 1 waited for a ZOMBIE child 2 status=2
*************************************
proc 1 running: Parent=0  child = [3 READY ]->[4 READY ]->NULL
input a char [f|s|q|w] :
```

图 3.3 修改后 MT 系统的输出示例

3.7 Unix/Linux 中的进程

了解了上述背景资料之后，现在，我们来介绍 Unix/Linux 中的进程（Bach 1990；Bovet 和 Cesati 2005；Silberschatz 等 2009；Love 2005）。读者可以将这些描述与 MT 系统中的进程联系起来，因为所有的术语和概念在两个系统中都同样适用。

3.7.1 进程来源

当操作系统启动时，操作系统内核的启动代码会强行创建一个 PID=0 的初始进程，即通过分配 PROC 结构体（通常是 proc[0]）进行创建，初始化 PROC 内容，并让**运行**指向 proc[0]。然后，系统执行初始进程 P0。大多数操作系统都以这种方式开始运行第一个进程。P0 继续初始化系统，包括系统硬件和内核数据结构。然后，它挂载一个根文件系统，使系统可以使用文件。在初始化系统之后，P0 复刻出一个子进程 P1，并把进程切换为以用户模式运行 P1。

3.7.2 INIT 和守护进程

当进程 P1 开始运行时，它将其执行映像更改为 INIT 程序。因此，P1 通常被称为 INIT **进程**，因为它的执行映像是 init 程序。P1 开始复刻出许多子进程。P1 的大部分子进程都是用来提供系统服务的。它们在后台运行，不与任何用户交互。这样的进程称为**守护进程**。守护进程的例子有

```
syslogd: log daemon process
inetd  : Internet service daemon process
httpd  : HTTP server daemon process
etc.
```

3.7.3 登录进程

除了守护进程之外，P1 还复刻了许多 LOGIN 进程，每个终端上一个，用于用户登录。每个 LOGIN 进程打开三个与自己的终端相关联的**文件流**。这三个文件流是用于标准输入的 stdin、用于标准输出的 stdout 和用于标准错误消息的 stderr。每个文件流都是指向进程堆区中 **FILE 结构体**的指针。每个 FILE 结构体记录一个文件描述符（数字），stdin 的文件描述符是 0，stdout 的是 1，stderr 的是 2。然后，每个 LOGIN 进程向 stdout 显示一个

```
login:
```

以等待用户登录。用户账户保存在 /etc/passwd 和 /etc/shadow 文件中。每个用户账户在表单的 /etc/passwd 文件中都有一行对应的记录，

```
name:x:gid:uid:description:home:program
```

其中 name 为用户登录名，x 表示登录时检查密码，gid 为用户组 ID, uid 为用户 ID, home 为用户主目录，program 为用户登录后执行的初始程序。其他用户信息保存在 /etc/shadow 文件中。shadow 文件的每一行都包含加密的用户密码，后面是可选的过期限制信息，如过期日期和时间等。当用户尝试使用登录名和密码登录时，Linux 将检查 /etc/passwd 文件和 /etc/shadow 文件，以验证用户的身份。

3.7.4 sh 进程

当用户成功登录时，LOGIN 进程会获取用户的 gid 和 uid，从而成为用户的进程。它将目录更改为用户的主目录并执行列出的程序，通常是命令解释程序 sh。现在，用户进程执行 sh，因此用户进程通常称为 sh 进程。它提示用户执行命令。一些特殊命令，如 cd(更改目录)、退出、注销等，由 sh 自己直接执行。其他大多数命令是各种 bin 目录（如 /bin、/sbin、/usr/bin、/usr/local/bin 等）中的可执行文件。对于每个（可执行文件）命令，sh 会复刻一个子进程，并等待子进程终止。子进程将其执行映像更改为命令文件并执行命令程序。子进程在终止时会唤醒父进程 sh，父进程会收集子进程终止状态、释放子进程 PROC 结构体并提示执行另一个命令等。除简单的命令之外，sh 还支持 **I/O 重定向**和通过**管道**连接的多个命令。

3.7.5 进程的执行模式

在 Unix/Linux 中，进程以两种不同的模式执行，即**内核模式**和**用户模式**，简称 Kmode 和 Umode。在每种执行模式下，一个进程有一个执行映像，如图 3.4 所示。

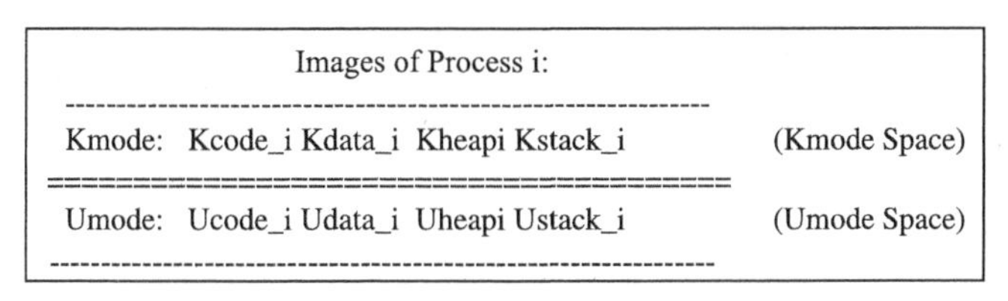

图 3.4 进程的执行映像

在图 3.4 中，索引 i 表示这些是进程 i 的映像。通常，进程在 Umode 下的映像都不相同。但是，在 Kmode 下，它们的 Kcode、Kdata 和 Kheap 都相同，这些都是操作系统内核的内容，但是每个进程都有自己的 Kstack。

在进程的生命周期中，会在 Kmode 和 Umode 之间发生多次迁移。每个进程都在 Kmode 下产生并开始执行。事实上，它在 Kmode 下执行所有相关操作，包括终止。在 Kmode 模式下，通过将 CPU 的状态寄存器从 K 模式更改为 U 模式，可轻松切换到 Umode。但是，一旦进入 Umode，就不能随意更改 CPU 的状态了，原因很明显。Umode 进程只能通过以下三种方式进入 Kmode：

（1）**中断**：中断是外部设备发送给 CPU 的信号，请求 CPU 服务。当在 Umode 下执行时，CPU 中断是启用的，因此它将响应任何中断。在中断发生时，CPU 将进入 Kmode 来处理中断，这将导致进程进入 Kmode。

（2）**陷阱**：陷阱是错误条件，例如无效地址、非法指令、除以 0 等，这些错误条件被 CPU 识别为异常，使得 CPU 进入 Kmode 来处理错误。在 Unix/Linux 中，内核陷阱处理程序将陷阱原因转换为信号编号，并将信号传递给进程。对于大多数信号，进程的默认操作是终止。

（3）**系统调用**：系统调用（简称 syscall）是一种允许 Umode 进程进入 Kmode 以执行内核函数的机制。当某进程执行完内核函数后，它将期望结果和一个返回值返回到 Umode，该值通常为 0（表示成功）或 -1（表示错误）。如果发生错误，外部全局变量 errno（在 errno.h 中）会包含一个 ERROR 代码，用于标识错误。用户可使用库函数

```
perror("error message");
```

来打印某个错误消息，该消息后面跟着一个描述错误的字符串。

每当一个进程进入 Kmode 时，它可能不会立即返回到 Umode。在某些情况下，它可能根本不会返回到 Umode。例如，_exit() syscall 和大多数陷阱会导致进程在内核中终止，这样它永远都不会再返回到 Umode。当某进程即将退出 Kmode 时，操作系统内核可能会切换进程以运行另一个具有更高优先级的进程。

3.8　进程管理的系统调用

在本节中，我们将讨论 Linux 中与进程管理相关的以下系统调用（Stallings 2011 ；Tanenbaum 和 Woodhull 2006）。

```
fork(), wait(), exec(), exit()
```

每个都是发出实际系统调用的库函数：

```
int syscall(int a, int b, int c, int d);
```

其中，第一个参数 a 表示系统调用号，b、c、d 表示对应核函数的参数。在基于 Intel x86 的 Linux 中，系统调用是由汇编指令 INT 0x80 实现的，使得 CPU 进入 Linux 内核来执行由系统调用号 a 标识的核函数。

3.8.1　fork()

```
Usage:   int pid = fork();
```

fork() 创建子进程并返回子进程的 pid，如果 fork() 失败则返回 -1。在 Linux 中，每个用户在同一时间只能有数量有限的进程。用户资源限制可在 /etc/security/limits.conf 文件中设置。用户可运行 ulimit -a 命令来查看各种资源限制值。图 3.5 显示了 fork() 操作，下面将进行解释。

```
             PROCi                 |          PROCj
             --------              |        -----------
Kmode :      Kcodei  kfork(){......}|     (When Pj runs:)---->---
             Kdatai      ^     =  |    |                        |
             kstacki     |     =  |    |      Kstackj           |
==================== | ===== | ================================= | ===
Umode :      Ucodei  pid=fork(); <-    |      Ucodej  pid=fork();<-
             Udatai                    |      Udataj
             Ustacki                   |      Ustackj
```

图 3.5　fork() 操作

（1）图 3.5 的左侧显示了进程 Pi 的映像，它在 Umode 下发出系统调用 pid=fork()。

（2）Pi 转到 Kmode 来执行内核中相应的 kfork() 函数，它在该函数中，创建一个有自己的 Kmode 堆栈和 Umode 映像的子进程 PROCj，如图右侧所示。Pj 的 Umode 映像与 Pi 的 Umode 映像完全相同。因此，Pj 的代码段也有这样的语句

```
pid=fork();
```

而且，kfork() 允许子进程继承父进程打开的所有文件。因此，父进程和子进程都可以从 stdin 获得输入，并显示到 stdout 和 stderr 的同一终端。

（3）在创建完子进程之后，Pi 返回到语句

```
pid = fork();      // parent return child PID
```

在它自己的 Umode 映像中，子进程的 pid = j。如果 fork() f 失败，会返回 -1，在这种情况下，不会创建子进程。

（4）当子进程 Pj 运行时，它会退出 Kmode 并返回到相同的语句

```
pid = fork();      // child returns 0
```

在它自己的 Umode 映像中，返回值为 0。

在 fork() 成功之后，父进程和子进程都执行它们自己的 Umode 映像，紧跟着 fork() 之后的映像是完全相同的。从程序代码的角度来看，借助返回的 pid 是判断当前正在执行进程的唯一方法。因此，程序代码应该写成：

```
int pid = fork();
if (pid){
   // parent executes this part
}
else{
   // child executes this part
}
```

我们通过一个例子来说明 fork()。

示例 3.1：示例程序 C3.1 演示了 fork()。

```
/********************** C3.1.c: fork() ************************/
#include <stdio.h>
int main()
{
     int pid;
     printf("THIS IS %d  MY PARENT=%d\n", getpid(), getppid());
(1). pid = fork();   // fork syscall; parent returns child pid,
     if (pid){ // PARENT EXECUTES THIS PART
(2).    printf("THIS IS PROCESS %d  CHILD PID=d\n", getpid(), pid);
     }
     else{           // child executes this part
(3).    printf("this is process %d  parent=%d\n", getpid(), getppid());
     }
}
```

该程序由 sh 的子进程运行。在示例代码中，getpid() 和 getppid() 是系统调用。getpid() 返回调用进程的 PID，getppid() 返回父进程的 PID。

- 第（1）行复刻一个子进程。
- 第（2）行打印（大写，更容易识别）正在执行进程的 PID 和新复刻子进程的 PID。
- 第（3）行打印（小写）子进程的 PID，应与第（2）行中子进程的 PID 相同，以及其父进程的 PID，应与第（2）行中进程的 PID 相同。

3.8.2 进程执行顺序

在 fork() 完成后，子进程与父进程和系统中所有其他进程竞争 CPU 运行时间。接下来

运行哪个进程取决于它们的调度优先级，优先级呈动态变化。下面的示例演示了进程可能的各种执行顺序。

示例3.2：示例程序C3.2演示了进程的执行顺序。

```
/***************** C3.2.c file ********************/
#include <stdio.h>
int main()
{
    int pid=fork();  // fork a child
    if (pid){        // PARENT
       printf("PARENT %d CHILD=%d\n", getpid(), pid);
(1).   // sleep(1);  // sleep 1 second ==> let child run next
       printf("PARENT %d EXIT\n", getpid());
    }
    else{            // child
       printf("child %d start my parent=%d\n", getpid(), getppid());
(2).   // sleep(2);  // sleep 2 seconds => let parent die first
       printf("child %d exit  my parent=%d\n", getpid(), getppid());
    }
}
```

在示例3.2的代码中，父进程复刻出一个子进程。在fork()完成后，接下来运行哪个进程取决于它们的优先级。子进程可能会先运行和终止，但也可能后运行和终止。如果进程执行非常长的代码，它们可能会轮流运行，因此它们的输出可能会交错显示在屏幕上。为了查看不同进程的执行顺序，读者可以进行以下实验。

（1）取消第（1）行注释，让父进程休眠1秒钟。然后，子进程先运行完成。

（2）取消第（2）行注释，但不取消第（1）行注释，让子进程休眠2秒钟。然后，父进程先运行完成。如果父进程先终止，子进程的ppid将会更改为1或其他PID号。后文将会解释这一原因。

（3）取消第（1）行和第（2）行注释。得到的结果应与第（2）种情况相同。

除了sleep(seconds)可以让调用进程延迟几秒之外，Unix/Linux还提供以下系统调用，可能会影响进程的执行顺序。

- nice(int inc)：nice()将进程优先级值增大一个指定值，这会降低进程调度优先级（优先级值越大，意味着优先级越低）。如果有优先级更高的进程，将会触发进程切换，首先运行优先级更高的进程。在非抢占式内核中，进程切换可能不会立即发生。它只在执行进程即将退出Kmode并返回Umode时发生。
- sched_yield(void)：sched_yield()使调用进程放弃CPU，允许优先级更高的其他进程先运行。但是，如果调用进程仍然具有最高优先级，它将继续运行。

3.8.3　进程终止

正如2.3.8节所指出，执行程序映像的进程可能以两种方式终止。

（1）**正常终止**：回顾前面的内容，我们知道，每个C程序的main()函数都是由C启动代码crt0.o调用的。如果程序执行成功，main()最终会返回到crt0.o，调用库函数exit(0)来终止进程。首先，exit(value)函数会执行一些清理工作，如刷新stdout、关闭I/O流等。然后，它发出一个_exit(value)系统调用，使进入操作系统内核的进程终止。退出值0通常表示正常终止。如果需要，进程可直接从程序内的任何位置调用 exit(value)，不必返回到 crt0.o。再直接一点，进程可能会发出 _exit(value) 系统调用立即执行终止，不必先进行清理工作。当内核中的某个进程终止时，它会将 _exit(value) 系统调用中的值记录为进程 PROC 结构体中的退出状态，并通知它的父进程并使该进程成为僵尸进程。父进程可通过系统调用找到僵尸子进程，获得其 pid 和退出状态

```
pid = wait(int *status);
```

它还会清空僵尸子进程 PROC 结构体，使该结构可被另一个进程重用。

（2）**异常终止**：在执行某程序时，进程可能会遇到错误，如非法指令、越权、除零等，这些错误会被 CPU 识别为**异常**。当某进程遇到异常时，它会阱入操作系统内核。内核的**异常处理程序**将陷阱错误类型转换为一个幻数，称为**信号**，将信号传递给进程，使进程终止。在这种情况下，进程非正常终止，僵尸进程的退出状态是信号编号。除了陷阱错误，信号也可能来自硬件或其他进程。例如，按下“Ctrl+C”组合键会产生一个硬件中断信号，它会向该终端上的所有进程发送信号 2 **中断信号**，使进程终止。或者，用户可以使用命令

```
kill -s signal_number pid         # signal_number=1 to 31
```

向通过 pid 识别的目标进程发送信号。对于大多数信号数值，进程的默认操作是终止。信号和信号处理将在后面第 6 章讨论。

在这两种情况下，当进程终止时，最终都会在操作系统内核中调用 kexit() 函数。kexit() 的一般算法见 3.5.1 节。唯一的区别是 Unix/Linux 内核将擦除终止进程的用户模式映像。

在 Linux 中，每个 PROC 都有一个 2 字节的退出代码（exitCode）字段，用于记录进程退出状态。如果进程正常终止，exitCode 的高位字节是 _exit(exitValue) 系统调用中的 exitValue。低位字节是导致异常终止的信号数值。因为一个进程只能死亡一次，所以只有一个字节有意义。

3.8.4 等待子进程终止

在任何时候，一个进程都可以使用

```
int pid = wait(int *status);
```

系统调用，等待僵尸子进程。如果成功，则 wait() 会返回僵尸子进程的 PID，而且 status 包含僵尸子进程的 exitCode。此外，wait() 还会释放僵尸子进程，以供重新使用。wait() 系统调用将调用内核中的 kwait() 函数。kwait() 的算法与 3.5.3 节中描述的算法完全相同。

示例 3.3：示例程序 C3.3 演示了等待和退出系统调用。

```
/************** C3.3.c: wait() and exit() ***************/
#include <stdio.h>
#include <stdlib.h>

int main()
{
    int pid, status;
    pid = fork();
    if (pid){ // PARENT:
       printf("PARENT %d WAITS FOR CHILD %d TO DIE\n", getpid(),pid);
```

```
        pid=wait(&status);  // wait for ZOMBIE child process
        printf("DEAD CHILD=%d, status=0x%04x\n", pid, status);
    }
    else{// child:
        printf("child %d dies by exit(VALUE)\n", getpid());
(1).    exit(100);
    }
}
```

在运行示例程序3.3时，子进程终止状态为0x6400，其中高位字节是子进程的退出值100。

为何wait()要等待任何僵尸子进程可解释如下。在复刻若干登录进程后，P1等待任何僵尸子进程。一旦用户从终端注销，P1一定会立即响应以复刻该终端上的另一个登录进程。由于P1不知道哪个登录进程会先终止，所以它必须等待任何僵尸登录子进程，而不是等待某个特定的进程。或者，某进程可以使用系统调用

```
int pid = waitpid(int pid, int *status, int options);
```

等待由pid参数指定的具有多个选项的特定僵尸子进程。例如，wait(&status)等于waitpid(-i, &status, 0)。读者可参考Linux手册页，以了解更多详细信息。

3.8.5 Linux中的subreaper进程

自内核3.4版本以来，Linux处理孤儿进程的方式略有不同。进程可以用系统调用将自己定义为subreaper：

```
prctl(PR_SET_CHILD_SUBREAPER);
```

这样，init进程P1将不再是孤儿进程的父进程。相反，标记为subreaper的最近活跃祖先进程将成为新的父进程。如果没有活跃的subreaper进程，孤儿进程仍然像往常一样进入INIT进程。实现该机制的原因如下：许多用户空间服务管理器（如upstart、systemd等）需要跟踪它们启动的服务。这类服务通常通过两次复刻来创建守护进程，但会让中间子进程立即退出，从而将孙子进程升级为P1的子进程。该方案的缺点是服务管理器不能再从服务守护进程接收SIGCHLD（death_of_child）信号，也不能等待任何僵尸子进程。当P1清理重定父级的进程时，将会丢失关于子进程的所有信息。使用subreaper机制，服务管理器可以将自己标记为“sub-init”，并且现在可以作为启动服务创建的所有孤儿进程的父进程。这也减少了P1的工作量，它不必处理系统中所有的孤儿进程。下面是一个很好的类比。当一个公司规模过大时，就需要拆分，以防止垄断，就像20世纪80年代早期AT&T的经历一样。在最初的Linux中，P1是唯一有权经营孤儿院的进程。subreaper机制打破了P1的垄断，允许任何进程通过宣布自己是subreaper（即使没有许可证！）来经营本地孤儿院。例如，在Ubuntu-15.10和后续版本中，每个用户init进程均标记为subreaper。它在Umode下运行，属于用户。读者可使用sh命令：

```
ps fxau | grep USERNAME | grep "/sbin/upstart"
```

显示subreaper进程的PID和信息。相反，它是用户所有孤儿进程的父进程，而不是P1。我们通过一个示例说明了Linux的subreaper进程的能力。

示例3.4：示例程序C3.4演示了Linux中的subreaper进程。

```
/************** C3.4.c: Subreaper Process ***************/
#include <stdio.h>
#include <unistd.h>
#include <wait.h>
#include <sys/prctl.h>
int main()
{
  int pid, r, status;
  printf("mark process %d as a subreaper\n", getpid());
  r = prctl(PR_SET_CHILD_SUBREAPER);
  pid = fork();
  if (pid){          // parent
    printf("subreaper %d child=%d\n", getpid(), pid);
    while(1){
       pid = wait(&status); // wait for ZOMBIE children
       if (pid>0)
          printf("subreaper %d waited a ZOMBIE=%d\n", getpid(), pid);
       else                  // no more children
           break;
    }
  }
  else{              // child
    printf("child %d parent=%d\n", getpid(), (pid_t)getppid());
    pid = fork();  // child fork a grandchild
    if (pid){        // child
       printf("child=%d start: grandchild=%d\n", getpid(), pid);
       printf("child=%d EXIT : grandchild=%d\n", getpid(), pid);
    }
    else{            // grandchild
       printf("grandchild=%d start: myparent=%d\n", getpid(),
               getppid());
       printf("grandchild=%d EXIT : myparent=%d\n", getpid(),
               getppid());
    }
  }
}
```

图 3.6 为运行示例 3.4 的程序的样本输出。

在示例 3.4 的程序中，进程（9620）首先将自己标记为 subreaper。然后，它复刻出一个子进程（9621），并使用 while 循环等待僵尸子进程，直到结束。该子进程复刻出自己的一个子进程，是第一个进程的孙子进程（9622）。当程序运行时，子进程（9621）或孙子进程（9622）可以先终止。如果孙子进程先终止，它的父进程会保持不变（9621）。但是，如果子进程先终止而且没有任何活跃的祖先进程被标记为 subreaper，孙子进程将会成为 P1 的子进程。因为第一个进程（9620）是 subreaper，如果孙子进程的父进程先死亡，它会将孙子进程作为孤儿进程来收养。输出表明，当孙子进程开始运行时，它的父进程是 9621，但是在退出时更改为 9620，因为原来的父进程已经死亡。输出还表明，subreaper 进程 9620 已将 9621 和 9622 作为僵尸子进程收养了。如果用户杀死每个用户的 init 进

```
mark process 9620 as a subreaper
subreaper 9620 child=9621
child 9621 parent=9620
child=9621 grandchild=9622
child=9621 EXIT: grandchild=9622
grandchild=9622 start: myparent=9621
subreaper 9620 waited a ZOMBIE=9621
grandchild=9622 EXIT : myparent=9620
subreaper 9620_waited a ZOMBIE=9622
```

图 3.6　示例 3.4 的样本输出

程，就相当于用户注销。在这种情况下，P1 将复刻出另一个用户 init 进程，要求用户再次登录。

3.8.6 exec()：更改进程执行映像

进程可以使用 exec() 将其 Umode 映像更改为不同的（可执行）文件。exec() 库函数有几个成员：

```
int execl( const char *path, const char *arg, ...);
int execlp(const char *file, const char *arg, ...);
int execle(const char *path, const char *arg,..,char *const envp[]);
int execv( const char *path, char *const argv[]);
int execvp(const char *file, char *const argv[]);
```

以上函数都是针对系统调用进行封装的库函数，准备参数并最终发出系统调用

```
int execve(const char *filename, char *const argv[ ], char *const envp[ ]);
```

在 execve() 系统调用中，第一个参数文件名与当前工作目录（CWD）或绝对路径名有关。参数 argv[ ] 是一个以 NULL 结尾的字符串指针数组，每个指针指向一个命令行参数字符串。按照惯例，argv[0] 是程序名，其他 argv[ ] 项是程序的命令行参数。例如，对于命令行：

```
a.out one two three
```

下图显示了 argv[ ] 的布局。

```
                0       1     2     3      4
argv[ ] = [ .     |  .   | .   | .    | NULL ]
            |        |     |     |
          "a.out" "one" "two" "three"
```

3.8.7 环境变量

环境变量是为当前 sh 定义的变量，由子 sh 或进程继承。当 sh 启动时，环境变量即在登录配置文件和 .bashrc 脚本文件中设置。它们定义了后续程序的执行环境。各环境变量定义为：

```
关键字 = 字符串
```

在 sh 会话中，用户可使用 env 或 printenv 命令查看环境变量。下面列出了一些重要的环境变量：

```
SHELL=/bin/bash
TERM=xterm
USER=kcw
PATH=/usr/local/bin:/usr/bin:/bin:/usr/local/games:/usr/games:./
HOME=/home/kcw
```

- SHELL：指定将解释任何用户命令的 sh。
- TERM：指定运行 sh 时要模拟的终端类型。
- USER：当前登录用户。

- PATH：系统在查找命令时将检查的目录列表。
- HOME：用户的主目录。在 Linux 中，所有用户主目录都在 /home 中。

在 sh 会话中，可以将环境变量设置为新的（字符串）值，如：

```
HOME=/home/newhome
```

可通过 EXPORT 命令传递给后代 sh，如：

```
export HOME
```

也可以将它们设置为空字符串来取消设置。在某个进程中，环境变量通过 env[ ] 参数传递给 C 程序，该参数是一个以 NULL 结尾的字符串指针数组，每个指针指向一个环境变量。

环境变量定义了后续程序的执行环境。例如，当 sh 看到一个命令时，它会在 PATH 环境变量的目录中搜索可执行命令（文件）。大多数全屏文本编辑器必须知道它们所运行的终端类型，该类型是在 TERM 环境变量中设置的。如果没有 TERM 信息，文本编辑器可能会出错。因此，命令行参数和环境变量都必须传递给正在执行的程序。这是所有 C 语言程序中 main() 函数的基础，可以写成：

```
int main(int argc, char *argv[ ], char *env[ ])
```

练习 3.1：在 C 程序的 main() 函数中，编写 C 代码来打印所有的命令行参数和环境变量。

练习 3.2：在 C 程序的 main() 函数中，找出 PATH 环境变量，形式为

```
PATH=/usr/local/bin:/usr/bin:/bin:/usr/local/games:/usr/games:./
```

然后编写 C 代码，将路径变量字符串标记到各个目录中。

如果成功，exec("filename", …) 将用可执行文件名中的新映像替换进程 Umode 映像。它仍然是同一个进程，只是多了一个新的 Umode 映像。旧的 Umode 映像被遗弃，因此除非 exec() 失败，否则永远不会返回，例如文件名不存在或不可执行。

通常，在 exec() 之后，进程中所有打开的文件都保持打开状态。被标记为 close-on-exec 的打开文件描述符将被关闭。进程的大部分信号被重置为默认值。如果可执行文件打开了 setuid 位，进程有效 uid/gid 将被更改为可执行文件的所有者，执行结束时将重置为保存的进程 uid/gid。

示例 3.5：示例程序 C3.5 演示了如何通过 execl() 更改进程映像，形式为

```
execl("a.out", "a.out", arg1, arg2,  ..., 0);
```

在调用 execve(“a.out” , argv[ ], env[ ]) 之前，库函数 execl() 先将参数汇编为 argv[ ] 形式。

```
/********** C3.5 program files ***************/
// (1). --------- b.c file: gcc to b.out ----------
#include <stdio.h>
int main(int argc, char *argv[])
{
  printf("this is %d in %s\n", getpid(), argv[0]);
}
// (2). --------- a.c file: gcc to a.out ----------
#include <stdio.h>
```

```
int main(int argc, char *argv[])
{
  printf("THIS IS %d IN %s\n", getpid(), argv[0]);
  int r = execl("b.out", "b.out", "hi", 0);
  printf("SEE THIS LINE ONLY IF execl() FAILED\n");
}
```

读者可先将 b.c 编译为 b.out。然后，再编译 a.c 并运行 a.out，将执行映像从 a.out 更改为 b.out，但进程 PID 不会更改，表明它仍是同一个进程。

示例 3.6：示例程序 C3.6 演示了如何通过 execve() 更改进程映像。

在这个示例程序中，我们将演示 execve() 如何在 /bin 目录中运行 Linux 命令。运行程序：

```
a.out  command  [options]
```

其中 command 是 /bin 目录下的任何 Linux 命令，而 [options] 是命令程序的可选参数，例如：

```
a.out  ls  -l;   a.out cat filename; etc.
```

程序将 command 和 [options] 汇编到 myargv[] 中，并发出系统调用：

```
execve("/bin/command", myargv,  env);
```

以执行 /bin/command 文件。如果 execve() 失败，进程将返回到旧映像。

```
/*********** C3.6.c file: compile and run a.out ***********/
#include <stdio.h>
#include <stdlib.h>
#include <string.h>

char *dir[64], *myargv[64]; // assume at most 64 parameters
char cmd[128];

int main(int argc, char *argv[], char *env[])
{
  int i, r;
  printf("THIS IS PROCESS %d IN %s\n", getpid(), argv[0]);
  if (argc < 2){
     printf("Usage: a.out command [options]\n");
     exit(0);
  }
  printf("argc = %d\n", argc);
  for (i=0; i<argc; i++)      // print argv[ ] strings
      printf("argv[%d] = %s\n", i, argv[i]);

  for (i=0; i<argc-1; i++)    // create myargv[ ]
      myargv[i] = argv[i+1];
  myargv[i] = 0;              // NULL terminated array

  strcpy(cmd, "/bin/");       // create /bin/command
  strcat(cmd, myargv[0]);
  printf(cmd = %s\n", cmd);   // show filename to be executed
  int r = execve(cmd, myargv, env);
```

```
    // come to here only if execve() failed
    printf("execve() failed: r = %d\n", r);
}
```

3.9 I/O 重定向

3.9.1 文件流和文件描述符

如前所述，sh 进程有三个用于终端 I/O 的文件流：stdin（标准输入）、stdout（标准输出）和 stderr（标准错误）。每个流都是指向执行映像堆区中 FILE 结构体的一个指针，如下文所示。

```
FILE *stdin -------> FILE structure
                     ---------------------------------
                     char fbuf[SIZE]
                     int counter, index, etc.
                     int fd = 0;  // fd[0] in PROC <== from KEYBOARD
                     ---------------------------------

FILE *stdout ------>  FILE structure
                     ---------------------------------
                     char fbuf[SIZE]
                     int counter, index, etc.
                     int fd = 1;  // fd[1] in PROC ==> to SCREEN
                     ---------------------------------

FILE *stderr ------>  FILE structure
                     ---------------------------------
                     char fbuf[SIZE]
                     int counter, index, etc.
                     int fd = 2;  // fd[2] in PROC ==> to SCREEN also
                     ---------------------------------
```

每个文件流对应 Linux 内核中的一个打开文件。每个打开文件都用一个**文件描述符**（数字）表示。stdin、stdout、stderr 的文件描述符分别为 0、1、2。当某个进程复刻出一个子进程时，该子进程会继承父进程的所有打开文件。因此，子进程也具有与父进程相同的文件流和文件描述符。

3.9.2 文件流 I/O 和系统调用

当进程执行库函数

```
scanf("%s", &item);
```

它会试图从 stdin 文件输入一个（字符串）项，指向 FILE 结构体。如果 FILE 结构体的 fbuf[] 为空，它会向 Linux 内核发出 read 系统调用，从文件描述符 0 中读取数据，映射到终端（/dev/ttyX）或伪终端（/dev/pts/#）键盘上。

3.9.3 重定向标准输入

如果我们用一个新打开的文件来替换文件描述符 0，那么输入将来自该文件而不是原始输入设备。因此，如果我们执行以下代码：

```
#include <fcntl.h>  // contains O_RDONLY, O_WRONLY,O_APPEND, etc
close(0);           // syscall to close file descriptor 0
int fd=open("filename", O_RDONLY);  // open filename for READ,
                                    // fd replace 0
```

close(0) 系统调用会关闭文件描述符 0，使 0 成为未使用的文件描述符。open() 系统调用会打开一个文件，并使用最小的未使用描述符数值作为文件描述符。在这种情况下，打开文件的文件描述符为 0。因此，原来的文件描述符 0 会被新打开的文件取代。或者，我们也可以使用：

```
int fd = open("filename", O_RDOMLY);  // get a fd first
close(0);                             // zero out fd[0]
dup(fd);                              // duplicate fd to 0
```

系统调用 dup(fd) 将 fd 复制到数值最小的未使用文件描述符中，允许 fd 和 0 都能访问同一个打开的文件。此外，系统调用

```
dup2(fd1, fd2)
```

将 fd1 复制到 fd2 中，如果 fd2 已经打开，则先关闭它。因此，Unix/Linux 提供了几种方法来替换 / 复制文件描述符。完成上述任何一项操作之后，文件描述符 0 将被替换或复制到打开文件中，以便每个 scanf() 调用都将从打开的文件中获取输入。

3.9.4 重定向标准输出

当进程执行库函数

```
printf("format=%s\n", items);
```

它试图将数据写入 stdout 文件 FILE 结构体中的 fbuf[ ]，这是缓冲行。如果 fbuf[ ] 有一个完整的行，它会发出一个 write 系统调用，将数据从 fbuf[ ] 写入文件描述符 1，映射到终端屏幕上。要想将标准输出重定向到一个文件，需执行以下操作。

```
close(1);
open("filename", O_WRONLY|O_CREAT, 0644);
```

更改文件描述符 1，指向打开的文件名。然后，stdout 的输出将会转到该文件而不是屏幕。同样，我们也可以将 stderr 重定向到一个文件。当某进程（在内核中）终止时，它会关闭所有打开的文件。

3.10 管道

管道是用于进程交换数据的单向进程间通信通道。管道有一个读取端和一个写入端。可从管道的读取端读取写入管道写入端的数据。自从管道在最初的 Unix 中首次出现以来，已经被用于几乎所有的操作系统中，有许多变体。一些系统允许双向管道，在双向管道上，数据可以双向传输。普通管道用于相关进程。**命名管道**是不相关进程之间的 FIFO 通信通道。读取和写入管道通常是同步、阻塞操作。一些系统支持管道的非阻塞、异步读 / 写操作。为简便起见，我们将管道视为一组相关进程之间的有限尺寸 FIFO 通信通道。管道的读、写进程按以下方式同步。当读进程从管道上读取数据时，如果管道上有数据，读进程会根据需要

读取（不超过管道大小）并返回读取的字节数。如果管道没有数据，但仍有写进程，读进程会等待数据。当写进程将数据写入管道时，它会唤醒等待的读进程，使它们继续读取。如果管道没有数据也没有写进程，读进程返回 0。如果管道仍然有写进程，读进程会等待数据，因此 0 返回值只能意味着管道没有数据也没有写进程。在这种情况下，读进程会停止从管道读取。当写进程写入管道时，如果管道有空间，它会根据需要尽可能多地写入，直至管道写满，即没有更多空间。如果管道没有空间，但仍有读进程，写进程会等待空间。当读进程从管道读取数据来释放更多空间时，它会唤醒等待的写进程，让它们继续写入。但是，如果管道不再有读进程，写进程必须将这种情况视为**管道中断**错误，并中止写入。

3.10.1 Unix/Linux 中的管道编程

在 Unix/Linux 中，一系列相关系统调用为管道提供支持。系统调用

```
int pd[2];              // array of 2 integers
int r = pipe(pd);       // return value r=0 if OK, -1 if failed
```

在内核中创建一个管道并在 pd[2] 中返回两个文件描述符，其中 pd[0] 用于从管道中读取，pd[1] 用于向管道中写入。然而，管道并非为单进程而创建。例如，在创建管道之后，如果进程试图从管道中读取 1 个字节，它将永远不会从读取的系统调用中返回。这是因为当进程试图从管道中读取数据时，管道中尚无数据，但是有一个写进程，所以它会等待数据。但是写进程是谁呢？它是进程本身。所以进程在等待自己，可以说是把自己锁起来了。相反，如果进程试图写入的大小超过管道大小（大多数情况下为 4KB），则当管道写满时，进程将再次等待自己。因此，进程只能是管道上的一个读进程或者一个写进程，但不能同时是读进程和写进程。管道的正确使用方式如下。在创建管道后，进程复刻一个子进程来共享管道。在复刻过程中，子进程继承父进程的所有打开文件描述符。因此，子进程也有 pd[0]（用于从管道中读取数据）和 pd[1]（用于向管道中写入数据）。用户必须将其中一个进程指定为管道的写进程，并将另一个进程指定为管道的读进程。只要指定每个进程只扮演一个角色，指定的顺序并不重要。假设父进程被指定为写进程，子进程被指定为读进程。各进程必须关闭它不需要的管道描述符，即写进程必须关闭 pd[0]，读进程必须关闭 pd[1]。然后，父进程可向管道写入数据，子进程可从管道读取数据。图 3.7 显示了管道操作的系统模型。

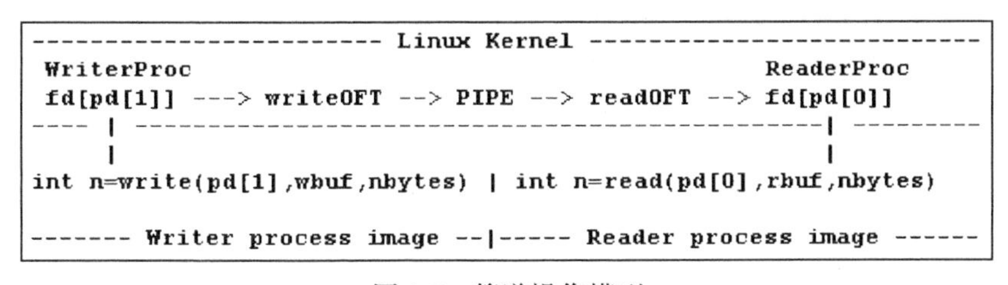

图 3.7　管道操作模型

在图 3.7 的左侧，一个写进程发出

```
write(pd[1], wbuf, nbytes)
```

系统调用，进入操作系统内核。它使用文件描述符 pd[1] 通过 writeOFT 来访问管道。它执行 write_pipe()，向管道的缓冲区写入数据，并在必要时等待空间。

在图 3.7 的右侧，一个读进程发出

```
read(pd[0],rbuf,nbytes)
```

系统调用，进入操作系统内核。它使用文件描述符 pd[0] 通过 readOFT 来访问管道。它执行 read_pipe()，从管道的缓冲区读取数据，并在必要时等待数据。

当写进程没有更多数据要写时，它可以先终止，在这种情况下，只要管道仍然有数据，读进程就可以继续读取。但是，如果读进程先终止，写进程须将这种情况视为**管道中断**错误，并随之终止。

注意，管道中断状况并不具有对称性。这是一种只有读进程没有写进程的通信通道。实际上，管道并未中断，因为只要管道有数据，读进程就仍可继续读取。下面的程序演示了 Unix/Linux 中的管道。

示例 3.7：示例程序 C3.7 演示了管道操作。

```
/*************** C3.7: Pipe Operations ************/
#include <stdio.h>
#include <stdlib.h>
#include <string.h>

int  pd[2], n, i;
char line[256];

int main()
{
   pipe(pd);            // create a pipe
   printf("pd=[%d, %d]\n", pd[0], pd[1]);
   if (fork()){          // fork a child to share the pipe
      printf("parent %d close pd[0]\n", getpid());
      close(pd[0]);     // parent as pipe WRITER
      while(i++ < 10){ // parent writes to pipe 10 times
         printf("parent %d writing to pipe\n", getpid());
         n = write(pd[1], "I AM YOUR PAPA", 16);
         printf("parent %d wrote %d bytes to pipe\n", getpid(), n);
      }
      printf("parent %d exit\n", getpid());
   }
   else{
      printf("child  %d close pd[1]\n", getpid());
      close(pd[1]);    // child as pipe READER
      while(1){        // child read from pipe
         printf("child  %d reading from pipe\n", getpid());
         if ((n = read(pd[0], line, 128))){ // try to read 128 bytes
            line[n]=0;
            printf("child read %d bytes from pipe: %s\n", n, line);
         }
         else // pipe has no data and no writer
            exit(0);
      }
   }
}
```

读进程可在 Linux 下编译和运行程序来观察它的行为。图 3.8 显示了运行示例 3.7 的程序的样本输出。

```
pd=[3, 4]
child  5101 close pd[1]
child  5101 reading from pipe
parent 5100 close pd[0]
parent 5100 writing to pipe
child read 16 bytes from pipe: I AM YOUR PAPA
child  5101 reading from pipe
parent 5100 wrote 16 bytes to pipe
parent 5100 writing to pipe
child read 16 bytes from pipe: I AM YOUR PAPA
child  5101 reading from pipe
parent 5100 wrote 16 bytes to pipe
parent 5100 writing to pipe
child read 16 bytes from pipe: I AM YOUR PAPA
```

图 3.8　管道程序的样本输出

在上面的管道程序中，父进程和子进程都会正常终止。读者可以修改程序来进行以下实验，并观察结果。

（1）将父进程指定为读进程，并将子进程指定为写进程。

（2）让写进程不停地写入，但只让读进程读取几次。

在第二种情况下，写进程会以“BROKEN_PIPE 错误”而终止。

3.10.2　管道命令处理

在 Unix/Linux 中，命令行

```
cmd1 | cmd2
```

包含一个管道符号“|”。sh 将通过一个进程运行 cmd1，并通过另一个进程运行 cmd2，它们通过一个管道连接在一起，因此 cmd1 的输出变成 cmd2 的输入。下文展示了管道命令的典型用法。

```
ps x | grep "httpd"      # show lines of ps x containing httpd
cat filename | more      # display one screen of text at a time
```

3.10.3　将管道写进程与管道读进程连接起来

（1）当 sh 获取命令行 cmd1|cmd2 时，会复刻出一个子进程 sh，并等待子进程 sh 照常终止。

（2）子进程 sh：浏览命令行中是否有 | 符号。在这种情况下，

```
cmd1 | cmd2
```

有一个管道符号 |。将命令行划分为头部 =cmd1，尾部 =cmd2

（3）然后，子进程 sh 执行以下代码段：

```
int pd[2];
pipe(pd);           // creates a PIPE
pid = fork();       // fork a child (to share the PIPE)
if (pid){           // parent as pipe WRITER
   close(pd[0]);    // WRITER MUST close pd[0]
   close(1);        // close 1
   dup(pd[1]);      // replace 1 with pd[1]
   close(pd[1]);    // close pd[1]
```

```
    exec(head);     // change image to cmd1
  }
  else{             // child as pipe READER
    close(pd[1]); // READER MUST close pd[1]
    close(0);
    dup(pd[0]);     // replace 0 with pd[0]
    close(pd[0]); // close pd[0]
    exec(tail);     // change image to cmd2
  }
```

管道写进程重定向其 fd=i 到 pd[i]，管道读进程重定向其 fd=0 到 pd[0]。这样，这两个进程就可以通过管道连接起来了。

3.10.4 命名管道

命名管道又叫作 FIFO。它们有“名称”，并在文件系统中以特殊文件的形式存在。它们会一直存在下去，直至用 rm 或 unlink 将其删除。它们可与非相关进程一起使用，并不局限于管道创建进程的子进程。

命名管道示例

（1）在 sh 中，通过 mknod 命令创建一个命名管道：

```
mknod mypipe p
```

（2）或者在 C 程序中，发出 mknod() 系统调用：

```
int r = mknod("mypipe", S_IFIFO, 0);
```

步骤（1）或（2）都可以在当前目录中创建一个名为 mypipe 的特殊文件。输入：

```
ls -l mypipe
```

将显示为：

```
prw-r—r— 1 root root  0  time  mypipe
```

其中文件类型 p 表示它是管道，链接数 =1，大小 =0。

（3）进程可像访问普通文件一样访问命名管道。但是，对命名管道的写入和读取是由 Linux 内核同步的。

下图显示了写进程和读进程通过 sh 命令在命名管道上的交互状态。从图上可以看出，如果没有读进程从管道上读取数据，写进程就会停止。如果管道上没有数据，读进程就会停止。

```
    写命令                               |  读命令
--------------------------------------------------------------------
echo test > mypipe (writer stops) |
                                  |  cat mypipe
(writer continues)                |  test        (data read from pipe)
                                  |  cat mypipe (reader stops)
echo test again > mypipe          |  test again (data read from pipe)
(writer continues)                |  (reader continues)
--------------------------------------------------------------------
```

除了 sh 命令之外，在 C 语言程序中创建的进程也可以使用命名管道。

示例 3.8：示例程序 C3.8 演示了命名管道操作。它显示了如何用不同进程打开命名管道来读 / 写。

```
/******* C3.8: Create and read/write named pipe ********/
```

(3).1 Writer process program:

```
#include <stdio.h>
#include <sys/stat.h>
#include <fcntl.h>
char *line = "tesing named pipe";

int main()
{
  int fd;
  mknod("mypipe", I_SFIFO, 0);      // create a named pipe
  fd = open("mypipe), O_WRONLY);   // open named pipe for write
  write (fd, line, strlen(line)); // write to pipe
  close (fd);
}
```

(3).2. Reader process program:

```
#include <stdio.h>
#include <sys/stat.h>
#include <fcntl.h>

int main()
{
  char buf[128];
  int fd = open("mypipe", O_RDONLY);
  read(fd, buf, 128);
  printf ("%s\n", buf);
  close (fd);
}
```

3.11 编程项目：sh 模拟器

本编程项目是编写一个 C 语言程序来模拟 Linux sh 实现对命令的执行。目标是让读者理解 Linux sh 如何工作。使用了 fork()、exec()、close()、exit()、pipe() 系统调用和字符串操作。建议读者尝试使用后续步骤实现该项目。

3.11.1 带有 I/O 重定向的单命令

（1）提示用户输入命令行，其形式为：

```
cmd arg1 arg2 arg3 ... argn
```

其中 cmd 是命令，arg1～argn 是 cmd 程序的命令行参数。有效命令包括“cd”“exit”和所有 Linux 可执行文件，如 echo、ls、date、pwd、cat、cp、mv、cc、emacs 等。简言之，sh 模拟器可运行与 Linux sh 相同的一系列命令。

（2）处理简单命令：

```
cmd = "cd"   :  chdir(arg1) OR chdir(HOME) if no arg1;
cmd = "exit" :  exit(0) to terminate;
```

（3）对于所有其他命令：

```
创建子进程;
等待子进程终止;
打印子进程的退出状态码;
继续执行步骤1;
```

（4）**子进程**：首先，假设命令行中没有管道。

（4）.1 处理 I/O 重定向：

```
cmd  arg1 arg2 ...  <  infile    // take inputs from infile
cmd  arg1 arg2 ...  >  outfile   // send outputs to outfile
cmd  arg1 arg2 ...  >> outfile   // APPEND outputs to outfile
```

（4）.2 通过 execve() 执行 cmd，传递参数

```
char *myargv[ ], char *env[ ]
```

至 cmd 文件，其中 myargv[ ] 是一个字符数组 *：

```
myargv[0]->cmd,
myargv[1]->arg1, .....,
End with a NULL pointer
```

sh 可以在 PATH 环境变量目录中搜索可执行命令，因此 sh 模拟器也必须这样做。为此，模拟器程序必须将 PATH 变量标记到用于定位命令文件的各个目录中。

3.11.2　带有管道的命令

（5）在验证了 mysh 适用于简单命令之后，将其扩展到处理管道。如果命令行有一个 | 符号，把它分为头部和尾部，如：

```
        cmd1 < infile | cmd 2 > outfile
head = "cmd < infile";   tail = "cmd 2 > outfile"
```

然后通过以下步骤实现管道。

创建管道
复刻出一个子进程来共享管道
安排一个进程作为管道的写进程，另一个进程作为管道的读进程。
然后，让各进程 execve() 其命令（可能带有 I/O 重定向）。

（6）**多管道**：如果命令行包含多个管道符号，则通过递归实现管道。如需了解递归算法，读者可参考文献（Wang 2015）第 13 章。

3.11.3　ELF 可执行文件与 sh 脚本文件

在 Linux 中，二进制可执行文件采用 ELF 文件格式。ELF 文件的前 4 个字节是 ELF 标识符“0x7F”ELF。sh 脚本是文本文件。大多数 sh 脚本文件的第一行包含：

```
#!  /bin/sh
```

有了这些提示，读者应该能够设计出一些方法来检验某个命令是 ELF 可执行文件还是 sh 脚本。如果是 sh 脚本文件，要在这些文件上运行 Linux 命令 /bin/bash。

3.11.4 示例解决方案

编程项目的示例解决方案可在线下载。读者可下载二进制可执行文件 kcsh.bin，然后在 Linux 下运行。如果提出要求，可向讲师提供项目源代码。图 3.9 显示了运行 sh 模拟器程序的样本输出。

```
root@wang:~/abc/360/F17/sh.simulator# kcsh.bin
************* Welcome to kcsh *************
show PATH: PATH=PATH=/usr/local/sbin:/usr/sbin:/sbin:/usr/local/bin:/usr/bin:/b
in:/usr/games:.:/usr/lib/java/bin:/usr/lib/java/jre/bin:/usr/lib/qt/bin
decompose PATH into dir strings:
/usr/local/sbin  /usr/sbin  /sbin  /usr/local/bin  /usr/bin  /bin  /usr/games
.  /usr/lib/java/bin  /usr/lib/java/jre/bin  /usr/lib/qt/bin
show HOME directory: HOME = /root
*********** kcsh processing loop **********
kcsh % : ls
32431 line=ls
32431 scan: head=ls  tail=(null)
32431 do_command: line=ls
32431 tries ls in each PATH dir
i=0   cmd=/usr/local/sbin/ls
i=1   cmd=/usr/sbin/ls
i=2   cmd=/sbin/ls
i=3   cmd=/usr/local/bin/ls
i=4   cmd=/usr/bin/ls
kcsh.bin  shve.c
parent kcsh 32430 forks a child process 32431
parent sh 32430 waits
child sh 32431 died : exit status = 0000
kcsh % : ▌
```

图 3.9 sh 模拟器程序的样本输出

3.12 习题

1. 在 Intel 32 位 x86 CPU 上，汇编指令 pushal 按顺序推送 CPU 寄存器 eax、ecx、ebx、old_esp、ebp、esi、edi。相应地，指令 popal 以相反的顺序弹出寄存器。假设 tswitch() 函数在 MT 系统中可写为：

```
tswitch:
SAVE:   pushal # save all CPU registers on stack
        pushfl
        movl   running,%ebx    # ebx -> PROC
        movl   %esp,4(%ebx)    # PORC.save_sp = esp
FIND:   call   scheduler
RESUME: movl   running,%ebx    # ebx -> PROC
        movl   4(%ebx),%esp    # esp = PROC>saved_sp
        popfl
        popal  # restore all saved CPU registers from stack
        ret
```

演示如何在 kfork() 中初始化新进程，以便它们可执行 body() 函数。重新编译和运行修改后的 MT 系统，并演示它的工作原理。

2. 假设 body() 函数在 MT 系统中可写为：

```
int body(int pid){ // use pid inside the body function }
```

其中 pid 是新任务的 PID。演示如何在 kfork() 中初始化新任务堆栈。提示：参数会在函数调用中以如第2章所述的方式传递。

3. 请参考示例程序 C3.3。如果将第（1）行替换为以下函数，子进程退出状态值是什么：

（1）exit(123）; Ans: 0x7B00 // 正常终止

（2）{ int *p=0; *p = 123; } Ans: 0x000B // 被信号 11 杀死

（3）{ int a,b=0; a = a/b; } Ans: 0x0008 // 被信号 8 杀死

4. 请参考示例程序 C3.5。解释在以下情况下会出现什么结果：

（1）在 a.c 文件中，用 execl(“a.out”, “a.out”, “hi”, 0) 更换 execl() 行。

（2）在 b.c 文件中，新增 execl(“a.out”, “a.out”, “again”, 0) 行。

运行修改后的程序来验证答案。

5. Linux 有一个 vfork() 系统调用，会创建一个子进程，就像 fork() 一样，但是比 fork() 更有效。

（1）请阅读 vfork() 的 Linux 手册页，了解为什么它比 fork() 更有效。

（2）将编程项目中的所有 fork() 系统调用替换为 vfork()。验证项目程序仍可正常工作。

参考文献

Bach, M. J., “The Design of the Unix operating system”, Prentice Hall, 1990
Bovet, D. P., Cesati, M., Understanding the Linux Kernel, O’Reilly, 2005
Silberschatz, A., P.A. Galvin, P.A., Gagne, G, “Operating system concepts, 8th Edition”, John Wiley & Sons, Inc. 2009
Love, R. Linux Kernel Development, 2nd Edition, Novell Press, 2005
Stallings, W. “Operating Systems: Internals and Design Principles (7th Edition)”, Prentice Hall, 2011
Tanenbaum, A.S., Woodhull, A.S., “Operating Systems, Design and Implementation, third Edition”, Prentice Hall, 2006
Wang, K.C., “Design and Implementation of the MTX Operating System”, Springer A.G. 2015

并发编程

摘要

本章论述了并发编程，介绍了并行计算的概念，指出了并行计算的重要性；比较了顺序算法与并行算法，以及并行性与并发性；解释了线程的原理及其相对于进程的优势；通过示例介绍了 Pthread 中的线程操作，包括线程管理函数，互斥量、连接、条件变量和屏障等线程同步工具；通过具体示例演示了如何使用线程进行并发编程，包括矩阵计算、快速排序和用并发线程求解线性方程组等方法；解释了死锁问题，并说明了如何防止并发程序中的死锁问题；讨论了信号量，并论证了它们相对于条件变量的优点；还解释了支持 Linux 中线程的独特方式。编程项目是为了实现用户级线程。它提供了一个基础系统来帮助读者开始工作。这个基础系统支持并发任务的动态创建、执行和终止，相当于在某个进程的同一地址空间中执行线程。读者可通过该项目实现线程同步的线程连接、互斥量和信号量，并演示它们在并发程序中的用法。该编程项目会让读者更加深入地了解多任务处理、线程同步和并发编程的原理及方法。

4.1 并行计算导论

在早期，大多数计算机只有一个处理组件，称为处理器或中央处理器（CPU）。受这种硬件条件的限制，计算机程序通常是为串行计算编写的。要求解某个问题，先要设计一种算法，描述如何一步步地解决问题，然后用计算机程序以串行指令流的形式实现该算法。在只有一个 CPU 的情况下，每次只能按顺序执行某算法的一个指令和步骤。但是，基于分治原则（如二叉树查找和快速排序等）的算法经常表现出高度的并行性，可通过使用并行或并发执行来提高计算速度。并行计算是一种计算方案，它尝试使用多个执行并行算法的处理器更快速地解决问题。过去，由于并行计算对计算资源的大量需求，普通程序员很少能进行并行计算。近年来，随着多核处理器的出现，大多数操作系统（如 Linux）都支持对称多处理（SMP）。甚至对于普通程序员来说，并行计算也已经成为现实。显然，计算的未来发展方向是并行计算。因此，迫切需要在计算机科学和计算机工程专业学生的早期学习阶段引入并行计算。在本章中，我们将介绍通过并发编程实现并行计算的基本概念和方法。

4.1.1 顺序算法与并行算法

在描述顺序算法时，常用的方法是用一个 begin-end 代码块列出算法，如下图左侧所示。

```
---        顺序算法          ---|---        并行算法         ---
        begin                   |       cobegin
          step_1                |         task_1
          step_2                |         task_2
          . . .                 |         . . .
          step_n                |         task_n
        end                     |       coend
        // next step            |       // next step
-----------------------------------------------------------
```

begin-end 代码块中的顺序算法可能包含多个步骤。所有步骤都是通过单个任务依次执行的，每次执行一个步骤。当所有步骤执行完成时，算法结束。相反，图的右侧为并行算法的描述，它使用 cobegin-coend 代码块来指定并行算法的独立任务。在 cobegin-coend 块中，所有任务都是并行执行的。紧接着 cobegin-coend 代码块的下一个步骤将只在所有这些任务完成之后执行。

4.1.2　并行性与并发性

通常，并行算法只识别可并行执行的任务，但是它没有规定如何将任务映射到处理组件。在理想情况下，并行算法中的所有任务都应该同时实时执行。然而，真正的并行执行只能在有多个处理组件的系统中实现，比如多处理器或多核系统。在单 CPU 系统中，一次只能执行一个任务。在这种情况下，不同的任务只能并发执行，即在逻辑上并行执行。在单 CPU 系统中，并发性是通过多任务处理来实现的，该内容已在第 3 章中讨论过。在本章的最后，我们将在一个编程项目中再次讲解和示范多任务处理的原理和方法。

4.2　线程

4.2.1　线程的原理

一个操作系统（OS）包含许多并发进程。在进程模型中，进程是独立的执行单元。所有进程均在内核模式或用户模式下执行。在内核模式下，各进程在唯一地址空间上执行，与其他进程是分开的。虽然每个进程都是一个独立的单元，但是它只有一个执行路径。当某进程必须等待某事件时，例如 I/O 完成事件，它就会暂停，整个进程会停止执行。线程是某进程同一地址空间上的独立执行单元。创建某个进程就是在一个唯一地址空间创建一个主线程。当某进程开始时，就会执行该进程的主线程。如果只有一个主线程，那么进程和线程实际上并没有区别。但是，主线程可能会创建其他线程。每个线程又可以创建更多的线程等。某进程的所有线程都在该进程的相同地址空间中执行，但每个线程都是一个独立的执行单元。在线程模型中，如果一个线程被挂起，其他线程可以继续执行。除了共享共同的地址空间之外，线程还共享进程的许多其他资源，如用户 id、打开的文件描述符和信号等。打个简单的比方，进程是一个有房屋管理员（主线程）的房子。线程是住在进程房子里的人。房子里的每个人都可以独立做自己的事情，但是他们会共用一些公用设施，比如同一个信箱、厨房和浴室等。过去，大多数计算机供应商都是在自己的专有操作系统中支持线程。不同系统之间的实现有极大的区别。目前，几乎所有的操作系统都支持 Pthread，它是 IEEE POSIX 1003.1c 的线程标准（POSIX 1995）。如需了解更多信息，读者可查阅更多关于 Pthread 编程的书籍（Buttlar 等 1996）和在线文章（Pthreads 2017）。

4.2.2　线程的优点

与进程相比，线程有许多优点。

（1）**线程创建和切换速度更快**：进程的上下文复杂而庞大。其复杂性主要来自管理进程映像的需要。例如，在具有虚拟内存的系统中，进程映像可能由叫作**页面**的许多内存单元组成。在执行过程中，有些页面在内存中，有些则不在内存中。操作系统内核必须使用多个页表和多个级别的硬件辅助来跟踪每个进程的页面。要想创建新的进程，操作系统必须为进程分配内存并构建页表。若要在某个进程中创建线程，操作系统不必为新的线程分配内存和创建页表，因为线程与进程共用同一个地址空间。所以，创建线程比创建进程更快。另外，由于以下原因，线程切换比进程切换更快。进程切换涉及将一个进程的复杂分页环境替换为另一个进程的复杂分页环境，需要大量的操作和时间。相比之下，同一个进程中的线程切换要简单得多，也快得多，因为操作系统内核只需要切换执行点，而不需要更改进程映像。

（2）**线程的响应速度更快**：一个进程只有一个执行路径。当某个进程被挂起时，整个进程都将停止执行。相反，当某个线程被挂起时，同一进程中的其他线程可以继续执行。这使得有多个线程的程序响应速度更快。例如，在一个多线程的进程中，当一个线程被阻塞以等待 I/O 时，其他线程仍可在后台进行计算。在有线程的服务器中，服务器可同时服务多个客户机。

（3）**线程更适合并行计算**：并行计算的目标是使用多个执行路径更快地解决问题。基于分治原则（如二叉树查找和快速排序等）的算法经常表现出高度的并行性，可通过使用并行或并发执行来提高计算速度。这种算法通常要求执行实体共享公用数据。在进程模型中，各进程不能有效共享数据，因为它们的地址空间都不一样。为了解决这个问题，进程必须使用**进程间通信**（IPC）来交换数据或使用其他方法将公用数据区包含到其地址空间中。相反，同一进程中的所有线程共享同一地址空间中的所有（全局）数据。因此，使用线程编写并行执行的程序比使用进程编写更简单、更自然。

4.2.3 线程的缺点

另一方面，线程也有一些缺点，其中包括：

（1）由于地址空间共享，线程需要来自用户的明确同步。

（2）许多库函数可能对线程不安全，例如传统 strtok() 函数将一个字符串分成一连串令牌。通常，任何使用全局变量或依赖于静态内存内容的函数，线程都不安全。为了使库函数适应线程环境，还需要做大量的工作。

（3）在单 CPU 系统上，使用线程解决问题实际上要比使用顺序程序慢，这是由在运行时创建线程和切换上下文的系统开销造成的。

4.3 线程操作

线程的执行轨迹与进程类似。线程可在内核模式或用户模式下执行。在用户模式下，线程在进程的相同地址空间中执行，但每个线程都有自己的执行堆栈。线程是独立的执行单元，可根据操作系统内核的调度策略，对内核进行系统调用，变为挂起、激活以继续执行等。为了利用线程的共享地址空间，操作系统内核的调度策略可能会优先选择同一进程中的线程，而不是不同进程中的线程。截至目前，几乎所有的操作系统都支持 POSIX Pthread，定义了一系列标准应用程序编程接口（API）来支持线程编程。下面，我们将讨论和演示 Linux 中的 Pthread 并发编程（Goldt 等 1995；IBM；Love 2005；Linux Man Page Progect 2017）。

4.4 线程管理函数

Pthread 库提供了用于线程管理的以下 API。

```
pthread_create(thread, attr, function, arg): create thread
pthread_exit(status)        : terminate thread
pthread_cancel(thread)      : cancel thread
pthread_attr_init(attr)     : initialize thread attributes
pthread_attr_destroy(attr): destroy thread attribute
```

4.4.1　创建线程

使用 pthread_create() 函数创建线程。

```
int pthread_create (pthread_t *pthread_id, pthread_attr_t *attr,
                    void *(*func)(void *),  void *arg);
```

如果成功则返回 0，如果失败则返回错误代码。pthread_create() 函数的参数为

- pthread_id 是指向 pthread_t 类型变量的指针。它会被操作系统内核分配的唯一线程 ID 填充。在 POSIX 中，pthread_t 是一种不透明的类型。程序员应该不知道不透明对象的内容，因为它可能取决于实现情况。线程可通过 pthread_self() 函数获得自己的 ID。在 Linux 中，pthread_t 类型被定义为无符号长整型，因此线程 ID 可以打印为 %lu。
- attr 是指向另一种不透明数据类型的指针，它指定线程属性，下面将对此进行更详细的说明。
- func 是要执行的新线程函数的入口地址。
- arg 是指向线程函数参数的指针，可写为：

```
void *func(void *arg)
```

其中，attr 参数最复杂。下面给出了 attr 参数的使用步骤。

（1）定义一个 pthread 属性变量 `pthread_attr_t attr`。

（2）用 `pthread_attr_init (&attr)` 初始化属性变量。

（3）设置属性变量并在 pthread_create() 调用中使用。

（4）必要时，通过 `pthread_attr_destroy (&attr)` 释放 attr 资源。

下面列出了使用属性参数的一些示例。每个线程在创建时都默认可与其他线程连接。必要时，可使用分离属性创建一个线程，使它不能与其他线程连接。下面的代码段显示了如何创建一个分离线程。

```
pthread_attr_t attr;                // define an attr variable
pthread_attr_init(&attr);           // initialize attr
pthread_attr_setdetachstate(&attr, PTHREAD_CREATE_DETACHED); // set attr
pthread_create(&thread_id, &attr, func, NULL);  // create thread with attr
pthread_attr_destroy(&attr);        // optional: destroy attr
```

每个线程都使用默认堆栈的大小来创建。在执行过程中，线程可通过函数找到它的堆栈大小：

```
size_t pthread_attr_getstacksize()
```

它可以返回默认的堆栈大小。下面的代码段显示了如何创建具有特定堆栈大小的线程。

```
pthread_attr_t attr;              // attr variable
size_t stacksize;                 // stack size
pthread_attr_init(&attr);         // initialize attr
stacksize = 0x10000;              // stacksize=16KB;
pthread_attr_setstacksize (&attr, stacksize);   // set stack size in attr
pthread_create(&threads[t], &attr, func, NULL); // create thread with stack
                                                   size
```

如果 attr 参数为 NULL，将使用默认属性创建线程。实际上，这是创建线程的建议方法，除非有必要更改线程属性，否则应该遵循这种方法。接下来，我们将 attr 设置为 NULL，就可始终使用默认属性。

4.4.2 线程 ID

线程 ID 是一种不透明的数据类型，取决于实现情况。因此，不应该直接比较线程 ID。如果需要，可以使用 pthread_equal() 函数对它们进行比较。

```
int pthread_equal (pthread_t t1, pthread_t t2);
```

如果是不同的线程，则返回 0，否则返回非 0。

4.4.3 线程终止

线程函数结束后，线程即终止。或者，线程可以调用函数

```
int pthread_exit (void *status);
```

进行显式终止，其中状态是线程的退出状态。通常，0 退出值表示正常终止，非 0 值表示异常终止。

4.4.4 线程连接

一个线程可以等待另一个线程的终止，通过：

```
int pthread_join (pthread_t thread, void **status_ptr);
```

终止线程的退出状态以 status_ptr 返回。

4.5 线程示例程序

4.5.1 用线程计算矩阵的和

示例 4.1：假设我们要计算一个 N×N 整数矩阵中所有元素的和。这个问题可通过使用线程的并发算法来解决。在本示例中，主线程会先生成一个 N×N 整数矩阵。然后，它会创建 N 个工作线程，将唯一行号作为参数传递给各工作线程，并等待所有工作线程终止。每个工作线程计算不同行的部分和，并将部分和存入全局数组 int sum[N] 的相应行中。当所有工作线程计算完成后，主线程继续进行计算。它将工作线程生成的部分和相加来计算总和。下面列出了示例程序 C4.1 的完整 C 代码。在 Linux 下，程序必须编译为：

```
gcc  C4.1.c  -pthread
```

```
/**** C4.1.c file: compute matrix sum by threads ***/
```

```
#include <stdio.h>
#include <stdlib.h>
#include <pthread.h>
#define  N   4
int A[N][N], sum[N];

void *func(void *arg)              // threads function
{
   int j, row;
   pthread_t tid = pthread_self(); // get thread ID number
   row = (int)arg;                 // get row number from arg
   printf("Thread %d [%lu] computes sum of row %d\n", row, tid, row);
   for (j=0; j<N; j++)      // compute sum of A[row]in global sum[row]
       sum[row] += A[row][j];
   printf("Thread %d [%lu] done: sum[%d] = %d\n",
                  row, tid, row, sum[row]);
   pthread_exit((void*)0); // thread exit: 0=normal termination
}

int main (int argc, char *argv[])
{
   pthread_t thread[N];      // thread IDs
   int i, j, r, total = 0;
   void *status;
   printf("Main: initialize A matrix\n");
   for (i=0; i<N; i++){
     sum[i] = 0;
     for (j=0; j<N; j++){
       A[i][j] = i*N + j + 1;
       printf("%4d ", A[i][j]);
     }
     printf("\n");
   }
   printf("Main: create %d threads\n", N);
   for(i=0; i<N; i++) {
      pthread_create(&thread[i], NULL, func, (void *)i);
   }
   printf("Main: try to join with threads\n");
   for(i=0; i<N; i++) {
      pthread_join(thread[i], &status);
      printf("Main: joined with %d [%lu]: status=%d\n",
                                 i, thread[i], (int)status);
   }
   printf("Main: compute and print total sum: ");
   for (i=0; i<N; i++)
       total += sum[i];
   printf("tatal = %d\n", total);
   pthread_exit(NULL);
}
```

图 4.1 显示了运行示例程序 C4.1 的输出。它显示了各个线程的执行情况及其计算出的部分和，还演示了线程连接操作。

```
Main: initialize A matrix
   1    2    3    4
   5    6    7    8
   9   10   11   12
  13   14   15   16
Main: create 4 threads
Thread 0 [3075390272] computes sum of row 0
Thread 0 [3075390272] done: sum[0] = 10
Thread 3 [3050212160] computes sum of row 3
Thread 3 [3050212160] done: sum[3] = 58
Thread 2 [3058604864] computes sum of row 2
Thread 2 [3058604864] done: sum[2] = 42
Thread 1 [3066997568] computes sum of row 1
Main: try to join with threads
Thread 1 [3066997568] done: sum[1] = 26
Main: joined with 0 [3075390272]: status=0
Main: joined with 1 [3066997568]: status=0
Main: joined with 2 [3058604864]: status=0
Main: joined with 3 [3050212160]: status=0
Main: compute and print total sum: tatal = 136
```

图 4.1　示例 4.1 的样本输出

4.5.2　用线程快速排序

示例 4.2：用并发线程快速排序。

在本示例中，我们将通过线程实现一个并行快速排序程序。当程序启动时，它作为进程的主线程运行。主线程调用 qsort(&arg)，其中 arg =[lowerbound=0, upperbound=N-1]。qsort() 函数实现一个 N 个整数数组的快速排序。在 qsort() 中，线程会选择一个基准元素，将数组分成两部分，这样左边部分的所有元素都小于基准元素，右边部分的所有元素都大于基准元素。然后，它会创建两个子线程来对这两部分进行排序，并等待子线程完成。每个子线程通过相同的递归算法对自己的部分进行排序。当所有的子线程工作完成后，主线程继续工作。它会打印排序后的数组并终止。众所周知，快速排序的排序步骤数取决于未排序数据的量级，这将影响 qsort 程序所需的线程数。

```
/****** C4.2.c: quicksort by threads *****/
#include <stdio.h>
#include <stdlib.h>
#include <pthread.h>

typedef struct{
  int upperbound;
  int lowerbound;
}PARM;

#define N 10
int A[N] = {5,1,6,4,7,2,9,8,0,3};   // unsorted data

int print()    // print current a[] contents
{
  int i;
  printf("[ ");
  for (i=0; i<N; i++)
    printf("%d ", a[i]);
  printf("]\n");
}
```

```
void *qsort(void *aptr)
{
  PARM *ap, aleft, aright;
  int pivot, pivotIndex, left, right, temp;
  int upperbound, lowerbound;

  pthread_t me, leftThread, rightThread;
  me = pthread_self();
  ap = (PARM *)aptr;
  upperbound = ap->upperbound;
  lowerbound = ap->lowerbound;
  pivot = a[upperbound];          // pick low pivot value
  left = lowerbound - 1;          // scan index from left side
  right = upperbound;             // scan index from right side
  if (lowerbound >= upperbound)
     pthread_exit(NULL);

  while (left < right) {          // partition loop
    do { left++;} while (a[left] < pivot);
       do { right--;} while (a[right] > pivot);
       if (left < right ) {
          temp = a[left];
          a[left] = a[right];
          a[right] = temp;
       }
    }
    print();
    pivotIndex = left;            // put pivot back
    temp = a[pivotIndex];
    a[pivotIndex] = pivot;
    a[upperbound] = temp;
    // start the "recursive threads"
    aleft.upperbound = pivotIndex - 1;
    aleft.lowerbound = lowerbound;
    aright.upperbound = upperbound;
    aright.lowerbound = pivotIndex + 1;
    printf("%lu: create left and right threads\n", me);
    pthread_create(&leftThread,  NULL, qsort, (void *)&aleft);
    pthread_create(&rightThread, NULL, qsort, (void *)&aright);
    // wait for left and right threads to finish
    pthread_join(leftThread, NULL);
    pthread_join(rightThread, NULL);
    printf("%lu: joined with left & right threads\n", me);
}

int main(int argc, char *argv[])
{
    PARM arg;
    int i, *array;
    pthread_t me, thread;
    me = pthread_self();
    printf("main %lu: unsorted array = ", me);
    print();
    arg.upperbound = N-1;
    arg.lowerbound = 0;
```

```
    printf("main %lu create a thread to do QS\n", me);
    pthread_create(&thread, NULL, qsort, (void *)&arg);
    // wait for QS thread to finish
    pthread_join(thread, NULL);
    printf("main %lu sorted array = ", me);
    print();
}
```

图 4.2 显示了运行示例程序 C4.2 的输出，通过并发线程演示了并行快速排序。

```
main 3075553024: unsorted array = [ 5 1 6 4 7 2 9 8 0 3 ]
main 3075553024 create a thread to do QS
[ 0 1 2 4 7 6 9 8 5 3 ]
3075550016: create left and right threads
[ 0 1 2 3 7 6 9 8 5 4 ]
3067157312: create left and right threads
[ 0 1 2 3 7 6 9 8 5 4 ]
3057646400: create left and right threads
[ 0 1 2 3 4 6 9 8 5 7 ]
3049253696: create left and right threads
[ 0 1 2 3 4 6 5 8 9 7 ]
3003116352: create left and right threads
[ 0 1 2 3 4 6 5 7 9 8 ]
2986330944: create left and right threads
[ 0 1 2 3 4 6 5 7 8 9 ]
2994723648: create left and right threads
3049253696: joined with left & right threads
3067157312: joined with left & right threads
2986330944: joined with left & right threads
2994723648: joined with left & right threads
3003116352: joined with left & right threads
3057646400: joined with left & right threads
3075550016: joined with left & right threads
main 3075553024 sorted array = [ 0 1 2 3 4 5 6 7 8 9 ]
```

图 4.2　并行快速排序程序的输出

4.6　线程同步

由于线程在进程的同一地址空间中执行，它们共享同一地址空间中的所有全局变量和数据结构。当多个线程试图修改同一共享变量或数据结构时，如果修改结果取决于线程的执行顺序，则称之为**竞态条件**。在并发程序中，绝不能有竞态条件。否则，结果可能不一致。除了连接操作之外，并发执行的线程通常需要相互协作。为了防止出现竞态条件并且支持线程协作，线程需要同步。通常，同步是一种机制和规则，用于确保共享数据对象的完整性和并发执行实体的协调性。它可以应用于内核模式下的进程，也可以应用于用户模式下的线程。下面，我们将讨论 Pthread 中线程同步的具体问题。

4.6.1　互斥量

最简单的同步工具是锁，它允许执行实体仅在有锁的情况下才能继续执行。在 Pthread 中，锁被称为**互斥量**，意思是**相互排斥**。互斥变量是用 pthread_mutex_t 类型声明的，在使用之前必须对它们进行初始化。有两种方法可以初始化互斥量。

（1）一种是静态方法，如：

```
pthread_mutex_t m =  PTHREAD_MUTEX_INITIALIZER;
```

定义互斥量 m，并使用默认属性对其进行初始化。

（2）另一种是动态方法，使用 pthread_mutex_init() 函数，可通过 attr 参数设置互斥属性，如：

```
pthread_mutex_init (pthread_mutex_t *m, pthread_mutexattr_t,*attr);
```

通常，attr 参数可以设置为 NULL，作为默认属性。

初始化完成后，线程可通过以下函数使用互斥量。

```
int pthread_mutex_lock (pthread_mutex_t *m);      // lock mutex
int pthread_mutex_unlock (pthread_mutex_t *m);    // unlock mutex
int pthread_mutex_trylock (pthread_mutex_t *m);   // try to lock mutex
int pthread_mutex_destroy (pthread_mutex_t *m);   // destroy mutex
```

线程使用互斥量来保护共享数据对象。互斥量的典型用法如下。线程先创建一个互斥量并对它进行一次初始化。新创建的互斥量处于解锁状态，没有所有者。每个线程都试图访问一个共享数据对象：

```
pthread_mutex_lock(&m);    // lock mutex
 access shared data object; // access shared data in a critical region
pthread_mutex_unlock(&m);  // unlock mutex
```

当线程执行 pthread_mutex_lock(&m) 时，如果互斥量被解锁，它将封锁互斥量，成为互斥量的所有者并继续执行。否则，它将被阻塞并在互斥量等待队列中等待。只有获取了互斥量的线程才能访问共享数据对象。一次只能由一个执行实体执行的一系列执行通常称为**临界区**（CR）。在 Pthread 中，互斥量用来保护临界区，以确保临界区内在任何时候最多只能有一个线程。当线程完成共享数据对象时，它通过调用 pthread_mutex_unlock(&m) 来解锁互斥量，从而退出临界区。封锁的互斥量只能由当前所有者解锁。在解锁某互斥量时，如果互斥量等待队列中没有阻塞的线程，它就会解锁该互斥量，这时，互斥量没有所有者。否则，它会从互斥量等待队列中解锁等待线程，解锁的线程成为新的所有者，互斥量继续保持封锁。当所有线程都完成后，如果是动态分配的互斥量，则可能会被销毁。下面我们通过一个例子用互斥量来演示线程同步。

示例 4.3：本示例为示例 4.1 的修改版本。和前面一样，我们用 N 个工作线程来计算一个 N × N 整数矩阵所有元素的和。每个工作线程计算每一行的部分和。各工作线程不是将部分和存储在全局 sum[ ] 数组中，而是试图将部分和加到全局变量 total 中来更新它。由于所有工作线程都试图更新同一个全局变量，因此必须对它们进行同步，防止出现竞态条件。这可以通过互斥量来实现，确保在临界区内每次只能有一个工作线程可以更新 total 变量。下面给出的是示例程序 C4.3。

```
/** C4.3.c: matrix sum by threads with mutex lock **/
#include <stdio.h>
#include <stdlib.h>
#include <pthread.h>
#define  N   4
int A[N][N];

int total = 0;                // global total
pthread_mutex_t *m;           // mutex pointer
void *func(void *arg)         // working thread function
{
   int i, row, sum = 0;
```

```
    pthread_t tid = pthread_self(); // get thread ID number
    row = (int)arg;                 // get row number from arg
    printf("Thread %d [%lu] computes sum of row %d\n", row, tid, row);
    for (i=0; i<N; i++)        // compute partial sum of A[row]in
        sum += A[row][i];
    printf("Thread %d [%lu] update total with %d : ", row, tid, sum);
    pthread_mutx_lock(m);
      total += sum;            // update global total inside a CR
    pthread_mutex_unlock(m);
    printf("total = %d\n", total);
}

int main (int argc, char *argv[])
{
    pthread_t thread[N];
    int i, j, r;
    void *status;
    printf("Main: initialize A matrix\n");
    for (i=0; i<N; i++){
      sum[i] = 0;
      for (j=0; j<N; j++){
        A[i][j] = i*N + j + 1;
        printf("%4d ", A[i][j]);
      }
      printf("\n");
    }
    // create a mutex m
    m = (pthread_mutex_t *)malloc(sizeof(pthread_mutex_t));
    pthread_mutex_init(m, NULL); // initialize mutex m
    printf("Main: create %d threads\n", N);
    for(i=0; i<N; i++) {
       pthread_create(&thread[i], NULL, func, (void *)i);
    }
    printf("Main: try to join with threads\n");
    for(i=0; i<N; i++) {
       pthread_join(thread[i], &status);
       printf("Main: joined with %d [%lu]: status=%d\n",
               i, thread[i], (int)status);
    }
    printf("Main: tatal = %d\n", total);
    pthread_mutex_destroy(m); // destroy mutex m
    pthread_exit(NULL);
}
```

图 4.3 显示的是运行示例 4.3 的程序的样本输出，演示了 Pthread 中的互斥量。

4.6.2 死锁预防

互斥量使用封锁协议。如果某线程不能获取互斥量，就会被阻塞，等待互斥量解锁后再继续。在任何封锁协议中，误用加锁可能会产生一些问题。最常见和突出的问题是死锁。**死锁是一种状态，在这种状态下，许多执行实体相互等待，因此都无法继续下去。**为了说明这个问题，我们假设某线程 T1 获取了互斥量 m1，并且试图加锁另一个互斥量 m2。另一个线程 T2 获取了互斥量 m2，并且试图加锁互斥量 m1，如下图所示。

```
Thread T1:          |     Thread T2:
--------------------|-----------------
 lock(m1);          |     lock(m2);
 . . . .            |     . . . .
 lock(m2);          |     lock(m1);
-------------------------------------
```

```
Main: initialize A matrix
   1    2    3    4
   5    6    7    8
   9   10   11   12
  13   14   15   16
Main: create thread 0
Main: create thread 1
Main: create thread 2
Thread 0 [3076299584] computes sum of row 0
thread 0 [3076299584] update total with 10 : Thread 0: total = 10
Main: create thread 3
Thread 1 [3067906880] computes sum of row 1
thread 1 [3067906880] update total with 26 : Thread 1: total = 36
Thread 2 [3059514176] computes sum of row 2
thread 2 [3059514176] update total with 42 : Thread 2: total = 78
Thread 3 [3051019072] computes sum of row 3
thread 3 [3051019072] update total with 58 : Thread 3: total = 136
Main: joined with 0 [3076299584]: status=0
Main: joined with 1 [3067906880]: status=0
Main: joined with 2 [3059514176]: status=0
Main: joined with 3 [3051019072]: status=0
Main: tatal = 136
```

图 4.3　示例 4.3 的程序的输出

在这种情况下，T1 和 T2 将永远相互等待，由于交叉加锁请求，它们处于死锁状态。与竞态条件类似，死锁决不能存在于并发程序中。有多种方法可以解决可能的死锁问题，其中包括**死锁预防**、死锁规避、死锁检测和恢复等。在实际系统中，唯一可行的方法是死锁预防，试图在设计并行算法时防止死锁的发生。一种简单的死锁预防方法是对互斥量进行排序，并确保每个线程只在一个方向请求互斥量，这样请求序列中就不会有循环。

但是，仅使用单向加锁请求来设计每个并行算法是不可能的。在这种情况下，可以使用条件加锁函数 pthread_mutex_trylock() 来预防死锁。如果互斥量已被加锁，则 trylock() 函数会立即返回一个错误。在这种情况下，调用线程可能会释放它已经获取的一些互斥量以便进行退避，从而让其他线程继续执行。在上面的交叉加锁示例中，我们可以重新设计其中一个线程，例如 T1，利用条件加锁和退避来预防死锁。

```
                Thread T1:
-------------------------------------------
 while(1){
    lock(m1);
    if (!trylock(m2)) // if trylock m2 fails
       unlock(m1);    // back-off and retry
    else
       break;
    // delay some random time before retry
 }
-------------------------------------------
```

4.6.3　条件变量

作为锁，互斥量仅用于确保线程只能互斥地访问临界区中的共享数据对象。条件变量提供了一种线程协作的方法。条件变量总是与互斥量一起使用。这并不奇怪，因为互斥是所有同步机制的基础。在 Pthread 中，使用类型 pthread_cond_t 来声明条件变量，而且必须在使用前进行初始化。与互斥量一样，条件变量也可以通过两种方法进行初始化。

（1）一种是静态方法，在声明时，如：

```
pthread_cond_t con = PTHREAD_COND_INITIALIZER;
```

定义一个条件变量 con，并使用默认属性对其进行初始化。

（2）另一种是动态方法，使用 pthread_cond_init() 函数，可通过 attr 参数设置条件变量。为简便起见，我们总是使用 NULL attr 参数作为默认属性。

下面的代码段展示了如何使用条件变量。

```
pthread_mutex_t con_mutex; // mutex for a condition variable
pthread_cond_t  con;       // a condition variable that relies on con_mutex
pthread_mutex_init(&con_mutex, NULL);  // initialize mutex
pthread_cond_init(&con, NULL);         // initialize con
```

当使用条件变量时，线程必须先获取相关的互斥量。然后，它在互斥量的临界区内执行操作，然后释放互斥量，如下所示：

```
pthread_mutex_lock(&con_mutex);
    modify or test shared data objects
    use condition variable con to wait or signal conditions
pthread_mutex_unlock(&con_mutex);
```

在互斥量的临界区中，线程可通过以下函数使用条件变量来相互协作。

pthread_cond_wait(condition, mutex)：该函数会阻塞调用线程，直到发出指定条件的信号。当互斥量被加锁时，应调用该例程。它会在线程等待时自动释放互斥量。互斥量将在接收到信号并唤醒阻塞的线程后自动锁定。

pthread_cond_signal(condition)：该函数用来发出信号，即唤醒正在等待条件变量的线程或解除阻塞。它应在互斥量被加锁后调用，而且必须解锁互斥量才能完成 pthread_cond_wait()。

pthread_cond_broadcast(condition)：该函数会解除被阻塞在条件变量上的所有线程阻塞。所有未阻塞的线程将争用同一个互斥量来访问条件变量。它们的执行顺序取决于线程调度。

我们通过一个示例使用条件变量来演示线程协作。

4.6.4 生产者 - 消费者问题

示例 4.4：在本示例中，我们将使用线程和条件变量来实现一个简化版的**生产者 - 消费者**问题，也称**有限缓冲**问题。生产者 - 消费者问题通常将进程定义为执行实体，可看作当前上下文中的线程。下面是该问题的定义。

一系列生产者和消费者进程共享数量有限的缓冲区。每个缓冲区每次有一个特定的项目。最开始，所有缓冲区都是空的。当一个生产者将一个项目放入一个空缓冲区时，该缓冲区就会变满。当一个消费者从一个满的缓冲区中获取一个项目时，该缓冲区就会变空。如果没有空缓冲区，生产者必须等待。同样，如果没有满缓冲区，则消费者必须等待。此外，当等待事件发生时，必须允许等待进程继续。

在示例程序中，假设每个缓冲区都有一个整数值。共享全局变量定义为：

```
// shared global variables
int buf[NBUF];          // circular buffers
int head, tail;         // indices
int data;               // number of full buffers
```

缓冲区用作一系列循环缓冲区。索引变量 head 用于将一个项目放入空缓冲区，tail 则用于从满缓冲区中取出一个项目。变量数据就是满缓冲区的数量。为支持生产者和消费者之间的协作，我们定义了一个互斥量和两个条件变量。

```
pthread_mutex_t mutex;        // mutex lock
pthread_cond_t empty, full;   // condition variables
```

其中，empty 表示所有空缓冲区的条件，full 表示所有满缓冲区的条件。当生产者发现没有空缓冲区时，它会等待 empty 条件变量，当消费者使用了一个满缓冲区时，就会发出信号。同样，当消费者发现没有满缓冲区时，它会等待 full 条件变量，当生产者将一个项放入空缓冲区时，就会发出信号。

该程序从主线程开始，将缓冲区控制变量和条件变量初始化。初始化完成后，它会创建一个生产者线程和一个消费者线程，并等待线程结合。缓冲区大小设置为 NBUF=5，如果生产者试图将 N=10 个项放入缓冲区，当所有缓冲区已满时，它就要等待。同样，如果消费者试图从缓冲区获取 N=10 个项，将会导致它在所有缓冲区都空时等待。在这两种情况下，当等待条件得到满足时，另一个线程会通知正在等待的线程。因此，这两个线程通过条件变量相互协作。下面是示例程序 C4.4 的程序代码，实现了一个简版的生产者 - 消费者问题，即只有一个生产者和一个消费者。

```
/* C4.4.c: producer-consumer by threads with condition variables */
#include <stdio.h>
#include <stdlib.h>
#include <pthread.h>

#define NBUF     5
#define N       10

// shared global variables
int buf[NBUF];          // circular buffers
int head, tail;         // indices
int data;               // number of full buffers
pthread_mutex_t mutex;        // mutex lock
pthread_cond_t empty, full;   // condition variables

int init()
{
  head = tail = data = 0;
  pthread_mutex_init(&mutex, NULL);
  pthread_cond_init(&fullBuf, NULL);
  pthread_cond_init(&emptyBuf, NULL);
}

void *producer()
{
```

```
  int i;
  pthread_t me = pthread_self();
  for (i=0; i<N; i++){ // try to put N items into buf[ ]
      pthread_mutex_lock(&mutex);     // lock mutex
      if (data == NBUF){
          printf ("producer %lu: all bufs FULL: wait\n", me);
          pthread_cond_wait(&empty, &mutex); // wait
      }
      buf[head++] = i+1;              // item = 1,2,..,N
      head %= NBUF;                   // circular bufs
      data++;                         // inc data by 1
      printf("producer %lu: data=%d value=%d\n", me, bp->data, i+1);
      pthread_mutex_unlock(&mutex);   // unlock mutex
      pthread_cond_signal(&full);     // unblock a consumer, if any
  }
  printf("producer %lu: exit\n", me);
}

void *consumer()
{
  int i, c;
  pthread_t me = pthread_self();
  for (i=0; i<N; i++) {
      pthread_mutex_lock(&mutex);       // lock mutex
      if (data == 0) {
         printf ("consumer %lu: all bufs EMPTY: wait\n", me);
         pthread_cond_wait(&full, &mutex); // wait
      }
      c = buf[tail++];                  // get an item
      tail %= NBUF;
      data--;                           // dec data by 1
      printf("consumer %lu: value=%d\n", me, c);
      pthread_mutex_unlock(&mutex);     // unlock mutex
      pthread_cond_signal(&empty);      // unblock a producer, if any
  }
  printf("consumer %lu: exit\n", me);
}

int main ()
{
  pthread_t pro, con;
  init();
  printf("main: create producer and consumer threads\n");
  pthread_create(&pro, NULL, producer, NULL);
  pthread_create(&con, NULL, consumer, NULL);
  printf("main: join with threads\n");
  pthread_join(pro, NULL);
  pthread_join(con, NULL);
  printf("main: exit\n");
}
```

图 4.4 显示了运行生产者 - 消费者示例程序的输出。

```
main: create producer and consumer threads
main: join with threads
producer 3076275008: data=1 value=1
producer 3076275008: data=2 value=2
producer 3076275008: data=3 value=3
producer 3076275008: data=4 value=4
producer 3076275008: data=5 value=5
producer 3076275008: all bufs FULL: wait
consumer 3067882304: value=1
consumer 3067882304: value=2
consumer 3067882304: value=3
consumer 3067882304: value=4
consumer 3067882304: value=5
consumer 3067882304: all bufs EMPTY: wait
producer 3076275008: data=1 value=6
producer 3076275008: data=2 value=7
producer 3076275008: data=3 value=8
producer 3076275008: data=4 value=9
consumer 3067882304: value=6
consumer 3067882304: value=7
consumer 3067882304: value=8
consumer 3067882304: value=9
consumer 3067882304: all bufs EMPTY: wait
producer 3076275008: data=1 value=10
producer 3076275008: exit
consumer 3067882304: value=10
consumer 3067882304: exit
main: exit
```

图 4.4　生产者 - 消费者程序的输出

4.6.5　信号量

信号量是进程同步的一般机制。(计数）信号量是一种数据结构

```
struct sem{
   int value;              // semaphore (counter) value;
   struct process *queue   // a queue of blocked processes
}s;
```

在使用信号量之前，必须使用一个初始值和一个空等待队列进行初始化。不论是什么硬件平台，即无论是在单 CPU 系统还是多处理器系统上，信号量的低级实现保证了每次只能由一个执行实体操作每个信号量，并且从执行实体的角度来看，对信号量的操作都是（不可分割的）**原子**操作或**基本**操作。读者可以忽略这些细节，将重点放在信号量的高级操作及其作为进程同步机制的使用上。最有名的信号量操作是 P 和 V（Dijkstra 1965），定义见下文。

```
-------------------------------------------------------
P(struct sempahore *s)       |   V(struct semaphore *s)
{                            |   {
  s->value--;                |      s->value++;
  if (s->value < 0)          |      if (s->value <= 0)
     BLOCK(s);               |         SIGNAL(s);
}                            |   }
-------------------------------------------------------
```

当 BLOCK 阻塞信号量等待队列中的调用进程时，SIGNAL 会从信号量等待队列中释放一个进程。

信号量不是原始 Pthreads 标准的一部分。然而，现在大多数 Pthreads 都支持 POSIX 1003.1b 的信号量。POSIX 信号量包含一些函数：

```
int sem_init(sem, value) : initialize sem with an initial value
int sem_wait(sem)        : similar to P(sem)
int sem_post(sem)        : similar to V(sem)
```

信号量和条件变量之间的主要区别是，前者包含一个计数器，可操作计数器，测试计数器值以做出决策等，所有这些都是临界区的原子操作或基本操作，而后者需要一个特定的互斥量来执行临界区。在 Pthreads 中，互斥量严格用于封锁，而条件变量可用于线程协作。相反，可以把使用初始值 1 计算信号量当作锁。带有其他初始值的信号量可用于协作。因此，信号量比条件变量更通用、更灵活。下面的示例说明了信号量相对于条件变量的优势。

示例 4.5：使用信号量可以更有效地解决生产者 - 消费者问题。在该示例中，empty=N 和 full=0 是生产者和消费者相互协作的信号量，mutex =1 是一个锁信号量，进程一次只能访问临界区中的一个共享缓冲区。下面显示了使用信号量的生产者 - 消费者问题的伪代码。

```
ITEM buf[N];           // N buffers of ITEM type
int head=0, tail=0;    // buffer indices
struct semaphore empty=N, full=0, mutex=1; // semaphores

--------- Producer ------------- Consumer------------
  while(1){                  |   while(1){
    produce an item;         |      ITEM item;
    P(&empty);               |      P(&full);
    P(&mutex);               |      P(&mutex)
      buf[head++]=item;      |        item=buf[tail++];
      head %= N;             |        tail %= N;
    V(&mutex);               |      V(&mutex);
    V(&full);                |      V(&empty);
   }                         |   }
-----------------------------------------------------
```

4.6.6　屏障

线程连接操作允许某线程（通常是主线程）等待其他线程终止。在等待的所有线程都终止后，主线程可创建新线程来继续执行并行程序的其余部分。创建新线程需要系统开销。在某些情况下，保持线程活动会更好，但应要求它们在所有线程都达到指定同步点之前不能继续活动。在 Pthreads 中，可以采用的机制是**屏障**以及一系列屏障函数。首先，主线程创建一个屏障对象

```
pthread_barrier_t barrier;
```

并且调用

```
pthread_barrier_init(&barrier NULL, nthreads);
```

用屏障中同步的线程数字对它进行初始化。然后，主线程创建工作线程来执行任务。工作线程使用

```
pthread_barrier_wait( &barrier)
```

在屏障中等待指定数量的线程到达屏障。当最后一个线程到达屏障时，所有线程重新开始执行。在这种情况下，屏障是线程的集合点，而不是它们的坟墓。我们用一个例子来说明屏障的使用。

4.6.7 用并发线程解线性方程组

我们通过一个例子来演示并发线程和线程连接以及屏障操作的应用。

示例 4.6：该示例用并发线程来解一个线性方程组。假设 AX = B 是一个线性方程组，其中 A 是一个 N×N 的实数矩阵，X 是 N 个未知数的列向量，B 是常数的列向量。问题是计算解向量 X。求解线性方程组最著名的算法是**高斯消元算法**。该算法包含两个主要步骤，即行简化和回代：**行简化**将组合矩阵 [A|B] 简化为一个上三角矩阵，然后进行**回代**，计算解向量 X。在行简化步骤中，可采用**部分选主元法**，确保用于简化其他行的首行元素有最大绝对值。部分选主元有助于提高数值计算的精确度。下面给出了部分选主元高斯消元算法。

/****** Gauss Elimination Algorithm with Partial Pivoting *******/**
Step 1: Row reduction: reduce [A|B] to upper triangular form

```
          for (i=0; i<N ; i++){          // for rows i = 0 to N-1
            do partial pivoting;         // exchange rows if needed
 (1).       // barrier
            for (j=i+1; j<=N; j++){      // for rows j = i+1 to N
              for (k=i+1; k<=N; k++){ // for columns k = i+1 to N
                  f = A[j,i]/A[i,i];     // reduction factor
                  A[j,k] -= A[j,k]*f; // reduce row j
              }
              A[j][i] = 0;               // A[j,i] = 0
            }
 (2).       // barrier
          }
 (3). // join
```

Step 2: Back Substitution: compute xN-1, xN-2, . . . , x0 in that order

高斯消元算法可进行以下并行化。算法从主线程开始，创建 N 个工作线程来执行 ge(thread_id) 函数，并等待所有工作线程加入。线程函数 ge() 可实现高斯消元算法的行简化步骤。在 ge() 函数中，对于 row = 0 至 N-2 的每个迭代，thread_ID=i 的线程在对应的行 i 中进行部分选元。所有其他线程都在 barrier（1）处等待，直到部分选元完成。然后，各工作线程在行号等于线程 ID 号的唯一行上进行行简化。由于所有工作线程都必须在下一行开始迭代之前完成当前行简化，所以它们在另一个 barrier（2）处等待。将矩阵 [A|B] 简化为上三角矩阵后，最后一步是计算解 x[N-i]，因为 i=1 到 N 的顺序是固有顺序。主线程必须等到所有的工作线程都结束后才能开始回代。这可通过主线程的连接操作来实现。下面列出了示例程序 C4.5 的完整代码。

```
/** C4.5.c: Gauss Elimination with Partial Pivoting **/
#include <stdio.h>
#include <stdlib.h>
#include <math.h>
#include <pthread.h>

#define N 4
double A[N][N+1];
pthread_barrier_t barrier;
```

```
int print_matrix()
{
   int i, j;
   printf("------------------------------------\n");
   for(i=0; i<N; i++){
       for(j=0;j<N+1;j++)
          printf("%6.2f  ",  A[i][j]);
       printf("\n");
   }
}

void *ge(void *arg) // threads function: Gauss elimination
{
  int i, j, prow;
  int myid = (int)arg;
  double temp, factor;
  for(i=0; i<N-1; i++){
     if (i == myid){
        printf("partial pivoting by thread %d on row %d: ", myid, i);
        temp = 0.0; prow = i;
        for (j=i; j<=N; j++){
            if (fabs(A[j][i]) > temp){
               temp = fabs(A[j][i]);
               prow = j;
            }
      }
      printf("pivot_row=%d  pivot=%6.2f\n", prow, A[prow][i]);
      if (prow != i){  // swap rows
          for (j=i; j<N+1; j++){
              temp = A[i][j];
              A[i][j] = A[prow][j];
              A[prow][j] = temp;
          }
      }
   }
   // wait for partial pivoting done
   pthread_barrier_wait(&barrier);
   for(j=i+1; j<N; j++){
       if (j == myid){
           printf("thread %d do row %d\n", myid, j);
           factor = A[j][i]/A[i][i];
           for (k=i+1; k<=N; k++)
                A[j][k] -= A[i][k]*factor;
           A[j][i] = 0.0;
       }
   }
   // wait for current row reductions to finish
   pthread_barrier_wait(&barrier);
   if (i == myid)
       print_matrix();
   }
}

int main(int argc, char *argv[])
{
```

```
    int i, j;
    double sum;
    pthread_t threads[N];

    printf("main: initialize matrix A[N][N+1] as [A|B]\n");
    for (i=0; i<N; i++)
      for (j=0; j<N; j++)
          A[i][j] = 1.0;
    for (i=0; i<N; i++)
        A[i][N-i-1] = 1.0*N;
    for (i=0; i<N; i++){
        A[i][N] = 2.0*N - 1;
    }
    print_matrix();  // show initial matrix [A|B]

    pthread_barrier_init(&barrier, NULL, N); // set up barrier

    printf("main: create N=%d working threads\n", N);
    for (i=0; i<N; i++){
       pthread_create(&threads[i], NULL, ge, (void *)i);
    }

      printf("main: wait for all %d working threads to join\n", N);
      for (i=0; i<N; i++){
           pthread_join(threads[i], NULL);
      }
      printf("main: back substitution : ");
      for (i=N-1; i>=0; i--){
          sum = 0.0;
          for (j=i+1; j<N; j++)
              sum += A[i][j]*A[j][N];
              A[i][N] = (A[i][N]- sum)/A[i][i];
      }
      // print solution
      printf("The solution is :\n");
      for(i=0; i<N; i++){
          printf("%6.2f  ", A[i][N]);
      }
      printf("\n");
}
```

图4.5给出了运行示例4.6的程序的样本输出，用部分选主元高斯消元算法求解一个N=4个未知数的方程组。

4.6.8 Linux中的线程

与许多其他操作系统不同，Linux不区分进程和线程。对于Linux内核，线程只是一个与其他进程共享某些资源的进程。在Linux中，进程和线程都是由clone()系统调用创建的，具有以下原型：

```
int clone(int (*fn)(void *), void *child_stack, int flags, void *arg)
```

可以看出，clone()更像是一个线程创建函数。它创建一个子进程来执行带有child_stack的函数fn(arg)。flags字段详细说明父进程和子进程共享的资源，包括：

```
main: initialize matrix A[N][N+1] as [A|B]
------------------------------------
  1.00     1.00     1.00     4.00     7.00
  1.00     1.00     4.00     1.00     7.00
  1.00     4.00     1.00     1.00     7.00
  4.00     1.00     1.00     1.00     7.00
main: create N=4 working threads
partial pivoting by thread 0 on row 0: pivot_row=3  pivot=  4.00
main: wait for all 4 working threads to join
thread 1 do row 1
thread 2 do row 2
thread 3 do row 3
------------------------------------
  4.00     1.00     1.00     1.00     7.00
  0.00     0.75     3.75     0.75     5.25
  0.00     3.75     0.75     0.75     5.25
  0.00     0.75     0.75     3.75     5.25
partial pivoting by thread 1 on row 1: pivot_row=2  pivot=  3.75
thread 3 do row 3
thread 2 do row 2
partial pivoting by thread 2 on row 2: pivot_row=2  pivot=  3.60
------------------------------------
  4.00     1.00     1.00     1.00     7.00
  0.00     3.75     0.75     0.75     5.25
  0.00     0.00     3.60     0.60     4.20
  0.00     0.00     0.60     3.60     4.20
thread 3 do row 3
------------------------------------
  4.00     1.00     1.00     1.00     7.00
  0.00     3.75     0.75     0.75     5.25
  0.00     0.00     3.60     0.60     4.20
  0.00     0.00     0.00     3.50     3.50
main: back substitution : The solution is :
  1.00     1.00     1.00     1.00
```

图 4.5　示例 4.6 的程序的样本输出

- CLONE_VM：父进程和子进程共享地址空间
- CLONE_FS：父进程和子进程共享文件系统信息，例如根节点、CWD
- CLONE_FILES：父进程和子进程共享打开的文件
- CLONE_SIGHAND：父进程和子进程共享信号处理函数和已屏蔽信号

如果指定了任何标志寄存器，两个进程会共享同一套资源，而不是单独的一套资源副本。如未指定标志寄存器，则子进程通常会获得一套单独的资源副本。在这种情况下，一个进程对资源的更改不会影响另一个进程的资源。Linux 内核保留了 fork() 作为系统调用，但它可以实现为一个库包装函数，使用适当的标志寄存器来调用 clone()。普通用户不必担心这些细节问题。可以说，Linux 有一种支持线程的有效方法。而且，当前大多数 Linux 内核都支持对称多处理系统（SMP）。在这类 Linux 系统中，进程（线程）计划在多处理器上并行运行。

4.7　编程项目：用户级线程

该编程项目的目的是实现**用户级线程**，以模拟 Linux 中的线程操作。该项目由 4 个部分组成。第 1 部分（4.7.1 节和 4.7.2 节）介绍了项目的基本代码，以帮助读者入门。基本代码实现了一个多任务系统，支持在 Linux 进程中独立执行各项任务。它与第 3 章中介绍的多任务处理系统相同，但适用于用户级线程环境。第 2 部分（4.7.3 节）对基本代码做了扩展，以支持任务连接操作。第 3 部分（4.7.4 节和 4.7.5 节）也对基本代码做了扩展，以支持互斥量操作。第 4 部分（4.7.6 节和 4.7.7 节）实现了计数信号量，以支持任务协作，并演示多任务处理系统中的信号量。

4.7.1　项目基本代码：一个多任务处理系统

编程项目第1部分介绍基本代码，下面给出了编程项目的基本代码，后面几节将对此进行解释。

（1）32位GCC汇编代码中的ts.s文件：

```
#------------ ts.s file ------------
.global tswitch, running, scheduler
tswitch:
SAVE:      pushal
           pushfl
           movl   running,%ebx
           movl   %esp, 4(%ebx)   # integers in GCC are 4 bytes
FIND:      call   scheduler
RESUME:    movl   running, %ebx
           movl   4(%ebx), %esp
           popfl
           popal
           ret
```

（2）type.h文件：该文件定义了系统常数和PROC结构体

```
/*********** type.h file *************/
#define NPROC    9
#define SSIZE 1024
// PROC status
#define FREE     0
#define READY    1
#define SLEEP    2
#define BLOCK    3
#define ZOMBIE   4

typedef struct proc{
  struct proc *next;     // next proc pointer
  int ksp;               // saved stack pointer
  int pid;               // proc PID
  int priority;          // proc scheduling priority
  int status;            // current status: FREE|READY, etc.
  int event;             // for sleep/wakeup
  int exitStatus;        // exit status
  int joinPid;           // join target pid
  struct proc *joinPtr;  // join target PROC pointer
  int stack[SSIZE];      // proc 4KB stack area
}PROC;
```

（3）queue.c文件，该文件实现了队列操作函数

```
// same as in MT system of Chapter 3
```

（4）t.c文件，该文件是程序的main文件

```
#include <stdio.h>
#include "type.h"
PROC proc[NPROC];        // NPROC proc structures
PROC *freeList;          // free PROC list
```

```
PROC *readyQueue;         // ready proc priority queue
PROC *sleepList;          // sleep PROC list
PROC *running;            // running proc pointer
#include "queue.c"        // enqueue(), dequeue(), printList()

int init()
{
  int i, j;
  PROC *p;
  for (i=0; i<NPROC; i++){
    p = &proc[i];
    p->pid = i;
    p->priority = 0;
    p->status = FREE;
    p->event = 0;
    p->joinPid = 0;
    p->joinPtr = 0;
    p->next = p+1;
  }
  proc[NPROC-1].next = 0;
  freeList = &proc[0];    // all PROCs in freeList
  readyQueue = 0;
  sleepList  = 0;
  // create P0 as initial running task
  running = p = dequeue(&freeList);
  p->status = READY;
  p->priority = 0;
  printList("freeList", freeList);
  printf("init complete: P0 running\n");
}

int texit(int value)
{
  printf("task %d in texit value=%d\n", running->pid, running->pid);
  running->status = FREE;
  running->priority = 0;
  enqueue(&freeList, running);
  printList("freeList", freeList);
  tswitch();
}

int do_create()
{
  int pid = create(func, running->pid); // parm = pid
}

int do_switch()
{
  tswitch();
}

int do_exit()
{
  texit(running->pid); // for simplicity: exit with pid value
}
```

```
void func(void *parm)
{
  int c;
  printf("task %d start: parm = %d\n", running->pid, (int)parm);
  while(1){
    printf("task %d running\n", running->pid);
    printList("readyQueue", readyQueue);
    printf("enter a key [c|s|q] : ");
    c = getchar(); getchar();
    switch (c){
       case 'c' : do_create(); break;
       case 's' : do_switch(); break;
       case 'q' : do_exit(); break;
    }
  }
}

int create(void (*f)(), void *parm)
{
  int i;
  PROC *p = dequeue(&freeList);
  if (!p){
    printf("create failed\n");
    return -1;
  }
  p->status = READY;
  p->priority = 1;
  p->joinPid = 0;
  p->joinPtr = 0;

  // initialize new task stack for it to resume to f(parm)
  for (i=1; i<13; i++)                    # zero out stack cells
      p->stack[SSIZE-i] = 0;
  p->stack[SSIZE-1] = (int)parm;      # function parameter
  p->stack[SSIZE-2] = (int)do_exit;   # function return address
  p->stack[SSIZE-3] = (int)f;         # function entry
  p->ksp = (int)&p->stack[SSIZE-12]; # ksp -> stack top
  enqueue(&readyQueue, p);
  printList("readyQueue", readyQueue);
  printf("task %d created a new task %d\n", running->pid, p->pid);
  return p->pid;
}

int main()
{
  printf("Welcome to the MT User-Level Threads System\n");
  init();
  create((void *)func, 0);
  printf("P0 switch to P1\n");
  while(1){
    if (readyQueue)
       tswitch();
  }
}
```

```
int scheduler()
{
  if (running->status == READY)
     enqueue(&readyQueue, running);
  running = dequeue(&readyQueue);
  printf("next running = %d\n", running->pid);
}
```

若要在 Linux 下编译和运行基本代码，可输入：

```
gcc -m32 t.c ts.s
```

然后运行 a.out。在运行程序时，读者可以输入以下命令

- ‘c’：创建新任务。
- ‘s’：切换到从 readyQueue 运行下一个任务。
- ‘q’：让正在运行的任务终止。

以测试和观察系统中的任务执行情况。图 4.6 为运行基本代码程序的样本输出。

```
Welcome to the MT User-Level Threads System
freeList = [1 0]->[2 0]->[3 0]->[4 0]->[5 0]->[6 0]->[7 0]->[8 0]->NULL
init complete
readyQueue = [1 1]->NULL
task 0 created a new task 1
P0 switch to P1
next running = 1
task 1 start: parm = 0
task 1 running
readyQueue = [0 0]->NULL
enter a key [c|s|q] : c
readyQueue = [2 1]->[0 0]->NULL
task 1 created a new task 2
task 1 running
readyQueue = [2 1]->[0 0]->NULL
enter a key [c|s|q] : s
next running = 2
task 2 start: parm = 1
task 2 running
readyQueue = [1 1]->[0 0]->NULL
enter a key [c|s|q] : q
task 2 in texit value=2
freeList = [3 0]->[4 0]->[5 0]->[6 0]->[7 0]->[8 0]->[2 0]->NULL
next running = 1
task 1 running
readyQueue = [0 0]->NULL
enter a key [c|s|q] : █
```

图 4.6　基本代码程序的样本输出

4.7.2　用户级线程

整个基本代码程序在用户模式下以 Linux 进程运行。Linux 进程内有独立的执行实体，相当于传统的线程。为了不与 Linux 进程或线程混淆，我们把执行实体称为任务。在该系统中，多个任务同时执行。图 4.7 为 Linux 进程内的并发任务模型。

```
----------- a.out = Linux Process ---------
|              (concurrent tasks)           |
|     task1      task2  . . . .     taskn   |
|       |          |                  |     |
|                                           |
|-------------------------------------------
```

图 4.7　Linux 进程中的并发任务

因为所有的任务都在 Linux 进程中执行，所以 Linux 内核并不知道它们的存在。对于 Linux 内核来说，整个多任务处理系统是一个 Linux 进程，但是我们将细分 CPU

的执行时间，以在Linux进程中运行不同的任务。因此，这些任务被称为**用户级线程**。该编程项目的目的是在Linux进程中创建一个支持用户级线程的多任务环境。该技术也可用于在任何操作系统下创建用户级线程。下面，我们将通过以下步骤更详细地解释基本代码。

（1）任务PROC结构体：进程和线程之间的一个主要区别是前者遵从父子关系，而后者不遵从。在线程模型中，所有线程都是相等的。线程之间的关系是对等的，并非父子关系，所以不需要父进程任务PID和任务家族树。因此，PROC结构体中删除了这些字段。新字段joinPid和joinPtr用于实现线程连接操作，稍后将对此进行讨论。

（2）init()：当系统启动时，main()调用init()来初始化系统。init()初始化PROC结构体，并将它们输入freeList中。它还将readyQueue初始化为空。然后使用proc[0]创建P0，作为初始运行任务。P0的优先级最低，为0。所有其他任务的优先级都是1，因此它们将轮流从readyQueue运行。

（3）P0调用create()来创建一个任务P1，以执行参数parm=0的func(parm)函数，并将其输入readyQueue中。然后P0调用tswitch()，以切换任务来运行P1。

（4）tswitch()：tswitch()函数实现任务上下文切换。它使用PUSHAL和POPAL指令，而不是pushl/popl单个CPU寄存器。tswitch()就像一个任务交换箱，一个进程进入时通常另一个进程会出现。它由3个独立的步骤组成，下面将详细解释这些步骤。

（4）.1 tswitch()中的SAVE函数：当某个任务调用tswitch()时，它会把返回地址保存在自己的堆栈上，并在汇编代码中进入tswitch()。在tswitch()中，SAVE函数将CPU寄存器保存到调用任务的堆栈中，并将堆栈指针保存到proc.ksp中。32位Intel x86 CPU有许多寄存器，但在用户模式下，只有eax、ebx、ecx、edx、esp、ebp、esi、edi和eflag对用户模式进程可见。因此，我们只需要在执行（Linux）进程时保存和恢复这些寄存器。下图显示了在执行tswitch()的SAVE函数后，调用任务的堆栈内容和保存的堆栈指针，其中xxx表示调用tswitch()之前的堆栈内容。

```
                                              proc.ksp
                                                 |
|xxx|retPC|eax|ecx|edx|ebx|old_esp|ebp|esi|edi|eflags|
```

在基于Intel x86的32位PC中，每个CPU寄存器的宽度为4字节，堆栈操作始终以4字节为单位。因此，我们可以将PROC堆栈定义为一个整数数组。

（4）.2 scheduler()：在执行了tswitch()中的SAVE函数之后，任务调用scheduler()来选择下一个正在运行的任务。在scheduler()中，如果调用任务仍然可以运行，则会调用enqueue()将自己按优先级放入readyQueue中。否则，任务不会在readyQueue中，因此也就无法运行。然后，它会调用dequeue()，将从readyQueue中删除的第一个就绪的PROC作为新运行任务返回。

（4）.3 tswitch()中的RESUME函数：当“执行”从scheduler()返回时，“运行”可能已经转而指向另一个任务的PROC。tswitch()中的RESUME函数将CPU的堆栈指针设置为当前运行任务的已保存堆栈指针。然后弹出保存的寄存器，接着是弹出RET，使当前运行的任务返回到之前调用tswitch()的位置。

（5）create()：函数create(func, parm)可创建一个新任务并将其输入readyQueue中。新任务将从指定的func()函数开始执行，参数为parm。虽然新任务以前从未存在过，但我们可以假装它不仅存在过，而且运行过。它现在不运行是因为它调用了tswitch()，所以提前放弃了使用 CPU。如果是这样的话，它的堆栈必须包含 tswitch() 中的 SAVE 函数保存的一个帧，而且它保存的 ksp 必须指向栈顶。此外，当一个新任务从 func() 开始执行时可写为：func(void *parm)，它必须在堆栈上有一个返回地址和一个参数 parm。当 func() 完成时，任务将"返回到先前调用 func() 的地方"。因此，在 create() 中，我们按以下方式初始化新任务的堆栈。

```
                                                      proc.ksp
                       |< ------- all saved registers = 0- -----> |
|parm|do_exit|func|eax|ecx|edx|ebx|oldesp|ebp|esi|edi|eflags|
  -1     -2     -3   -4  -5  -6  -7    -8    -9 -10 -11   -12
```

其中索引 -i 表示 SSIZE-i。这些通过 create() 中的以下代码段完成。

```
for (i=1; i<13; i++)              # zero out stack cells
    p->stack[SSIZE-i] = 0;
p->stack[SSIZE-1] = (int)parm;    # function parameter
p->stack[SSIZE-2] = (int)do_exit; # function return address
p->stack[SSIZE-3] = (int)func;    # function entry
p->ksp = (int)&p->stack[SSIZE-12]; # ksp -> stack top
```

当新任务开始运行时，它首先执行 tswitch() 中的 RESUME 函数，使它返回 func() 函数的入口地址。当"执行"进入 func() 时，栈顶包含一个指向 do_exit() 的指针和一个参数 parm，就像从 do_exit() 的入口地址调用 func(parm) 一样。实际上，任务函数很少返回。如果返回，会返回到 do_texit()，导致任务终止。

（6）func()：为便于演示，所有创建的任务都执行同一个 func(parm) 函数。为方便起见，函数参数可设置为创建者任务的 pid，但是读者可将任何参数（如结构体指针）传递给函数。在 func() 中执行时，任务提示输入 char = [c | s | q] 命令，其中：

- 'c'：创建一个新任务来执行 func(parm)，parm= 调用者 pid。
- 's'：切换任务。
- 'q'：终止进程，并将 PROC 以 FREE 状态返回 freeList 中。

（7）空闲任务 P0：P0 的特殊之处在于它在所有任务中具有最低的优先级。在系统初始化之后，P0 创建 P1 并切换到运行 P1。当且仅当没有可运行任务时，P0 将会再次运行。在这种情况下，P0 会一直循环。读者可按下"Ctrl+C"组合键来终止（Linux）进程。注意，P0 的堆栈实际上是 Linux 进程的用户模式堆栈。

4.7.3 线程连接操作的实现

编程项目第 2 部分扩展基本代码，以支持任务连接操作。扩展代码按以下顺序列出。

（1）tsleep(int event)：当某个任务必须等待的事件未发生时，它就会在某个事件值上调用 tsleep() 进入休眠状态，该事件值表示休眠的原因。tsleep() 的算法如下：

```
/************ Algorithm of tsleep(int event) **************/
  1. record event value in proc.event:  running->event = event;
  2. change status to SLEEP:             running->status = SLEEP;
  3. for ease of maintenance, enter caller into a (global) PROC *sleepList:
                                         enqueue(&sleepList, running);
  4. give up CPU:                        tswitch();
```

休眠的任务在被另一个任务唤醒之前不可运行。因此，在 tsleep() 中，任务可调用 tswitch() 来放弃 CPU。

（2）int twakeup(int event)：twakeup() 可唤醒在指定事件值上休眠的所有任务。如果在该事件值上没有任务休眠，就不执行任何操作。twakeup() 的算法如下：

```
/*********** Algorithm of twakeup(int event) **********/
for each PROC *p in sleepList do{ // assume sleepers are in a global sleepList
     if (p->event == event){
        delete p from sleepList;
        p->status = READY;          // make p READY to run again
        enqueue(&readyQueue, p);    // enter p into readyQueue by priority
    }
 }
```

（3）texit(int status)：基本代码中修改的 texit() 用于终止任务。在进程模型中，每个进程都有一个唯一父进程（可能是 INIT 进程 P1），始终等待该进程终止。因此，一个终止进程可变成一个僵尸进程并唤醒它的父进程，父进程会处理僵尸子进程。如果终止进程有子进程，必须要在成为僵尸进程之前将子进程分发给 INIT 进程或 subreaper 进程。在线程模型中，所有线程都是平等的。线程没有父线程或子线程。当一个线程终止时，如果没有其他线程等待它终止，那么它必须以 FREE 状态退出。如果有线程等待它终止，它会成为僵尸线程，唤醒所有这类线程，让其中一个线程找到并处理僵尸线程。在线程模型中，线程终止算法如下：

```
/************ Algorithm of texit(int status) ************/
1.   try to find a (any) task which wants to join with this task;
2.   if no task wants to join with this task: exit as FREE
     //  some task is waiting for this task to terminate
3.   record status value in proc.exitStatus: running->exitStatus = status;
4.   become a ZOMBIE:                         running->status = ZOMBIE;
5.   wake up joining  tasks:                  twakeup(running->pid)
6.   give up CPU:                             tswitch();
```

（4）join(int targetPid, int *status)：连接操作等待带有 targetPid 的线程终止。在进程模型中，每个进程只能由唯一父进程等待，而一个进程永远不会等待自己的父进程终止。所以等待序列始终是单向的。在线程模型中，一个线程可与其他任何线程连接。这可能会导致两个问题。第一，当某线程试图与目标线程连接时，目标线程可能已经不存在了。在这种情况下，正在连接的线程会返回一个错误。如果目标线程存在但尚未终止，则连接线程会在目标线程 PID 上休眠，等待目标线程终止。第二，由于线程不遵从父子关系，因此一系列连接线程可能导致死锁以等待循环。为防止出现这类死锁，各连接线程必须在其 PROC 结构体中记录目标线程的身份，以允许其他连接线程检查连接请求是否会导致死锁。当目标线程终止时，它会检查是否有线程等待它终止。如果有，它会变成僵尸线程，并唤醒所有这些连接线程。被唤醒后，连接线程会再次执行连接操作，因为僵尸进程已经被另一个连接线程释放。因此，线程连接操作的算法如下。

```
/********* Algorithm of join(int targetPid, int *status) *********/
while(1)
{
  1.  if (no task with targetPid) return NOPID error;
```

```
  2. if (targetPid's joinPtr list leads to this task) return DEADLOCK error;
  3. set running->joinPid = targetPid, running->joinPtr ->targetPid's PROC;
  4. if (found targetPid's ZOMBIE proc *p){
       *status = p->exitStatus;  // extract ZOMBIE exit status
       p->status =  FREE; p->priority = 0;
       enqueue(&freeList, p); // release p to freeList
       return p->pid;
     }
  5. tsleep(targetPid); // sleep on targetPID until woken up by target task
}
```

join()算法的第2步可防止因交叉或循环连接请求而导致的死锁，可通过遍历targetPid的joinPtr指针列表的方法检查这些请求。各连接任务的joinPtr指向它打算连接的一个目标进程。如果列表向后指向当前运行的任务，它会是一个循环等待列表，由于可能造成死锁，所以必须拒绝循环等待列表。

编程项目的第2部分为读者实现了tsleep()、twakeup()和join()函数。然后，通过下面的程序代码使用编程项目的任务创建和连接操作来测试得到的系统。

```
/************ PART 2 of Programming Project ***********/
#include <stdio.h>
#include <stdlib.h>
#include "type.h"

PROC proc[NPROC];
PROC *freeList;
PROC *sleepList;
PROC *readyQueue;
PROC *running;

/****** implement these functions ******/
int tsleep(int event){ }
int twakeup(int event){ }
int texit(int status){ }
int join(int pid, int *status){ }
/****** end of implementations *********/

int init(){ }  // SAME AS in PART 1

int do_exit()
{
  // for simplicity: exit with pid value as status
  texit(running->pid);
}

void task1(void *parm) // task1: demonstrate create-join operations
{
  int pid[2];
  int i, status;
  //printf("task %d create subtasks\n", running->pid);
  for (i=0; i<2; i++){   // P1 creates P2, P3
      pid[i] = create(func, running->pid);
  }
  join(5, &status);      // try to join with targetPid=5
```

```
  for (i=0; i<2; i++){  // try to join with P2, P3
      pid[i] = join(pid[i], &status);
      printf("task%d joined with task%d: status = %d\n",
             running->pid, pid[i], status);
  }
}

void func(void *parm)  // subtasks: enter q to exit
{
  char c;
  printf("task %d start: parm = %d\n", running->pid, parm);
  while(1){
    printList("readyQueue", readyQueue);
    printf("task %d running\n", running->pid);
    printf("enter a key [c|s|q|j]: ");
    c = getchar(); getchar(); // kill \r
    switch (c){
       case 'c' : do_create(); break;
       case 's' : do_switch(); break;
       case 'q' : do_exit();   break;
       case 'j' : do_join();   break;
    }
  }
}

int create(void (*f)(), void *parm)
{
  int i;
  PROC *p = dequeue(&freeList);
  if (!p){
    printf("create failed\n");
    return -1;
  }
  p->status = READY;
  p->priority = 1;
  p->joinPid = 0;
  p->joinPtr = 0;
  for (i=1; i<13; i++)
      p->stack[SSIZE-i] = 0;
  p->stack[SSIZE-1] = (int)parm;
  p->stack[SSIZE-2] = (int)do_exit;
  p->stack[SSIZE-3] = (int)f;
  p->ksp = &p->stack[SSIZE-12];

  enqueue(&readyQueue, p);
  printList("readyQueue", readyQueue);
  printf("task%d created a new task%d\n", running->pid, p->pid);
  return p->pid;
}

int main()
{
  int i, pid, status;
  printf("Welcome to the MT User-Threads System\n");
  init();
```

```
    create((void *)task1, 0);
    printf("P0 switch to P1\n");
    tswitch();
    printf("All tasks ended: P0 loops\n");
    while(1);
}

int scheduler()
{
    if (running->status == READY)
        enqueue(&readyQueue, running);
    running = dequeue(&readyQueue);
}
```

在测试程序中，P0 创建了一个新的任务 P1，来执行函数 task1()。当 P1 运行时，它创建两个新任务 P2 和 P3 来执行函数 func()。P1 先尝试连接 targetPid =5，然后再尝试与 P2 和 P3 连接。图 4.8 显示了正在运行项目程序第 2 部分的样本输出，演示了任务连接操作。如图所示，当 P1 尝试与 P5 进行连接时，由于 targetPid 无效，会产生连接错误。然后 P1 尝试与任务 2 连接，会导致它等待。当 P2 运行时，它试图与 P1 连接，但 P1 拒绝连接，因为这会导致死锁。然后 P2 退出，将任务切换到运行 P3。当 P3 退出时，P1 仍在等待与 P2 连接，因此 P3 以 FREE 状态退出，允许 P1 完成与 P2 的连接操作。当 P1 试图与 P3 连接时，由于 P3 已经不存在，因此会得到一个无效 targetPid 错误。读者可以先不让 P3 退出，而是让 P3 切换到 P1，处理僵尸进程 P2，然后再尝试与 P3 连接。在这种情况下，当 P3 退出时，P1 将成功地完成与 P3 的连接操作。读者可修改测试程序来创建更多的任务，并输入不同的命令序列来测试系统。

```
Welcome to the MT User-Level Threads System
freeList = 1 --> 2 --> 3 --> 4 --> 5 --> 6 --> 7 --> 8 --> NULL
init complete
task0 create a new task 1
P0 switch to P1
task 1 running
task1 create a new task 2
task1 create a new task 3
task1 try to join with task5: join error: no such pid 5
task1 try to join with task2: sleepList = 1 --> NULL
readyQueue = 3 --> 0 --> NULL
task 2 running: enter a key [c|s|q|j]: j
enter a pid to join with : 1
task2 try to join with task1: join error: DEADLOCK
readyQueue = 3 --> 0 --> NULL
task 2 running: enter a key [c|s|q|j]: q
readyQueue = 3 --> 1 --> 0 --> NULL
task 2 exited with status = 2
readyQueue = 1 --> 0 --> NULL
task 3 running: enter a key [c|s|q|j]: q
task 3: no joiner=>exit as FREE: task 3 exited with status = 3
task1 joined with task2 status = 2
task1 try to join with task3: join error: no such pid 3
task 1: no joiner=>exit as FREE: task 1 exited with status = 1
All tasks ended: P0 loops
```

图 4.8 任务连接程序的样本输出

4.7.4 互斥量操作的实现

编程项目第 3 部分扩展项目的第 2 部分以支持互斥量操作。扩展代码如下文所述。

互斥量结构体

```
typedef struct mutex{
  int  lock;      // mutex lock state: 0 for unlocked, 1 for locked
  PROC *owner;    // pointer to owner of mutex; may also use PID
  PROC *queue;    // FIFO queue of BLOCKED waiting PROCs
}MUTEX;

MUTEX *mutex_create() // create a mutex and initialize it
{
  MUTEX *mp = (MUTEX *)malloc(sizeof(MUTEX));
  mp->lock = mp->owner = mp->queue = 0;
  return mp;
}

void mutex_destroy(MUTEX *mp){ free(mp); }
int mutex_lock(MUTEX *mp){  // implement mutex locking operation }
int mutex_unlock(MUTEX *mp){// implement mutex unlocking operation }

/******** Algorithm of mutex_lock() ************/
(1). if (mutex is in unlocked state){
        change mutex to locked state;
        record caller as mutex owner;
     }
(2). else{
        BLOCK caller in mutex waiting queue;
        switch task;
     }

/******* Algorithm of mutex_unlock() *********/
(1). if (mutex is unlocked ||(mutex is locked && caller NOT owner))
        return error;
(2). // mutex is locked && caller is owner
     if (no waiter in mutex waiting queue){
        change mutex to unlocked state;
        clear mutex owner field to 0;
     }
(3). else{ // mutex has waiters
        PROC *p = dequeue a waiter from mutex waiting queue;
        change mutex owner to p; // mutex remains locked
        enter p into readyQueue;
     }
```

4.7.5　用并发程序测试有互斥量的项目

编程项目第3部分实现多任务处理系统的mutex_lock()和mutex_unlock()函数。用下面的程序测试互斥量操作。测试程序为使用Pthreads的示例4.3的修改版。它通过多任务处理系统的并发任务和互斥量来计算矩阵元素的和。在测试程序中，P0要初始化一个4×4矩阵，并创建P1来执行task1()。P1要创建4个工作任务来执行函数func(parm)。每个工作任务计算行的部分和，并使用互斥量更新全局变量total以实现同步。输出如图4.9所示。

```
/************ test program for mutex operations ************/
#include <stdio.h>
#include <stdlib.h>
```

```
#include "type.h"

PROC proc[NPROC], *freeList, *sleepList, *readyQueue, *running;

#include "queue.c"  // queue operation functions
#include "wait.c"   // tsleep/twakeup/texit/join functions
#include "mutex.c"  // mutex operation functions

#define N 4
int A[N][N];         // matrix A
MUTEX *mp;           // mutex for task synchronization
int total;           // global total

int init()
{
  int i, j;
  PROC *p;
  for (i=0; i<NPROC; i++){
    p = &proc[i];
    p->pid = i;
    p->ppid = 1;
    p->priority = 0;
    p->status = FREE;
    p->event = 0;
    p->next = p+1;
  }
  proc[NPROC-1].next = 0;
  freeList = &proc[0];
  readyQueue = 0;
  sleepList  = 0;
  p = running = dequeue(&freeList);
  p->status = READY;
  p->priority = 0;
  printList("freeList", freeList);

  printf("P0: initialize A matrix\n");
  for (i=0; i<N; i++){
     for (j=0; j<N; j++){
       A[i][j] = i*N + j + 1;
     }
  }

  for (i=0; i<N; i++){   // show the matrix
     for (j=0; j<N; j++){
       printf("%4d ", A[i][j]);
     }
     printf("\n");
  }
  mp = mutex_create();   // create a mutex
  total = 0;
  printf("init complete\n");
}

int myexit()             // for task exit
{
```

```
  texit(0);
}

void func(void *arg)
{
  int i, row, s;
  int me = running->pid;
  row = (int)arg;
  printf("task %d computes sum of row %d\n", me, row);
  s = 0;
  for (i=0; i < N; i++){
     s += A[row][i];
  }
  printf("task %d update total with %d\n", me, s);
  mutex_lock(mp);
   total += s;
   printf("[total = %d] ", total);
  mutex_unlock(mp);
}

void task1(void *parm)
{
  int pid[N];
  int i, status;
  int me = running->pid;
  printf("task %d: create working tasks : ", me);
  for(i=0; i < N; i++) {
     pid[i] = create(func, (void *)i);
     printf("%d ", pid[i]);

  }
  printf(" to compute matrix row sums\n");
  for(i=0; i<N; i++) {
    printf("task %d tries to join with task %d\n",
                 running->pid, pid[i]);
    join(pid[i], &status);
  }
  printf("task %d : total = %d\n", me, total);
}

int create(void (*f)(), void *parm)
{
  int i;
  PROC *p = dequeue(&freeList);
  if (!p){
    printf("fork failed\n");
    return -1;
  }
  p->ppid = running->pid;
  p->status = READY;
  p->priority = 1;
  p->joinPid = 0;
  p->joinPtr = 0;
```

```
    for (i=1; i<12; i++){
      p->stack[SSIZE-i] = 0;
      p->stack[SSIZE-1] = (int)parm;
      p->stack[SSIZE-2] = (int)myexit;
      p->stack[SSIZE-3] = (int)f;
      p->ksp = &p->stack[SSIZE-12];
    }
    enqueue(&readyQueue, p);
    return p->pid;
}

int main()
{
    printf("Welcome to the MT User-Level Threads System\n");
    init();
    create((void *)task1, 0);
    //printf("P0 switch to P1\n");
    tswitch();
    printf("all task ended: P0 loops\n");
    while(1);
}

int scheduler()
{
    if (running->status == READY)
       enqueue(&readyQueue, running);
    running = dequeue(&readyQueue);
}
```

图 4.9 显示了运行互斥量测试程序的输出样本。

```
Welcome to the MT User-Level Threads System
freeList = 1 --> 2 --> 3 --> 4 --> 5 --> 6 --> 7 --> 8 --> NULL
P0: initialize A matrix
   1   2   3   4
   5   6   7   8
   9  10  11  12
  13  14  15  16
init complete
task 1: create working tasks : 2 3 4 5  to compute matrix row sums
task 1 tries to join with task 2
sleepList = 1 --> NULL
task 2 computes sum of row 0
task 2 update total with 10
task 2 mutex_lock() [total = 10] task 2 mutex_unlock()
task 3 computes sum of row 1
task 3 update total with 26
task 3 mutex_lock() [total = 36] task 3 mutex_unlock()
task 4 computes sum of row 2
task 4 update total with 42
task 4 mutex_lock() [total = 78] task 4 mutex_unlock()
task 5 computes sum of row 3
task 5 update total with 58
task 5 mutex_lock() [total = 136] task 5 mutex_unlock()
task 1 tries to join with task 3
task 1 tries to join with task 4
task 1 tries to join with task 5
task 1 : tatal = 136
all task ended: P0 loops
```

图 4.9　通过并发任务和互斥量计算矩阵和

4.7.6　信号量的实现

编程项目第4部分实现4.6.5节中所述的计数信号量。通过下面的测试程序演示信号量，示例4.5中的生产者－消费者问题将信号量用于多任务处理系统。

4.7.7　使用信号量实现生产者－消费者问题

```
/********** test program for semaphore operations ********/
#include <stdio.h>
#include "type.h"
PROC proc[NPROC], *freeList, *sleepList, *readyQueue, *running;
#include "queue.c"    // queue operation function
#include "wait.c"     // tsleep/twakeup/texit/join functions

#define NBUF  4
#define N     8
int buf[NBUF], head, tail;    // buffers for producer-consumer

typedef struct{
   int value;
   PROC *queue;
}SEMAPHORE;
SEMAPHORE full, empty, mutex; // semaphores
int P(SEMAPHORE *s)
{  // implement P function }

int V(SEMAPHORE *s)
{  // implement V function }

void producer()              // produce task code
{
  int i;
  printf("producer %d start\n", running->pid);
  for (i=0; i<N; i++){
    P(&empty);
    P(&mutex);
     buf[head++] = i+1;
     printf("producer %d: item = %d\n", running->pid, i+1);
     head %= NBUF;
    V(&mutex);
    V(&full);
  }
  printf("producer %d exit\n", running->pid);
}

void consumer()              // consumer task code
{
  int i, c;
  printf("consumer %d start\n", running->pid);
  for (i=0; i<N; i++) {
    P(&full);
    P(&mutex);
     c = buf[tail++];
     tail %= NBUF;
```

```
    printf("consumer %d: got item = %d\n", running->pid, c);
    V(&mutex);
    V(&empty);
  }
  printf("consumer %d exit\n", running->pid);
}

int init()
{
  int i, j;
  PROC *p;
  for (i=0; i<NPROC; i++){
    p = &proc[i];
    p->pid = i;
    p->ppid = 1;
    p->priority = 0;
    p->status = FREE;
    p->event = 0;
    p->next = p+1;
  }
  proc[NPROC-1].next = 0;
  freeList = &proc[0];
  readyQueue = 0;
  sleepList  = 0;
  p = running = dequeue(&freeList);
  p->status = READY;
  p->priority = 0;
  printList("freeList", freeList);

  // initialize semaphores full, empty, mutex
  head = tail = 0;
  full.value = 0;  full.queue = 0;
  empty.value=NBUF; empty.queue = 0;
  mutex.value = 1; mutex.queue = 0;
  printf("init complete\n");
}
int myexit(){  texit(0); }

int task1()
{
  int status;
  printf("task %d creates producer-consumer tasks\n", running->pid);
  create((void *)producer, 0);
  create((void *)consumer, 0);
  join(2, &status);
  join(3, &status);
  printf("task %d exit\n", running->pid);
}

int create(void (*f)(), void *parm)
{
  int i;
  PROC *p = dequeue(&freeList);
  if (!p){
    printf("fork failed\n");
```

```
    return -1;
  }
  p->ppid = running->pid;
  p->status = READY;
  p->priority = 1;
  p->joinPid = 0;
  p->joinPtr = 0;
  for (i=1; i<12; i++){
    p->stack[SSIZE-i] = 0;
    p->stack[SSIZE-1] = (int)parm;
    p->stack[SSIZE-2] = (int)myexit;
    p->stack[SSIZE-3] = (int)f;
    p->ksp = &p->stack[SSIZE-12];
  }
  enqueue(&readyQueue, p);
  return p->pid;
}

int main()
{
  printf("Welcome to the MT User-Level Threads System\n");
  init();
  create((void *)task1, 0);
  printf("P0 switch to P1\n");
  tswitch();
  printf("all task ended: P0 loops\n");
  while(1);
}

int scheduler()
{
  if (running->status == READY)
     enqueue(&readyQueue, running);
  running = dequeue(&readyQueue);
}
```

图 4.10 显示了使用信号量实现生产者 - 消费者问题的示例输出。

编程项目方案可在线下载。如有需求，可面向讲师提供编程项目源代码。

4.8　习题

1. 修改示例程序 C4.1，通过并发线程找到一个 N × N 矩阵的最大元素值。
2. 在示例程序 C4.2 的 qsort() 函数中，选择未排序区间中最右边的元素作为基准元素。修改示例程序 C4.2，使用最左边的元素作为基准元素，比较两种情况下所需的排序步骤数。
3. 修改示例程序 C4.4，以创建多个生产者和消费者线程。
4. 使用信号量实现生产者 - 消费者问题，如示例 4.5 所示。
5. 在示例程序 C4.5 中，工作线程数量 N 等于 A 矩阵的维数。如果 N 值较大，可能该维数不可取。修改程序，以使用 NTHREADS 工作线程，其中 NTHREADS <=N。例如，当 N=8 时，NTHREADS 可以是 2、4 或任何 <=N 的数值。
6. 修改示例程序 C4.1，通过多任务处理系统中的并发任务计算矩阵中所有元素的和。结果应与示例 4.1 相同。
7. 假设在用户级线程多任务处理系统中，每个任务都与某个任务相连接。当系统结束时，应该不会剩下僵尸任务。实现 texit() 和 join() 来满足以上要求。

```
Welcome to the MT User-Level Threads System
freeList = 1 --> 2 --> 3 --> 4 --> 5 --> 6 --> 7 --> 8 --> NULL
init complete
P0 switch to P1
task 1 creates producer-consumer tasks
sleepList = 1 --> NULL
producer 2 start
producer 2: item = 1
producer 2: item = 2
producer 2: item = 3
producer 2: item = 4
task 2 block  on sem=0x80541e8
consumer 3 start
consumer 3: got item = 1
task 3 V up 2 on sem=0x80541e8
consumer 3: got item = 2
consumer 3: got item = 3
consumer 3: got item = 4
task 3 block  on sem=0x805420c
producer 2: item = 5
task 2 V up 3 on sem=0x805420c
producer 2: item = 6
producer 2: item = 7
producer 2: item = 8
producer 2 exit
consumer 3: got item = 5
consumer 3: got item = 6
consumer 3: got item = 7
consumer 3: got item = 8
consumer 3 exit
task 1 exit
all task ended: P0 loops
■
```

图 4.10 使用信号量实现生产者 – 消费者问题

参考文献

Buttlar, D, Farrell, J, Nichols, B., “PThreads Programming, A POSIX Standard for Better Multiprocessing”, O’Reilly Media, 1996

Dijkstra, E.W., Co-operating Sequential Processes, Programming Languages, Academic Press, 1965

Goldt, S,. van der Meer, S. Burkett, S. Welsh, M. The Linux Programmer’s Guide, Version 0.4. March 1995

IBM MVS Programming Assembler Services Guide, Oz/OS V1R11.0, IBM

Love, R. Linux Kernel Development, 2nd Edition, Novell Press, 2005

POSIX.1C, Threads extensions, IEEE Std 1003.1c, 1995

Pthreads: https://computing.llnl.gov/tutorials/pthreads, 2017

The Linux Man page Project: https://www.kernel.org/doc/man-pages, 2017

定时器及时钟服务

摘要

本章讨论了定时器和定时器服务；介绍了硬件定时器的原理和基于 Intel x86 的 PC 中的硬件定时器；讲解了 CPU 操作和中断处理；描述了 Linux 中与定时器相关的系统调用、库函数和定时器服务命令；探讨了进程间隔定时器、定时器生成的信号，并通过示例演示了进程间隔定时器。编程项目的目的是要在一个多任务处理系统中实现定时器、定时器中断和间隔定时器。多任务处理系统作为一个 Linux 进程运行，该系统是 Linux 进程内并发任务的一个虚拟 CPU。Linux 进程的实时模式间隔定时器被设计为定期生成 SIGALRM 信号，充当虚拟 CPU 的定时器中断，虚拟 CPU 使用 SIGALRM 信号捕捉器作为定时器的中断处理程序。该项目可让读进程通过定时器队列实现任务间隔定时器，还可让读进程使用 Linux 信号掩码来实现临界区，以防止各项任务和中断处理程序之间出现竞态条件。

5.1 硬件定时器

定时器是由时钟源和可编程计数器组成的硬件设备。时钟源通常是一个晶体振荡器，会产生周期性电信号，以精确的频率驱动计数器。使用一个倒计时值对计数器进行编程，每个时钟信号减 1。当计数减为 0 时，计数器向 CPU 生成一个**定时器中断**，将计数值重新加载到计数器中，并重复倒计时。计数器周期称为定时器刻度，是系统的基本计时单元。

5.2 个人计算机定时器

基于 Intel x86 的个人计算机有数个定时器（Bovet 和 Cesati 2005）。

（1）实时时钟（RTC）：RTC 由一个小型备用电池供电。即使在个人计算机关机时，它也能连续运行。它用于实时提供时间和日期信息。当 Linux 启动时，它使用 RTC 更新系统时间变量，以与当前时间保持一致。在所有类 Unix 系统中，时间变量是一个长整数，包含从 1970 年 1 月 1 日起经过的秒数。

（2）可编程间隔定时器（PIT）(Wang 2015）：PIT 是与 CPU 分离的一个硬件定时器。可对它进行编程，以提供以毫秒为单位的定时器刻度。在所有 I/O 设备中，PIT 可以最高优先级 IRQ0 中断。PIT 定时器中断由 Linux 内核的定时器中断处理程序来处理，为系统操作提供基本的定时单元，例如进程调度、进程间隔定时器和其他许多定时事件。

（3）多核 CPU 中的本地定时器（Intel 1997；Wang 2015）：在多核 CPU 中，每个核都是一个独立的处理器，它有自己的本地定时器，由 CPU 时钟驱动。

（4）高分辨率定时器：大多数电脑都有一个时间戳定时器（TSC），由系统时钟驱动。它的内容可通过 64 位 TSC 寄存器读取。由于不同系统主板的时钟频率可能不同，TSC 不适合作为实时设备，但它可提供纳秒级的定时器分辨率。一些高端个人计算机可能还配备有专用高速定时器，以提供纳秒级定时器分辨率。

5.3 CPU 操作

每个 CPU 都有一个程序计数器（PC），也称为指令指针（IP），以及一个标志或状态寄存器（SR）、一个堆栈指针（SP）和几个通用寄存器，当 PC 指向内存中要执行的下一条指令时，SR 包含 CPU 的当前状态，如操作模式、中断掩码和条件码，SP 指向当前堆栈栈顶。堆栈是 CPU 用于特殊操作（如 push、pop 调用和返回等）的一个内存区域。CPU 操作可通过无限循环进行建模。

```
while(power-on){
  (1). fetch instruction: load *PC as instruction, increment PC to point to the
       next instruction in memory;
  (2). decode instruction: interpret the instruction's operation code and
       generate operands;
  (3). execute instruction: perform operation on operands, write results to
       memory if needed; execution may use the stack, implicitly change PC,
       etc.
  (4). check for pending interrupts; may handle interrupts;
}
```

在以上各步骤中，由于无效地址、非法指令、越权等原因，可能会出现一个错误状态，称为**异常**或陷阱。当 CPU 遇到异常时，它会根据内存中预先安装的指针来执行软件中的异常处理程序。在每条指令执行结束时，CPU 会检查挂起的中断。**中断**是 I/O 设备或协处理器发送给 CPU 的外部信号，请求 CPU 服务。如果有挂起的中断请求，但是 CPU 未处于接受中断的状态，即它的状态寄存器已经屏蔽了中断，CPU 会忽略中断请求，继续执行下一条指令。否则，它将直接执行中断处理。在中断处理结束时，它将恢复指令的正常执行。中断处理和异常处理都在操作系统内核中进行。在大多数情况下，用户级程序无法访问它们，但它们是理解操作系统（如 Linux）定时器服务和信号的关键。

5.4 中断处理

外部设备（如定时器）的中断被馈送到**中断控制器**的预定义输入行（Intel 1990；Wang 2015），按优先级对中断输入排序，并将具有最高优先级的中断作为中断请求（IRQ）路由到 CPU。在每条指令执行结束时，如果 CPU 未处于接受中断的状态，即在 CPU 的状态寄存器中屏蔽了中断，它将忽略中断请求，使其处于挂起状态，并继续执行下一条指令。如果 CPU 处于接受中断状态，即中断未被屏蔽，那么 CPU 将会转移它正常的执行顺序来进行中断处理。对于每个中断，可以编程中断控制器以生成一个唯一编号，叫作**中断向量**，标识中断源。在获取中断向量号后，CPU 用它作为内存中**中断向量表**（AMD64 2011）中的条目索引，条目包含一个指向**中断处理程序**入口地址的指针来实际处理中断。当中断处理结束时，CPU 恢复指令的正常执行。

5.5 时钟服务函数

在几乎所有的操作系统（OS）中，操作系统内核都会提供与时钟相关的各种服务。时钟服务可通过系统调用、库函数和用户级命令调用。在本节中，我们将介绍 Linux 的一些基本时钟服务函数。

5.5.1 gettimeofday-settimeofday

```
#include <sys/time.h>
int gettimeofday(struct timeval *tv, struct timezone *tz);
int settimeofday(const struct timeval *tv, const struct timezone *tz);
```

这些是对 Linux 内核的系统调用。第一个参数 tv 指向一个 timeval 结构体。

```
struct timeval {
      time_t          tv_sec;          /* seconds */
      suseconds_t     tv_usec;         /* microseconds */
};
```

第二个参数 timezone 已过期，应设置为 NULL。gettimeofday() 函数用于返回当前时间（当前秒的秒和微秒）。settimeofday() 函数用于设置当前时间。在 Unix/Linux 中，时间表示自 1970 年 1 月 1 日 00:00:00 起经过的秒数。它可以通过库函数 ctime(&time) 转换为日历形式。下面给出了 gettimeofday() 函数和 settimeofday() 函数的示例。

1. gettimeofday 系统调用

示例 5.1：通过 gettimeofday() 获取系统时间。

```
/********* gettimeofday.c file *********/
#include <stdio.h>
#include <stdlib.h>
#include <sys/time.h>

struct timeval t;

int main()
{
  gettimeofday(&t, NULL);
  printf("sec=%ld usec=%d\n", t.tv_sec, t.tv_usec);
  printf((char *)ctime(&t.tv_sec));
}
```

程序应以秒、微秒显示当前时间，并以日历形式显示当前日期和时间，如：

```
sec=1515624303 usec=860772
Wed Jan 10 14:45:03 2018
```

2. settimeofday 系统调用

示例 5.2：通过 settimeofday() 设置系统时间。

```
/********* settimeofday.c file *********/
#include <stdio.h>
#include <stdlib.h>
#include <sys/time.h>
#include <time.h>

struct timeval t;

int main()
{
```

```
  int r;
  t.tv_sec = 123456789;
  t.tv_usec= 0;
  r = settimeofday(&t, NULL);
  if (!r){
     printf("settimeofday() failed\n");
     exit(1);
  }
  gettimeofday(&t, NULL);
  printf("sec=%ld usec=%ld\n", t.tv_sec, t.tv_usec);
  printf("%s", ctime(&t.tv_sec)); // show time in calendar form
}
```

程序的输出显示如下：

```
sec=123456789 usec=862
Thu Nov 29 13:33:09 1973
```

根据打印日期和年份（1973）来看，settimeofday()操作似乎已经成功。但是，在某些Linux系统中，如在Ubuntu 15.10中，产生的结果可能只是暂时的。如果读进程再次运行gettimeofday程序，结果将会显示Linux已经将系统时间更改回正确的实时时间。这表明Linux内核能够使用实时时钟（和其他时间同步协议）来纠正系统时间与实时时间之间的任何偏差。

5.5.2 time系统调用

示例5.3：time系统调用。

time_t time(time_t *t)

以秒为单位返回当前时间。如果参数t不是NULL，还会将时间存储在t指向的内存中。time系统调用具有一定的局限性，只提供以秒为单位的分辨率，而不是以微秒为单位。该示例说明了如何获取以秒为单位的系统时间。

```
/************ time.c file ***********/
#include <stdio.h>
#include <time.h>

time_t start, end;

int main()
{
  int i;
  start = time(NULL);
  printf("start=%ld\n", start);
  for (i=0; i<123456789; i++); // delay to simulate computation
  end = time(NULL);
  printf("end  =%ld time=%ld\n", end, end-start);
}
```

输出应打印开始时间、结束时间以及从开始到结束的秒数。

5.5.3 times 系统调用

times 系统调用

clock_t times(struct tms *buf);

可用于获取某进程的具体执行时间。它将进程时间存储在 struct tms buf 中，即：

```
struct tms{
        clock_t tms_utime;   // user mode time
        clock_t tms_stime;   // system mode time
        clock_t tms_cutime;  // user time of children
        clock_t tms_cstime;  // system time of children
};
```

以时钟计时单元报告所有时间。这可以为分析某个正在执行的进程提供信息，包括其子进程的时间（如有）。

5.5.4 time 和 date 命令

- date：打印或设置系统日期和时间。
- time：报告进程在用户模式和系统模式下的执行时间和总时间。
- hwclock：查询并设置硬件时钟（RTC），也可以通过 BIOS 来完成。

5.6 间隔定时器

Linux 为每个进程提供了三种不同类型的间隔计时器，可用作进程计时的虚拟时钟。间隔定时器由 setitimer() 系统调用创建。getitimer() 系统调用返回间隔定时器的状态。

```
int getitimer(int which, struct itimerval *curr_value);
int setitimer(int which, const struct itimerval *new_value,
              struct itimerval *old_value);
```

各间隔定时器在参数 which 指定的不同时间域中工作。当间隔定时器定时到期时，会向进程发送一个信号，并将定时器重置为指定的间隔值（如果是非零数）。一个**信号**就是发送给某个进程进行处理的一个数字（1 到 31）。信号和信号处理将在第 6 章中讨论。有 3 类间隔定时器，分别是：

（1）ITIMER_REAL：实时减少，在到期时生成一个 SIGALRM（14）信号。

（2）ITIMER_VIRTUAL: 仅当进程在用户模式下执行时减少，在到期时生成一个 SIGVTALRM（26）信号。

（3）ITIMER_PROF: 当进程正在用户模式和系统（内核）模式下执行时减少。这类间隔定时器与 ITIMER_VIRTUAL 结合使用，通常用于分析应用程序在用户模式和内核模式下花费的时间。它在到期时生成一个 SIGPROF（27）信号。

间隔定时器的值用以下结构体（在 <sys/time.h> 中）定义：

```
struct itimerval {
   struct timeval it_interval;   /* interval for periodic timer */
   struct timeval it_value;      /* time until next expiration */
};
struct timeval {
```

```
    time_t      tv_sec;          /* seconds */
    suseconds_t tv_usec;         /* microseconds */
};
```

函数 getitimer() 用当前值填充 curr_value 指向的结构体，即参数 which（ITIMER_REAL、ITIMER_VIRTUAL 或 ITIMER_PROF 三者之一）指定的定时器在下次到期之前剩余的时间。将 it_value 字段的子字段设置为定时器上的剩余时间，如果定时器被禁用，则设置为 0。将 it_interval 字段设置为定时器间隔（周期）；如果该字段中的返回值（两个子字段）为 0，则表明这是 singleshot 定时器。

函数 setitimer() 将指定定时器设置为 new_value 中的值。如果 old_value 为非 NULL，定时器的原来值，即 getitimer() 返回的信息，会存储在那里。

周期定时器从 it_value 逐渐减小到 0，生成一个信号，并重置为 it_interval。设置为 0 的定时器（it_value 为 0 或定时器到期以及 it_interval 为 0）可将定时器停止。tv_sec 和 tv_usec 对确定定时器的持续时间都有重要影响。下面我们通过示例来演示进程间隔定时器。

示例 5.4：该示例展示了如何设置 VIRTUAL 模式的间隔定时器，该定时器仅在进程以用户模式执行时才减少计时。该定时器设置为完成最初 100 毫秒计时后开始计时。然后，它以 1 秒为周期运行。当定时器计时减少为 0 时，它会向进程发出一个 SIGVTALRM（26）信号。如果进程未安装该信号的捕捉器，将会对该信号进行默认处理，即终止。在这种情况下，进程将以信号数 26 终止。如果进程安装了信号捕捉器，Linux 内核会让进程执行信号捕捉器，以用户模式处理信号。在间隔时间开始之前，程序通过以下代码安装 SIGVTALRM 信号的信号捕捉器：

```
void timer_handler(int sig){ . . . .}
signal(SIGALRM, timer_handler)
```

安装信号捕捉器后，程序启动定时器，然后在 while(1) 循环中执行。当在循环中执行时，每个硬件中断（例如来自硬件定时器的中断）都会导致 CPU 以及在 CPU 上执行的进程进入 Linux 内核来处理中断。当进程处于内核模式时，会检查待处理信号。如有待处理信号，它会试图先处理信号再返回用户模式。在这种情况下，SIGVTALRM 信号将导致进程在用户模式下执行信号捕捉器。由于信号定时器程序设计为每秒生成一个信号，进程将每秒执行一次 timer_handler()，使打印消息像脉冲星一样每秒显示一次。信号捕捉函数 timer_handler() 可计算定时器的时间结束次数。当计数达到规定值（例如 8）时，它用定时器值 0 来取消 setitimer() 设置的间隔定时器。虽然定时器已经停止，但进程仍在无限 while(1) 循环中执行。在这种情况下，从键盘按下“Ctrl+C”组合键，可以使进程以 SIGINT（2）信号终止。信号和信号处理的具体内容将在第 6 章中讨论。下面列出了示例 5.5 的 setitimer 程序代码。

```
/*********** setitimer.c file *********/
#include <signal.h>
#include <stdio.h>
#include <sys/time.h>
int count = 0;
struct itimerval t;

void timer_handler(int sig)
```

```
{
  printf("timer_handler: signal=%d count=%d\n", sig, ++count);
  if (count>=8){
     printf("cancel timer\n");
     t.it_value.tv_sec = 0;
     t.it_value.tv_usec = 0;
     setitimer(ITIMER_VIRTUAL, &t, NULL);
  }
}

int main()
{
  struct itimerval timer;
  // Install timer_handler as SIGVTALRM signal handler
  signal(SIGVTALRM, timer_handler);
  // Configure the timer to expire after 100 msec
  timer.it_value.tv_sec = 0;
  timer.it_value.tv_usec = 100000; // 100000 nsec
  // and every 1 sec afterward
  timer.it_interval.tv_sec = 1;
  timer.it_interval.tv_usec = 0;
  // Start a VIRTUAL itimer
  setitimer(ITIMER_VIRTUAL, &timer, NULL);
  printf("looping: enter Control-C to terminate\n");
  while(1);
}
```

图 5.1 显示运行 setitimer 程序的输出。

```
looping: enter Control-C to terminate
timer_handler: signal=26 count=1
timer_handler: signal=26 count=2
timer_handler: signal=26 count=3
timer_handler: signal=26 count=4
timer_handler: signal=26 count=5
timer_handler: signal=26 count=6
timer_handler: signal=26 count=7
timer_handler: signal=26 count=8
cancel timer
^C
```

图 5.1　settitimer 程序的输出

5.7 REAL 模式间隔定时器

VIRTUAL 和 PROF 模式下的间隔计时器仅在执行进程时才有效。这类定时器的信息可保存在各进程的 PROC 结构体中。（硬件）定时器中断处理程序只需要访问当前运行进程的 PROC 结构体，就可以减少定时器计时，在定时结束时重新加载定时器计时，并向进程生成一个信号。操作系统内核不必使用额外的数据结构来处理进程的 VIRTUAL 和 PROF 定时器。但是，REAL 模式间隔定时器各不相同，因为无论进程是否正在执行，它们都必须由定时器中断处理程序来更新。因此，操作系统内核必须使用额外的数据结构来处理进程的 REAL 模式定时器，并在定时器到期或被取消时采取措施。在大多数操作系统内核中，使用的数据结构都是定时器队列。我们将在本章末尾解释编程项目中的定时器队列。

5.8 编程项目

本编程项目将在一个有并发执行任务的多任务处理系统中实现定时器、定时器中断和间隔定时器。编程项目包含四个步骤。第 1 步（5.8.1 节）是提供多任务系统的基本代码以启动读进程。第 2 步（5.8.2 节）是向基本系统添加定时器和定时器中断。第 3 步（5.8.3 节）是实现任务间隔定时器。第 4 步（5.8.4 节）是通过时间片在系统和任务调度中实现临界区。该项目的目的是不仅让读者学习如何使用定时器，还要学习如何在操作系统内核中实现定时器。

5.8.1 系统基本代码

多任务处理系统的基本代码与第 4 章的用户级线程编程项目大致相同。为便于参考完整代码，我们在这里再次列出基本代码。下面列出了多任务处理系统的基本代码。它由 32 位汇编中的 ts.s 文件和 C 语言中的 t.c 文件组成。

```
#------------ ts.s file -------------
.global tswitch, scheduler, running
tswitch:
SAVE:    pushal
         pushfl
         movl running, %ebx
         movl %esp, 4(%ebx)
FIND:    call scheduler
RESUME:  movl running, %ebx
         movl 4(%ebx), %esp
         popfl
         popal
         ret

/********* t.c file *********/
#include <stdio.h>
#include <stdlib.h>
#include <signal.h>
#include <string.h>
#include <sys/time.h>

#define NPROC    9
#define SSIZE 1024
// PROC status
#define FREE     0
#define READY    1
#define SLEEP    2
#define BLOCK    3
#define PAUSE    4
#define ZOMBIE   5

typedef struct proc{
  struct proc *next;
  int ksp;             // saved sp when NOT running
  int pid;             // task PID
  int priority;        // task priority
  int status;          // status=FREE|READY, etc.
  int event;           // sleep event
  int exitStatus;
  int joinPid;
  struct proc joinPtr;
  int time;            // time slice in ticks
  int pause;           // pause time in seconds
  int stack[SSIZE];    // per task stack
}PROC;
```

```
PROC proc[NPROC];      // task PROCs
PROC *freeList, *readyQueue, *running;
PROC *sleepList;       // list of SLEEP tasks
PROC *pauseList;       // list of PAUSE tasks

#include "queue.c"     // same queue.c file as before
#include "wait.c"      // tsleep, twakeup, texit, join functions
int menu() // command menu: to be expanded later
{
  printf("*********** menu ***********\n");
  printf("* create  switch  exit  ps *\n");
  printf("****************************\n");
}

int init()
{
  int i, j;
  PROC *p;
  for (i=0; i<NPROC; i++){
    p = &proc[i];
    p->pid = i;
    p->priority = 1;
    p->status = FREE;
    p->event = 0;
    p->next = p+1;
  }
  proc[NPROC-1].next = 0;
  freeList = &proc[0];      // all PROCs in freeList
  readyQueue = 0;
  sleepList  = 0;
  pauseList  = 0;
  // create P0 as initial running task
  running = dequeue(&freeList);
  running->status = READY;
  running->priority = 0;  // P0 has lowest priority 0
  printList("freeList", freeList);
  printf("init complete: P0 running\n");
}

int do_exit()  // task exit as FREE
{
  printf("task %d exit: ", running->pid);
  running->status = FREE;
  running->priority = 0;
  enqueue(&freeList, running);
  printList("freeList", freeList);
  tswitch();
}

int do_ps()    // print task status
{
  printf("--------- ps ----------\n");
  printList("readyQueue", readyQueue);
  printList("sleepList ", sleepList);
  printf("----------------------\n");
```

```
}

int create(void (f)(), void *parm) // create a new task
{
  int i;
  PROC *p = dequeue(&freeList);
  if (!p){
     printf("create failed\n");
     return -1;
  }
  p->ppid = running->pid;
  p->status = READY;
  p->priority = 1;
  for (i=1; i<12; i++)
      p->stack[SSIZE-i] = 0;
  p->stack[SSIZE-1] = (int)parm;
  p->stack[SSIZE-2] = (int)do_exit;
  p->stack[SSIZE-3] = (int)f;
  p->ksp = &p->stack[SSIZE-12];
  enqueue(&readyQueue, p);
  printf("%d created a new task %d\n", running->pid, p->pid);
  return p->pid;
}

void func(void *parm) // task function
{
  char line[64], cmd[16];
  printf("task %d start: parm = %d\n", running->pid, parm);
  while(1){
    printf("task %d running\n", running->pid);
    menu();
    printf("enter a command line: ");
    fgets(line, 64, stdin);
    line[strlen(line)-1] = 0; // kill \n at end of line
    sscanf(line, "%s", cmd);
    if (strcmp(cmd, "create")==0)
       create((void *)func, 0);
    else if (strcmp(cmd, "switch")==0)
       tswitch();
    else if (strcmp(cmd, "exit")==0)
       do_exit();
    else if (strcmp(cmd, "ps")==0)
       do_ps();
  }
}

int main()
{
  int i;
  printf("Welcome to the MT multitasking system\n");
  init();
  for (i=1; i<5; i++)     // create tasks
      create(func, 0);
  printf("P0 switch to P1\n");
  while(1){
```

```
    if (readyQueue)
       tswitch();
  }
}

int scheduler()
{
  if (running->status == READY)
     enqueue(&readyQueue, running);
  running = dequeue(&readyQueue);
  printf("next running = %d\n", running->pid);
}
```

读者可查看第 4 章中的 create() 函数，该函数可创建一个新任务来执行指定函数。若要在 Linux 下编译并运行基本系统，可输入：

gcc –m32 t.c ts.s # assembly code ts.s is for 32-bit Linux

然后运行 a.out。当任务运行时，它会显示一个命令菜单，其中的命令有：

- create：创建新任务。
- switch：切换任务。
- exit：任务退出。
- ps：显示任务状态。

图 5.2 显示了运行多任务处理系统基本代码的样本输出。

```
Welcome to the MT multitasking system
freeList = 1 -> 2 -> 3 -> 4 -> 5 -> 6 -> 7 -> 8 -> NULL
init complete
task 0 created new task 1
task 0 created new task 2
task 0 created new task 3
task 0 created new task 4
P0 switch to P1
task 1 start: parm = 0
task 1 running
*********** menu ***********
* create  switch  exit  ps *
****************************
enter a command line: switch
task 2 start: parm = 0
task 2 running
*********** menu ***********
* create  switch  exit  ps *
****************************
enter a command line: █
```

图 5.2 基本多任务处理系统的样本输出

5.8.2 定时器中断

整个基本系统在一个虚拟 CPU 上运行，它是一个 Linux 进程。定时器向 Linux 进程发出的信号可看作是对基本系统虚拟 CPU 的中断。为 Linux 进程创建一个 REAL 模式间隔定时器。编写间隔定时器程序，每 10 毫秒生成一个 SIGALRM 信号。安装一个 SIGALRM 信号捕捉器，作为虚拟 CPU 的定时器中断处理程序。在定时器中断处理程序中，记录经过的秒数、分钟数和小时数。下面给出了所需的扩展代码段。

```
void thandler(int sig)
{
  // count the number of timer ticks; update ss, mm, hh
  // print a message every second
}
signal(SIGALRM, thandler);          // install SIGALRM catcher
struct itimerval t;                 // configure timer
t.it_value.tv_sec = 0;
t.it_value.tv_usec = 10000;         // start in 10 msec
t.it_interval.tv_sec = 0;
t.it_interval.tv_usec = 10000;      // period = 10 msec
setitimer(ITIMER_REAL, &t, NULL);   // start REAL mode timer
```

第 2 步的输出应与图 5.2 类似，不同的是它每秒钟显示一条消息。

5.8.3 定时器队列

在基本系统中为任务添加间隔定时器支持。添加命令：

- pause t：任务暂停 t 秒。
- timer t：任务设置一个间隔为 t 秒的（REAL 模式）间隔定时器。

pause 命令会让一个任务休眠指定的秒数，在暂停时间到期时再唤醒该任务。设置间隔定时器后，任务可以继续执行，当它的定时器到期时，信号将通知该任务。由于多任务处理系统还不能生成和发送信号，我们假设任务在执行定时器命令后将进入休眠状态，当定时器过期时，该任务将被唤醒。

正如 5.7 节中所指出，系统必须使用定时器队列来记录任务的 REAL 模式定时器状态。定时器队列由定时器队列元素（TQE）的条目组成，格式为：

```
typedef struct tqe{
   struct tqe *next;
   PROC *proc;         // pointer to requesting process
   int   time;         // expiration time count
   void (action)()     // action function = twakeup
}TQE;
TQE *timerQueue = 0; // initialized to NULL
```

定时器队列元素可从定时器队列元素池中分配和取消分配。由于每个进程只能有一个 REAL 模式定时器，所以定时器队列元素的最大数量等于进程数量。必要时，可以把它们包含到 PROC 结构体中。这里，我们假设定时器队列元素是从一个空闲定时器队列元素池中动态分配或取消分配的。当某进程请求 REAL 模式间隔定时器时，分配一个定时器队列元素来记录请求的进程、到期时间和定时器到期时所采取的操作。然后，它会将定时器队列元素输入 timerQueue 中。最简单的 timerQueue 是按到期时间从小到大排序的链表。如果有多个 REAL 模式定时器请求，则该 timerQueue 会如下图所示，其中的数字表示定时器队列元素的到期时间。

```
timerQueue = TQE1 ->TQE2 ->TQE3
               2      7      15
```

（硬件）定时器中断处理程序每秒减少1个定时器队列元素。当定时器队列元素的时间减为0时，将会从timerQueue中删除，并调用动作函数。默认操作是向进程生成SIGALRM（14）信号，但也可能是其他类型的操作。通常，一个（硬件）中断处理程序应尽快运行完。上述timerQueue组织方式的缺点是定时器中断处理程序每次只能减少1个定时器队列元素，非常耗时。更好的方法是将定时器队列元素保存在一个**累积**队列中，其中，每个定时器队列元素的到期时间都与前面所有定时器队列元素的到期时间累积和有关。下面给出了一个累积timerQueue，和原来的timerQueue相同。

```
timerQueue = TQE1 ->TQE2 ->TQE3
               2      5      8
```

使用这种tiemrQueue组织方式，定时器中断处理程序只需要减少第一个定时器队列元素并处理所有时间到期的定时器队列元素，这大大加快了定时器中断处理速度。

该项目第3步是为Linux进程添加一个REAL模式间隔定时器，为多任务处理系统中的任务实现间隔定时器。

图5.3所示为运行该这类系统的样本输出。如图所示，task 1请求一个间隔6秒的定时器。Task 2请求一个间隔2秒的定时器，比task 1晚4秒。此时，累积timerQueue的内容如下所示：

timerQueue = [2, 2] =>[1, 2]

在每个定时器队列元素中，第一个数字是任务PID，第二个数字是到期时间。定时器中断处理程序每秒只减少task 2的第一个定时器队列元素。当task 2的定时器队列元素到期时，它会从timerQueue中删除该定时器队列元素，并唤醒task 2。之后，task 1的定时器队列元素成为timerQueue中的第一个元素。下一秒，它会减少task 1的定时器队列元素。同样地，pause命令会让一个任务休眠t秒，在暂停时间到期时再唤醒该任务。

```
Welcome to the multitasking system
freeList = 1 -> 2 -> 3 -> 4 -> 5 -> 6 -> 7 -> 8 -> NULL
P0 creates tasks
readyQueue = 1 -> 2 -> 3 -> 4 -> NULL
task 1 running
**************** menu ******************
* create  switch  exit  pause  ps  timer *
*****************************************
enter a command line: timer 6
timerQueue = [1, 6] =>
task 2 running
**************** menu ******************
* create  switch  exit  pause  ps  timer *
*****************************************
enter a command line: timerQueue = [1, 5] =>
timer timerQueue = [1, 4] =>
2
timerQueue = [2, 2] => [1, 2] =>
task 3 running
**************** menu ******************
* create  switch  exit  pause  ps  timer *
*****************************************
enter a command line: timerQueue = [2, 1] => [1, 2] =>
timerQueue = [2, 0] => [1, 2] =>
timer wakeup task 2
timerQueue = [1, 1] =>
timerQueue = [1, 0] =>
timer wakeup task 1
■
```

图5.3　任务间隔定时器的样本输出

5.8.4 临界区

在基本代码系统中，只有一种执行实体，即任务，一次只执行一个任务。某任务在收到切换命令、进入休眠或退出之前，会一直执行下去。此外，任务切换只会发生在操作结束时，而不会发生在任何操作过程中。因此，任务之间没有竞争，因此在基本代码系统中没有**临界区**。但是，一旦我们将中断引入系统，情况就会改变。有两种类型的实体来执行中断，分别是任务和中断处理程序，它们可能会争夺系统中的同一（共享）数据对象。例如，当某任务请求间隔定时器时，必须将请求作为定时器队列元素输入 timerQueue 中。当某任务修改 timerQueue 时，如果出现定时器中断，它将转移任务以执行中断处理程序，可能会改动同一 timerQueue，造成竞态条件。因此，timerQueue 是临界区，必须对它进行保护，以确保它一次只能由一个执行实体访问。同样，当某进程在 sleep() 函数过程中执行时，可能被转移到执行中断处理程序，即可执行 wakeup()，以试图在进程完成休眠操作之前唤醒它，从而导致另一个竞态条件。所以，问题是如何防止任务和中断处理程序相互干扰。

下面是临界区的实现。

当某中断处理程序执行时，任务在逻辑上不能执行，这样任务就无法干扰中断处理程序，但反之则不然。当某任务执行时，可能会发生定时器中断，将该任务转移到执行中断处理程序，这可能会干扰任务。为了防止出现这种情况，执行任务只需屏蔽临界区中的中断即可。如前文所述，多任务处理系统在一个虚拟 CPU 上运行，该 CPU 是一个 Linux 进程。对于多任务处理系统而言，定时器中断是发送给 Linux 进程的 SIGALRM 信号。在 Linux 中，除 SIGKILL（9）和 SIGSTOP（19）之外的信号都可通过以下方法进行阻塞或解锁。

```
(1). sigset_t sigmask, oldmask;           // define signal mask sets
(2). sigemptyset(&sigmask);               // initialize signal mask set to empty
(3). sigaddset(&sigmask, SIGALRM);  // add signal numbers to the set
(4). To block signals specified in the mask set, issue the system call
     sigprocmask(SIG_BLOCK, &sigmask, &oldmask);
(5). To unblock signals specified in the mask set, issue the system call
     sigprocmask(SIG_UNBLOCK, &sigmask, &oldmask);
(6). For convenience, the reader may use the following functions to block/unblock
     signals
     int_off(){ sigprocmask(SIG_BLOCK, &sigmask, &oldmask); }
     int_on(){ sigprocmask(SIG_UNBLOCK, &oldmask, &sigmask); }
```

该编程项目的第 4 步是使用 Linux 的 sigprocmask() 在多任务处理系统临界区阻塞或取消阻塞定时器中断，以确保任务和定时器中断处理程序之间不存在竞态条件。

5.8.5 高级主题

作为高级研究的一个主题，读者可尝试通过时间片来实现任务调度，并讨论其影响。在基于时间片的任务调度中，每个任务都要运行有一个固定数的定时器刻度，称为时间片。当任务运行时，它的时间片由定时器中断处理程序减小。当运行任务的时间片减为 0 时，系统将切换任务，以运行另一个任务。这些听起来简单轻松，但却关系重大。鼓励读者们思考问题并找到可能的解决方案。

5.9　习题

1. 库函数 ctime(&seconds) 将以秒为单位的时间转换成日历形式，即一个字符串。修改示例 5.1 的程序，以日历形式打印当前日期和时间。
2. 修改示例 5.1 的程序，使其包含一个延迟循环，以模拟冗长的程序计算。使用 gettimeofday() 获取循环前后的当前时间。然后打印这两个时间的差值。我们可以用这种方法测量程序的总执行时间吗？请证明你的答案。
3. 按以下方式修改示例 5.1 中的程序。

   （1）安装 SIGPROF（27）信号的信号捕捉器。

   （2）将另一个有相同周期的 PROF 模式间隔定时器启动为 VIRTUAL 模式定时器。

   （3）令 pcount=PROF 定时器产生的信号数，vcount=VIRTUAL 定时器产生的信号数。

   在信号处理器中同时打印 pcount 和 vcount。当任一计数达到 100 时，停止定时器并打印计数。

   （4）在主程序的 while(1) 循环中，添加系统调用 getpid() 和 getppid()，这样，进程将在系统模式中花费一些时间。

   编译和运行程序，并回答以下问题。

   （5）哪个定时器产生信号更快，为什么？

   （6）如何确定用户模式和系统模式下的进程执行时间？

   （7）以 time a.out 运行程序。它将打印 a.out 在实时模式、用户模式和系统模式下的执行时间。解释"时间 time 是如何实现的？"
4. 修改示例 5.1 中的程序，以设置一个 REAL 模式间隔定时器。
5. 在累积定时器队列中，每个定时器队列元素的到期时间都与前面所有定时器队列元素的到期时间累积和有关。设计一个算法，将一个到期时间为 t 的定时器队列元素插入 timerQueue 中。设计一个算法，当定时器请求在到期之前被取消时，删除一个定时器队列元素。

参考文献

AMD64 Architecture Programmer's manual Volume 2: System Programming, 2011

Bovet, D.P., Cesati, M., "Understanding the Linux Kernel, Third Edition", O'Reilly, 2005

Intel i486 Processor Programmer's Reference Manual, 1990

Intel MultiProcessor Specification, v1.4, 1997

Wang, K. C., "Design and Implementation of the MTX Operating System", Springer A.G., 2015

信号和信号处理

摘要

本章讲述了信号和信号处理；介绍了信号和中断的统一处理，有助于从正确的角度看待信号；将信号视为进程中断，将进程从正常执行转移到信号处理；解释了信号的来源，包括来自硬件、异常和其他进程的信号；然后举例说明了信号在Unix/Linux中的常见用法；详细解释了Unix/Linux中的信号处理，包括信号类型、信号向量位、信号掩码位、进程PROC结构体中的信号处理程序以及信号处理步骤；用示例展示了如何安装信号捕捉器来处理程序异常，如用户模式下的段错误；还讨论了将信号用作进程间通信（IPC）机制的适用性。读者可借助该编程项目，使用信号和管道来实现用于进程交换信息的进程间通信机制。

6.1 信号和中断

“中断”是从I/O设备或协处理器发送到CPU的外部请求，它将CPU从正常执行转移到中断处理。与发送给CPU的中断请求一样，“信号”是发送给进程的请求，将进程从正常执行转移到中断处理。在讨论信号和信号处理之前，我们先来回顾中断的概念和机制，这有助于正确看待信号。

（1）首先，我们将进程的概念概括为：一个“进程”（引号中）就是一系列活动。广义的“进程”包括

- 从事日常事务的人。
- 在用户模式或内核模式下运行的Unix/Linux进程。
- 执行机器指令的CPU。

（2）“中断”是发送给“进程”的事件，它将“进程”从正常活动转移到其他活动，称为“中断处理”。“进程”可在完成“中断”处理后恢复正常活动。

（3）“中断”一词可应用于任何“进程”，并不仅限于计算机中的CPU。例如，我们可能会提到以下几种“中断”。

（3）.1 人员中断。

当我在办公室读书、评分、做白日梦时，可能会发生一些真实事件，比如：

真实事件	ID	动作函数
大楼着火了	1	马上离开
电话响了	2	拿起电话和打电话的人聊天
有人敲门	3	回答请进（或假装不在）
切到自己的手指	4	贴创可贴
……		

所有这些事件都叫作人员中断，因为他们把人从正常活动转向“应对或处理中断”。处理完中断后，此人可以继续此前的活动（如果这个人还活着而且仍然记得自己之前的活动）。

每个中断都分配有一个唯一的ID识别号，并有一个预先安装的动作函数，人可在收到中断请求时“执行”动作函数。根据来源，中断可分为三类：

- 来自硬件的中断：大楼着火，闹钟响了等。
- 来自其他人的中断：电话响了，有人敲门等。
- 自己造成的中断：切到手指，吃得太多等。

按照紧急程度，中断可分为以下几类：

- 不可屏蔽（NMI）：大楼着火！
- 可屏蔽：有人敲门等。

人员的每个动作函数都是通过本能或经验实现的。由于人员中断的种类太多，所以不能在上表中全部列出，但是思路应该清晰。

（3）.2 进程中断。

这类中断是发送给进程的中断。当某进程正在执行时，可能会收到来自3个不同来源的中断：

- 来自硬件的中断：终端、间隔定时器的“Ctrl+C”组合键等。
- 来自其他进程的中断：kill(pid，SIG#)、death_of_child等。
- 自己造成的中断：除以0、无效地址等。

每个进程中断都被转换为一个唯一ID号，发送给进程。与多种类的人员中断不同，我们始终可限制在一个进程中的中断的数量。Unix/Linux中的进程中断称为信号，编号为1到31。进程的PROC结构体中有对应每个信号的动作函数，进程可在收到信号后执行该动作函数。与人员类似，进程也可屏蔽某些类型的信号，以推迟处理。必要时，进程还可能会修改信号动作函数。

（3）.3 硬件中断。

这类中断是发送给处理器或CPU的信号。它们也有三个可能的来源：

- 来自硬件的中断：定时器、I/O设备等。
- 来自其他处理器的中断：FFP、DMA、多处理器系统中的其他CPU。
- 自己造成的中断：除以0、保护错误、INT指令。

每个中断都有唯一的中断向量号。动作函数是中断向量表中的中断处理程序。前面说过，CPU始终执行一个进程。CPU不会导致任何自己造成的中断（除非出错）。这种中断是由于进程正在使用或在大多数情况下误用CPU造成的。前一种情况包括INT n或等效指令，使CPU从用户模式切换到内核模式。后一种情况包括CPU识别为异常的所有陷阱错误。因此，我们可以排除CPU自身造成的中断，只留下CPU外部的中断。

（3）.4 进程的陷阱错误。

进程可能会自己造成中断。这些中断是由被CPU识别为异常的错误引起的，例如除以0、无效地址、非法指令、越权等。当进程遇到异常时，它会陷入操作系统内核，将陷阱原因转换为信号编号，并将信号发送给自己。如果在用户模式下发生异常，则进程的默认操作是终止，并使用一个可选的内存转储进行调试。我们会在后面学习到，进程可以用信号捕捉器代替默认动作函数，允许它在用户模式下处理信号。如果在内核模式下发生陷阱，原因一定是硬件错误，或者很可能是内核代码中的漏洞，在这种情况下，内核无法处理。在Unix/Uinux中，内核只打印一条PANIC错误消息，然后就停止了。希望在下一个内核版本中可以跟踪并修复这个问题。

6.2 Unix/Linux 信号示例

（1）按“Ctrl+C”组合键通常会导致当前运行的进程终止。原因如下。“Ctrl+C”组合键会生成一个键盘硬件中断。键盘中断处理程序将“Ctrl+C”组合键转换为 SIGINT（2）信号，发送给终端上的所有进程，并唤醒等待键盘输入的进程。在内核模式下，每个进程都要检查和处理未完成的信号。进程对大多数信号的默认操作是调用内核的 kexit(exitValue) 函数来终止。在 Linux 中，exitValue 的低位字节是导致进程终止的信号编号。

（2）用户可使用 nohup a.out & 命令在后台运行一个程序。即使在用户退出后，进程仍将继续运行。nohup 命令会使 sh 像往常一样复刻子进程来执行程序，但是子进程会忽略 SIGHUP（1）信号。当用户退出时，sh 会向与终端有关的所有进程发送一个 SIGHUP 信号。后台进程在接收到这一信号后，会忽略它并继续运行。为防止后台进程使用终端进行 I/O，后台进程通常会断开与终端的连接（通过将其文件描述符 0、1、2 重定向到 /dev/null），使其完全不受任何面向终端信号的影响。

（3）也许几天后，用户再次登录时会发现（通过 ps -u LTD）后台进程仍在运行。用户可以使用 sh 命令

```
kill pid   (or  kill -s 9 pid)
```

杀死该进程。方法如下。执行杀死的进程向 pid 标识的目标进程发送一个 SIGTERM（15）信号，请求它死亡。目标进程将会遵从请求并终止。如果进程选择忽略 SIGTERM 信号，它可能拒绝死亡。在这种情况下，我们可以使用 kill -s 9 pid，肯定能杀死它。因为进程不能修改对 9 号信号的动作。读者可能会问，为什么是 9 号信号呢？在最初的 Unix 中，只有 9 个信号。9 号信号被保留为终止进程的终极手段。虽然后来的 Unix/ Linux 系统将信号编号扩展到了 31，但是信号编号 9 的含义仍然保留了下来。

6.3 Unix/Linux 中的信号处理

6.3.1 信号类型

Unix/Linux 支持 31 种不同的信号，每种信号在 signal.h 文件中都有定义。

```
#define SIGHUP          1
#define SIGINT          2
#define SIGQUIT         3
#define SIGILL          4
#define SIGTRAP         5
#define SIGABRT         6
#define SIGIOT          6
#define SIGBUS          7
#define SIGFPE          8
#define SIGKILL         9
#define SIGUSR1        10
#define SIGSEGV        11
#define SIGUSR2        12
#define SIGPIPE        13
#define SIGALRM        14
#define SIGTERM        15
#define SIGSTKFLT      16
#define SIGCHLD        17
```

```
#define SIGCONT         18
#define SIGSTOP         19
#define SIGTSTP         20
#define SIGTTIN         21
#define SIGTTOU         22
#define SIGURG          23
#define SIGXCPU         24
#define SIGXFSZ         25
#define SIGVTALRM       26
#define SIGPROF         27
#define SIGWINCH        28
#define SIGPOLL         29
#define SIGPWR          30
#define SIGSYS          31
```

每种信号都有一个符号名，如 SIGHUP（1）、SIGEMT（2）、SIGKILL（9）、SIGSEGV（11）等。

6.3.2 信号的来源

- **来自硬件中断的信号**：在进程执行过程中，一些硬件中断被转换为信号发送给进程。硬件信号示例是
  - 中断键（Ctrl+C），它产生一个 SIGINT（2）信号。
  - 间隔定时器，当它的时间到期时，会生成一个 SIGALRM（14）、SIGVTALRM（26）或 SIGPROF（27）信号。
  - 其他硬件错误，如总线错误、IO 陷阱等。
- **来自异常的信号**：当用户模式下的进程遇到异常时，会陷入内核模式，生成一个信号，并发送给自己。常见的陷阱信号有 SIGFPE（8），表示浮点异常（除以 0），最常见也是最可怕的是 SIGSEGV（11），表示段错误，等等。
- **来自其他进程的信号**：进程可使用 kill(pid, sig) 系统调用向 pid 标识的目标进程发送信号。读者可以尝试以下实验。在 Linux 下，运行简单的 C 程序

  ```
  main(){ while(1); }
  ```

  使进程无限循环。从另一个（X-window）终端，使用 ps -u 查找循环进程 pid。然后输入 sh 命令

  ```
  kill -s 11 pid
  ```

  循环进程会因为段错误而死亡。读者可能会问：这怎么可能呢？所有进程都在一个 while(1) 循环中执行，它是如何产生段错误的呢？答案是：这并不重要。当某进程被某个信号终止时，它的 exitValue 就包含这个信号编号。父进程 sh 只是将死亡子进程的信号编号转换为一个错误字符串，不管它是什么。

6.3.3 进程 PROC 结构体中的信号

每个进程 PROC 都有一个 32 位向量，用来记录发送给进程的信号。在位向量中，每一位（0 位除外）代表一个信号编号。此外，它还有一个信号 MASK 位向量，用来屏蔽相应的信号。可使用一系列系统调用，如 sigmask、sigsetmask、siggetmask、sigblock 等设置、清

除和检查 MASK 位向量。待处理信号只在未被屏蔽的情况下才有效。这样可以让进程延迟处理被屏蔽的信号，类似于 CPU 屏蔽某些中断。

6.3.4 信号处理函数

每个进程 PROC 都有一个信号处理数组 int sig[32]。sig[32] 数组的每个条目都指定了如何处理相应的信号，其中 0 表示 DEFault（默认），1 表示 IGNore（忽略），其他非零值表示用户模式下预先安装的信号捕捉（处理）函数。图 6.1 给出了信号位向量、屏蔽位向量和信号处理函数。

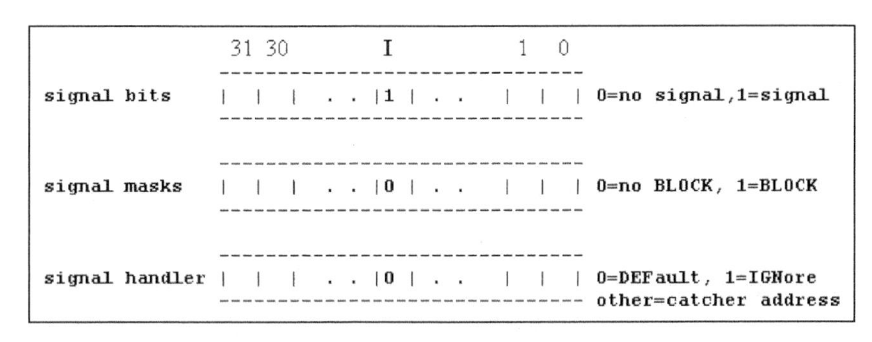

图 6.1 信号处理示意图

如果信号位向量中的位 I 为 1，则会**生成**一个信号 I 或将其发送给进程。如果屏蔽位向量的位 I 为 1，则信号会被**阻塞**或屏蔽。否则，信号**未被阻塞**。只有当信号存在并且未被阻塞时，信号才会生效或**传递**给进程。当内核模式下的进程发现一个未阻塞信号时，会将信号位清除为 0，并尝试通过信号处理数组中的处理函数来处理该信号。0 表示 DEFault，1 表示 IGNore，其他数值表示用户空间内预先安装的捕捉函数。

6.3.5 安装信号捕捉函数

进程可使用系统调用：

```
int r = signal(int signal_number, void *handler);
```

来修改选定信号编号的处理函数，SIGKILL（9）和 SIGSTOP（19）除外，它们不能修改。已安装的处理函数（若不是 0 或 1）一定是以下形式用户空间中信号捕捉函数的入口地址：

```
void catcher(int signal_number){.............}
```

signal() 系统调用在所有类 Unix 系统中均可用，但它有一些不理想的特点。

（1）在执行已安装的信号捕捉函数之前，通常将信号处理函数重置为 DEFault。为捕捉下次出现的相同信号，必须重新安装捕捉函数。这可能会导致下一个信号和信号处理函数重新安装之间出现竞态条件。相反，sigaction() 在执行当前捕捉函数时会自动阻塞下一个信号，因此不会出现竞态条件。

（2）signal() 不能阻塞其他信号。必要时，用户必须使用 sigprocmask() 显式地阻塞或解锁其他信号。相反，sigaction() 可以指定要阻塞的其他信号。

（3）signal() 只能向捕捉函数发送一个信号编号。sigaction() 可以传输关于信号的其他信息。

（4）signal() 可能不适用于多线程程序中的线程。sigaction() 适用于线程。

（5）不同 Unix 版本的 signal() 可能会有所不同。sigaction() 采用的是 POISX 标准，可移植性更好。

由于这些原因，signal() 已经被 POSIX sigaction() 函数所代替。在 Linux（Bovet 和 Cesati 2005）中，sigaction() 是一个系统调用。它的原型是：

```
   int sigaction (int signum, const struct sigaction *act, struct sigaction
*oldact);
```

sigaction 结构体的定义为：

```
struct sigaction{
   void     (*sa_handler)(int);
   void     (*sa_sigaction)(int, siginfo_t *, void *);
   sigset_t sa_mask;
   int      sa_flags;
   void     (*sa_restorer)(void);
};
```

其中最重要的字段是：

- sa_handler：该字段是指向处理函数的指针，该函数与 signal() 的处理函数有相同的原型。
- sa_sigaction：该字段是运行信号处理函数的另一种方法。它的信号编号旁边有两个额外参数，其中 siginfo t * 提供关于所接收信号的更多信息。
- sa_mask：可在处理函数执行期间设置要阻塞的信号。
- sa_flags：可修改信号处理进程的行为。若要使用 sa_sigaction 处理函数，必须将 sa_flags 设置为 SA_SIGINFO。

如需了解有关 sigaction 结构体字段的详细描述，请参见 sigaction 手册页。下面是一个使用 sigaction() 系统调用的简单示例。

示例 6.1：sigaction() 的使用示例。

```
/****** sigaction.c file *******/
#include <stdio.h>
#include <unistd.h>
#include <signal.h>

void handler(int sig, siginfo_t *siginfo, void *context)
{
  printf("handler: sig=%d from PID=%d UID=%d\n",
                   sig, siginfo->si_pid, siginfo->si_uid);
}

int main(int argc, char *argv[])
{
  struct sigaction act;
  memset(&act, 0, sizeof(act));
  act.sa_sigaction = &handler;
  act.sa_flags = SA_SIGINFO;
  sigaction(SIGTERM, &act, NULL);
  printf("proc PID=%d looping\n");
```

```
    printf("enter kill PID to send SIGTERM signal to it\n", getpid());
    while(1)(
      sleep(10);
    }
  }
```

在程序运行时，读者可从另一个 X 终端输入 kill PID，向进程发送一个 SIGTERM（15）信号。任何信号都会唤醒进程，即使它处于睡眠状态，也能够处理该信号。在信号处理函数中，我们从 siginfo 参数中读取两个字段来显示信号发送者的 PID 和 UID，如果信号处理函数是由 signal() 系统调用安装的，则这两个字段不可用。

进程可使用信号调用

```
int r = kill(pid, signal_number);
```

向 pid 标识的另一个进程发送信号。sh 命令

```
kill -s signal_number pid
```

使用 kill 系统调用。通常，只有相关进程，例如有相同 uid 的进程，才会相互发送信号。但是，超级用户进程（uid=0）可以向任意进程发送信号。kill 系统调用使用一个无效 pid 来表示传递信号的方式不同。例如，pid=0 表示向同一进程组中的所有进程发送信号，pid=-1 表示向 pid > 1 的所有进程发送信号等。读者可参考 Linux 手册页，以了解关于 signal/kill 的更多详细信息。

6.4 信号处理步骤

（1）当某进程处于内核模式时，会检查信号并处理未完成的信号。如果某信号有用户安装的捕捉函数，该进程会先清除信号，获取捕捉函数地址，对于大多数陷阱信号，则将已安装的捕捉函数重置为 DEFault。然后，它会在用户模式下返回，以执行捕捉函数，以这种方式篡改返回路径。当捕捉函数结束时，它会返回到最初的中断点，即它最后进入内核模式的地方。因此，该进程会先迂回执行捕捉函数，然后再恢复正常执行。

（2）重置用户安装的信号捕捉函数：用户安装的陷阱相关信号捕捉函数用于处理用户代码中的陷阱错误。由于捕捉函数也在用户模式下执行，因此可能会再次出现同样的错误。如果是这样，该进程最终会陷入无限循环，一直在用户模式和内核模式之间跳跃。为了防止这种情况，Unix 内核通常会在允许进程执行捕捉函数之前先将处理函数重置为 DEFault。这意味着用户安装的捕捉函数只对首次出现的信号有效。若要捕捉再次出现的同一信号，则必须重新安装捕捉函数。但是，用户安装的信号捕捉函数的处理方法并不都一样，在不同 Unix 版本中会有所不同。例如，在 BSD Unix 中，信号处理函数不会被重置，但是该信号在执行信号捕捉函数时会被阻塞。感兴趣的读者可参考关于 Linux 信号和 sigaction 函数的手册页，以了解更多详细信息。

（3）信号和唤醒：在 Unix/Linux 内核中有两种 SLEEP 进程；深度休眠进程和浅度休眠进程。前一种进程不可中断，而后一种进程可由信号中断。如果某进程处于不可中断的 SLEEP 状态，到达的信号（必须来自硬件中断或其他进程）不会唤醒进程。如果它处于可中断的 SLEEP 状态，到达的信号将会唤醒它。例如，当某进程等待终端输入时，它会以低优先级休眠，这种休眠是可中断的，SIGINT 这类信号即可唤醒它。

6.5 信号与异常

Unix 信号最初设计用于以下用途。

- 作为进程异常的统一处理方法：当进程遇到异常时，它会陷入内核模式，将陷阱原因转换为信号编号，并将信号发送给自己。如果在内核模式下发生异常，内核只打印一条 PANIC 错误消息，然后就停止了。如果在用户模式下发生异常，则进程通常会终止，并以内存转储进行调试。
- 让进程通过预先安装的信号捕捉函数处理用户模式下的程序错误。这类似于 MVS [IBM MVS] 中的 ESPIE 宏。
- 在特殊情况下，它会让某个进程通过信号杀死另一个进程。注意，这里所说的杀死并不是直接杀死某个进程，而只是向目标进程发出“死亡”请求。为什么我们不直接杀死某个进程呢？我们鼓励读者思考其中的原因。（提示：瑞士银行有大量无人认领的匿名账户。）

6.6 信号用作 IPC

在许多操作系统的书籍中，信号被归类为进程间的通信机制。基本原理是一个进程可以向另一个进程发送信号，使它执行预先安装的信号处理函数。由于以下原因，这种分类即使不算不恰当也颇具争议。

- 该机制并不可靠，因为可能会丢失信号。每个信号由位向量中的一个位表示，只能记录一个信号的一次出现。如果某个进程向另一个进程发送两个或多个相同的信号，它们可能只在接收 PROC 中出现一次。实时信号被放入队列，并保证按接收顺序发送，但操作系统内核可能不支持实时信号。
- 竞态条件：在处理信号之前，进程通常会将信号处理函数重置为 DEFault。要想捕捉同一信号的再次出现，进程必须在该信号再次到来之前重新安装捕捉函数。否则，下一个信号可能会导致该进程终止。在执行信号捕捉函数时，虽然可以通过阻塞同一信号来防止竞态条件，但是无法防止丢失信号。
- 大多数信号都有预定义的含义。不加区别地任意使用信号不仅不能达到通信的目的，反而会造成混乱。例如，向循环进程发送 SIGSEGV（11）段错误信号，就像对水里游泳的人大喊：“你的裤子着火了！”

因此，试图将信号用作进程间通信手段实际上是对信号预期用途的过度延伸，应避免出现这种情况。

示例 6.2：段错误捕捉函数　正如第 2 章中所指出的，C 语言程序中最常见的段错误的原因是解除空指针或无效指针关联、数组越界等。当某进程遇到无效内存异常时，它会陷入操作系统内核，生成 SIGSEGV（11）信号，并发送给自己。SIGSEGV 信号的默认处理函数是 0，将导致进程终止。如果进程忽略该信号，它会再次返回同一错误指令，导致无限循环。若用户已经安装了 SIGSEGV 信号的捕捉函数，进程会执行信号捕捉函数，但是在执行结束后仍会返回同一错误指令。在任何段错误情况下，进程的唯一选择似乎只能是异常终止。本示例说明了如何使用信号捕捉函数和 long jump 来绕过导致段错误的程序代码，使程序继续执行或正常终止。

```
/********** t7.c file **********/
#include <stdio.h>
#include <stdlib.h>
#include <unistd.h>
#include <signal.h>
#include <setjmp.h>  // for long jump

jmp_buf env; // for saving longjmp environment
int count = 0;

void handler(int sig, siginfo_t *siginfo, void *context)
{
  printf ("handler: sig=%d from PID=%d UID=%d count=%d\n",
           sig, siginfo->si_pid, siginfo->si_uid, ++count);
  if (count >= 4) // let it occur up to 4 times
     longjmp(env, 1234);
}

int BAD()
{
  int *ip = 0;
  printf("in BAD(): try to dereference NULL pointer\n");
  *ip = 123;            // dereference a NULL pointer
  printf("should not see this line\n");
}

int main (int argc, char *argv[])
{
  int r;
  struct sigaction act;
  memset (&act, 0, sizeof(act));
  act.sa_sigaction = &handler;
  act.sa_flags = SA_SIGINFO;
  sigaction(SIGSEGV, &act, NULL); // install SIGSEGV catcher
  if ((r = setjmp(env)) == 0)     // call setjmp(env)
     BAD();                       // call BAD()
  else
     printf("proc %d survived SEGMENTATION FAULT: r=%d\n",getpid(), r);

  printf("proc %d looping\n");
  while(1);
}
```

图 6.2 显示了 segfault.c 程序的输出。

练习 6.1：修改示例 6.1 的程序，以捕捉除以零异常。

```
in BAD(): try to dereference NULL pointer
handler: sig=11 from PID=0 UID=4 count=1
handler: sig=11 from PID=0 UID=4 count=2
handler: sig=11 from PID=0 UID=4 count=3
handler: sig=11 from PID=0 UID=4 count=4
proc 2855 survived SEGMENTATION FAULT: r=1234
proc 2855 looping
```

图 6.2 段错误捕捉函数示意图

6.7 Linux 中的 IPC

IPC 是指用于进程间通信的机制。在 Linux 中，IPC 包含以下组成部分。

6.7.1 管道和 FIFO

管道和管道编程在 3.10.1 节中有过讨论。一个管道有一个读取端和一个写入端。管道的主要用途是连接一对管道写进程和读进程。管道写进程可将数据写入管道，读进程可从管道中读取数据。管道控制机制要对管道读写操作进行同步控制。未命名管道供相关进程使用。命名管道是FIFO的，可供不相关进程使用。在Linux中的管道读取操作为同步和阻塞。如果管道仍有写进程但没有数据，读进程会进行等待。

必要时，可通过对管道描述符的fcntl系统调用将管道操作更改为非阻塞。

6.7.2　信号

进程可使用kill系统调用向其他进程发送信号，其他进程使用信号捕捉函数处理信号。将信号用作IPC的一个主要缺点是信号只是用作通知，不含任何信息内容。

6.7.3　System V IPC

Linux支持System V IPC，包括共享内存、信号量和消息队列。在Linux中，多种System V IPC函数，例如用于添加/移除共享内存的shmat/shmdt、用于获取/操作信号量的semget/semop和用于发送/接收消息的msgsnd/msgrcv，都是库包装函数，它们都会向Linux内核发出一个ipc()系统调用。ipc()的实现是Linux所特有的，不可移植。

6.7.4　POSIX消息队列

POSIX标准（IEEE 1003.1-2001）以消息队列为基础定义了IPC机制。它们类似于System V IPC的消息队列，但更通用且可移植。

6.7.5　线程同步机制

Linux不区分进程和线程。在Linux中，进程是共享某些公共资源的线程。如果是使用有共享地址空间的clone()系统调用创建的进程，它们可使用互斥量和条件变量通过共享内存进行同步通信。另外，常规进程可添加到共享内存，使它们可作为线程进行同步。

6.7.6　套接字

套接字是用于跨网络进程通信的IPC机制。套接字和网络编程将在第13章中讨论。

6.8　编程项目：实现一个消息IPC

正如前文所指出，管道可用于进程交换信息，但是如果使用阻塞协议，进程必须等待管道内容，如果使用非阻塞协议，进程必须反复检查管道内容的可用性。当某些事件发生时，信号可用作通知，但不传递其他信息。读者可借助该编程项目，结合使用信号和管道来实现用于进程交换信息的进程间通信机制。在本项目中，消息定义为固定长度的文本字符串，例如64字符字符串。进程不必等待消息。当某个进程通过向管道写入来发送消息时，它会向目标进程发送一个信号，向进程通知到达的消息。接收进程可对信号做出响应，从管道中读取消息，不需要检查，也不会被阻塞。该项目分为两个部分。第1部分介绍了项目的基本代码，读者可从这一部分开始阅读，具体内容见下文。

编程项目第1部分：项目的基本代码。

```
/******* Base Code of Programming Project *******/
#include <stdio.h>
```

```
#include <signal.h>
#include <fcntl.h>
#include <string.h>

#define LEN 64
int ppipe[2];                    // pipe descriptors
int pid;                         // child pid
char line[LEN];
int parent()
{
  printf("parent %d running\n", getpid());
  close(ppipe[0]);               // parent = pipe writer
  while(1){
    printf("parent %d: input a line : \n", getpid());
    fgets(line, LEN, stdin);
    line[strlen(line)-1] = 0; // kill \n at end
    printf("parent %d write to pipe\n", getpid());
    write(ppipe[1], line, LEN); // write to pipe
    printf("parent %d send signal 10 to %d\n", getpid(), pid);
    kill(pid, SIGUSR1);         // send signal to child process
  }
}

void chandler(int sig)
{
  printf("\nchild %d got an interrupt sig=%d\n", getpid(), sig);
  read(ppipe[0], line, LEN);    // read pipe
  printf("child %d get a message = %s\n", getpid(), line);
}

int child()
{
  char msg[LEN];
  int parent = getppid();
  printf("child %d running\n", getpid());
  close(ppipe[1]);              // child is pipe reader
  signal(SIGUSR1, chandler);    // install signal catcher
  while(1);
}

int main()
{
  pipe(ppipe);                 // create a pipe
  pid = fork();                // fork a child process
  if (pid) // parent
    parent();
  else
    child();
}
```

在基本代码中，父进程可创建管道，复刻子进程，并充当管道的写进程。子进程是管道的读进程。它会安装一个 SIGUSR1 信号捕捉函数，并进行循环。当父进程运行时，会获取一个输入字符串，将该字符串作为一条消息写入管道，向子进程发送一个 SIGUSR1 信号，在信号捕捉函数中接收和显示消息。然后，它会重复刚才的循环。图 6.3 为运行基本代码程序的样本输出。

编程项目第2部分：修改项目基本代码以实现以下操作。

（1）发送回复：发送者会在发送消息后，先等待回复，然后再发送下一条消息。

相应地，接收者会在接收到消息后，向发送者发送一条回复。例如，回复内容可以是转换成大写的原始信息。

（2）时间戳：在发送消息或回复时，按当前时间向消息添加时间戳。在收到回复后，计算并显示消息的往返时间。

（3）超时和重发：假设父进程不可靠。它可能会在收到消息后随机决定不发送回复，这将导致发送者无限期等待。按以下方法添加超时和重发。在发送消息后，将实时模式间隔定时器设置为10ms。如果在定时器到期之前收到回复，则取消定时器。否则，需要重新发送消息（使用新的时间戳），直到收到回复。

```
parent 28931 running
parent 28931: input a line :
child  28932 running
message one
parent 28931 write to pipe
parent 28931 send signal 10 to 28932
parent 28931: input a line :
child  28932 got an interrupt sig=10
child  28932 got a message = message one
message two
parent 28931 write to pipe
parent 28931 send signal 10 to 28932
parent 28931: input a line :
child  28932 got an interrupt sig=10
child  28932 got a message = message two
■
```

图 6.3 项目基本代码的样本输出

提示：请参考第5章关于间隔定时器的相关内容。

6.9 习题

1. 复习题：
（1）用自己的话来定义“中断”。
（2）什么是硬件中断、进程中断？
（3）INT n 指令通常称为软件中断。INT n 和进程中断有什么区别？
（4）一个进程如何获取进程中断信号？
（5）一个进程如何处理进程中断？
2. 信号9的作用是什么？为什么需要它？
3. CPU 通常会在当前指令结束时检查未处理的中断。进程会在什么时间、什么地点检查未处理的信号？
4. 在处理信号之前，内核通常会将信号的处理函数重置为 DEFault。为什么必须进行该操作？
5. 在 Linux 中，按下“Ctrl+C”组合键将会导致进程终止。但是，当主进程 sh 在伪终端运行时，按下“Ctrl+C”组合键并不会导致主进程 sh 终止。主进程 sh 是如何做到的？
6. 在 C 语言编程中，回调函数（指针）可作为一个参数传递给被调函数，以此执行回调函数。回调函数与信号捕捉函数有什么区别？
7. 假设某进程已经安装了 SIGALRM（14）的信号捕捉函数。如果该进程执行到另一个映像中，会出现什么结果？如何处理这类问题？

参考文献

Bovet, D.P., Cesati, M., “Understanding the Linux Kernel, Third Edition”, O’Reilly, 2005
IBM MVS Programming Assembler Services Guide, Oz/OS V1R11.0, IBM
Linux: http://www.linux.org signal(7), Linux Programmer’s Manual: descriptions of Posix/Linux signals, including real-time signals.

文件操作

摘要

本章讨论了多种文件系统；解释了操作系统中的各种操作级别，包括为文件存储准备存储设备、内核中的文件系统支持函数、系统调用、文件流上的I/O库函数、用户命令和各种操作的sh脚本；系统性概述了各种操作，包括从用户空间的文件流读/写到内核空间的系统调用，直到底层的设备I/O驱动程序级别；描述了低级别的文件操作，包括磁盘分区、显示分区表的示例程序、文件系统的格式化分区以及挂载磁盘分区；介绍了Linux系统的EXT2文件系统，包括EXT2文件系统的系统数据结构、显示超级块、组描述符、块和索引节点位图以及目录内容的示例程序。编程项目将本章中讨论的EXT2/3文件系统和编程技术集中到一个程序中，将路径名转换为索引节点并打印它们的信息。

7.1 文件操作级别

文件操作分为五个级别，按照从低到高的顺序排列如下。

（1）**硬件级别**：硬件级别的文件操作包括：

- fdisk：将硬盘、U盘或SDC盘分区。
- mkfs：格式化磁盘分区，为系统做好准备。
- fsck：检查和维修系统。
- 碎片整理：压缩文件系统中的文件。

其中大多数是针对系统的实用程序。普通用户可能永远都不需要它们，但是它们是创建和维护系统不可缺少的工具。

（2）**操作系统内核中的文件系统函数**：每个操作系统内核均可为基本文件操作提供支持。下文列出了类Unix系统内核中的一些函数，其中前缀k表示内核函数。

```
kmount(), kumount()           (mount/umount file systems)
kmkdir(), krmdir()            (make/remove directory)
kchdir(), kgetcwd()           (change directory, get CWD pathname)
klink(),  kunlink()           (hard link/unlink files)
kchmod(), kchown(), kutime()  (change r|w|x permissions,owner,time)
kcreat(), kopen()             (create/open file for R,W,RW,APPEND)
kread(),  kwrite()            (read/write opened files)
klseek(); kclose()            (lseek/close file descriptors)
ksymlink(), kreadlink()       (create/read symbolic link files)
kstat(),  kfstat(), klstat()  (get file status/information)
kopendir(), kreaddir()        (open/read directories)
```

（3）**系统调用**：用户模式程序使用系统调用来访问内核函数。例如，下面的程序可读取文件的第二个1024字节。

```
#include <fcntl.h>
int main(int argc, char *argv[ ]) // run as a.out filename
{
   int fd, n;
   char buf[1024];
   if ((fd = open(argv[1], O_RDONLY)) < 0) // if open() fails
       exit(1);
   lseek(fd, 1024, SEEK_SET);      // lseek to byte 1024
   n = read(fd, buf, 1024);        // try to read 1024 bytes
   close(fd);
}
```

open()、read()、lseek() 和 close() 函数都是 C 语言库函数。每个库函数都会发出一个系统调用，使进程进入内核模式来执行相应的内核函数，例如 open 可进入 kopen()，read 可进入 kread() 函数，等等。当进程结束执行内核函数时，会返回到用户模式，并得到所需的结果。在用户模式和内核模式之间切换需要大量的操作（和时间）。因此，内核和用户空间之间的数据传输成本昂贵。虽然可以发出 read(fd, buf, 1) 系统调用来只读取一个字节的数据，但是这种做法是不明智的，因为一个字节也会带来可怕的高成本。我们在每次必须进入内核时，都要尽可能不虚此行。对于读 / 写文件，最好的方法是匹配内核的功能。内核会按数据块大小（从 1 KB 到 8KB）来读取 / 写入文件。例如，在 Linux 中，硬盘的默认数据块大小是 4KB，软盘的是 1KB。因此，每个读 / 写系统调用还要尝试一次传输一个数据块。

（4）**I/O 库函数**：系统调用可让用户读 / 写多个数据块，这些数据块只是一系列字节。它们不知道，也不关心数据的意义。用户通常需要读 / 写单独的字符、行或数据结构记录等。如果只有系统调用，用户模式程序则必须自己从缓冲区执行这些操作。大多数用户会认为这非常不方便。为此，C 语言库提供了一系列标准的 I/O 函数，同时也提高了运行效率。I/O 库函数包括：

```
FILE mode I/O: fopen(),fread();  fwrite(),fseek(),fclose(),fflush()
char mode I/O: getc(), getchar() ugetc(); putc(),putchar()
line mode I/O: gets(), fgets();  puts(), fputs()
formatted I/O: scanf(),fscanf(),sscanf(); printf(),fprintf(),sprintf()
```

除了读 / 写内存位置的 sscanf()/sprintf() 函数之外，所有其他 I/O 库函数都建立在系统调用之上，也就是说，它们最终会通过系统内核发出实际数据传输的系统调用。

（5）**用户命令**：用户可以使用 Unix/Linux 命令来执行文件操作，而不是编写程序。用户命令的示例如下：

```
mkdir, rmdir, cd, pwd, ls, link, unlink, rm, cat, cp, mv, chmod, etc.
```

每个用户命令实际上是一个可执行程序（cd 除外），通常会调用库 I/O 函数，而库 I/O 函数再发出系统调用来调用相应的内核函数。用户命令的处理顺序为：

```
Command => Library I/O function => System call => Kernel Function
```

或者

```
Command ======================== > System call => Kernel Function
```

（6）**sh 脚本**：虽然比系统调用方便得多，但是必须要手动输入命令，如果使用的是 GUI，必须要拖放文件图标和点击指向设备来输入，操作烦琐而且耗时。sh 脚本是用 sh 编程语言编写的程序，可通过命令解释程序 sh 来执行。sh 语言包含所有的有效 Unix/Linux 命令。它还支持变量和控制语句，如 if、do、for、while、case 等。实际上，sh 脚本广泛用于 Unix/Linux 系统编程。除 sh 之外，Perl 和 Tcl 等其他许多脚本语言也使用广泛。

7.2 文件 I/O 操作

图 7.1 为文件 I/O 操作示意图。

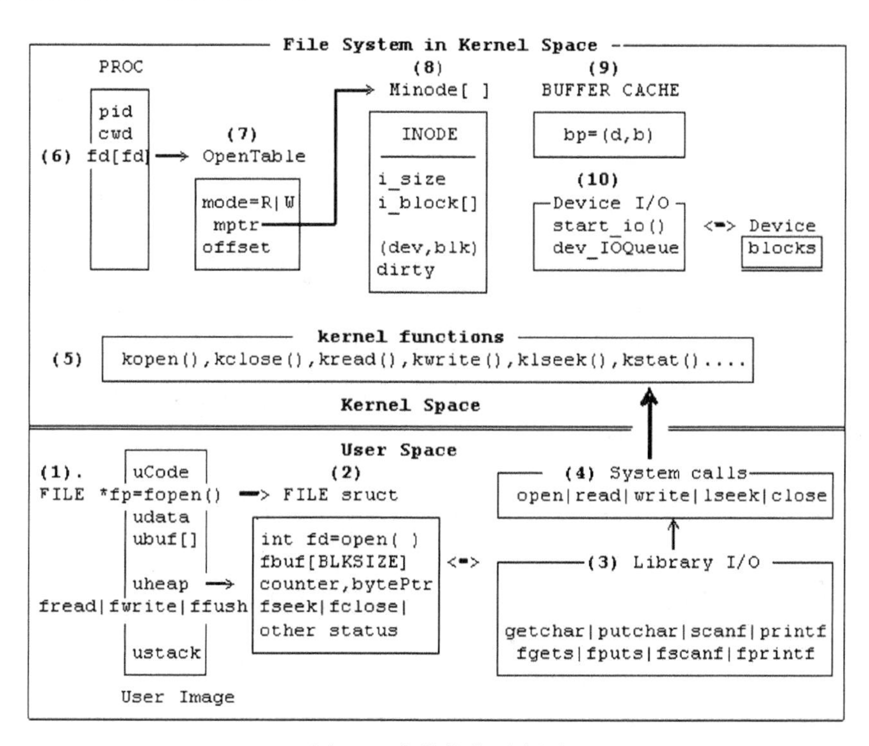

图 7.1 文件操作示意图

在图 7.1 中，双线上方的上半部分表示内核空间，下半部分表示进程的用户空间。该图显示了进程读 / 写文件流时的操作序列。控制流用标签（1）到（10）标识，说明如下。

下面的步骤（1）～（46）是用户模式下的操作，步骤（5）～（10）是内核模式下的操作。

（1）用户模式下的程序执行操作

```
FILE *fp = fopen("file", "r"); or FILE *fp = fopen("file", "w");
```

可以打开一个读 / 写文件流。

（2）fopen() 在用户（heap）空间中创建一个 FILE 结构体，包含一个文件描述符 fd、一个 fbuf [BLKSIZE] 和一些控制变量。它会向内核中的 kopen() 发出一个 fd = open("file", flags=READ or WRITE) 系统调用，构建一个 OpenTable 来表示打开文件示例。OpenTable 的 mptr 指向内存中的文件 INODE。对于非特殊文件，INODE 的 i_block 数组指向存储设备上的数据块。成功后，fp 会指向 FILE 结构体，其中 fd 是 open() 系统调用返回的文件描述符。

（3）fread(ubuf, size, nitem, fp)：将 nitem 个 size 字节读取到 ubuf 上，通过：

- 将数据从 FILE 结构体的 fbuf 上复制到 ubuf 上，若数据足够，则返回。
- 如果 fbuf 没有更多数据，则执行（4a）。

（4a）发出 read(fd, fbuf, BLKSIZE) 系统调用，将文件数据块从内核读取到 fbuf 上，然后将数据复制到 ubuf 上，直到数据足够或者文件无更多数据可复制。

（4b）fwrite(ubuf, size, nitem, fp)：将数据从 ubuf 复制到 fbuf。

- 若（fbuf 有空间）：将数据复制到 fbuf 上，并返回。
- 若（fbuf 已满）：发出 write(fd, fbuf, BLKSIZE) 系统调用，将数据块写入内核，然后再次写入 fbuf。

这样，fread()/fwrite() 会向内核发出 read()/write() 系统调用，但仅在必要时发出，而且它们会以块集大小来传输数据，提高效率。同样，其他库 I/O 函数，如 fgetc/fputc、fgets/lputs、fscanf/fprintf 等也可以在用户空间内的 FILE 结构体中对 fbuf 进行操作。

（5）内核中的文件系统函数：

假设非特殊文件的 read(fd, fbuf[ ], BLKSIZE) 系统调用。

（6）在 read() 系统调用中，fd 是一个打开的文件描述符，它是运行进程的 fd 数组中的一个索引，指向一个表示打开文件的 OpenTable。

（7）OpenTable 包含文件的打开模式、一个指向内存中文件 INODE 的指针和读 / 写文件的当前字节偏移量。从 OpenTable 的偏移量，

- 计算逻辑块编号 lbk。
- 通过 INODE.i_block[ ] 数组将逻辑块编号转换为物理块编号 blk。

（8）Minode 包含文件的内存 INODE。EMODE.i_block[ ] 数组包含指向物理磁盘块的指针。文件系统可使用物理块编号从磁盘块直接读取数据或将数据直接写入磁盘块，但将会导致过多的物理磁盘 I/O。

（9）为提高磁盘 I/O 效率，操作系统内核通常会使用一组 I/O 缓冲区作为高速缓存，以减少物理 I/O 的数量。磁盘 I/O 缓冲区管理将在第 12 章中讨论。

（9a）对于 read(fd, buf, BLKSIZE) 系统调用，要确定所需的（dev, blk）编号，然后查询 I/O 缓冲区高速缓存，以执行以下操作：

```
.get a buffer = (dev, blk);
.if (buffer's data are invalid){
    start_io on buffer;
    wait for I/O completion;
 }
.copy data from buffer to fbuf;
.release buffer to buffer cache;
```

（9b）对于 write(fd, fbuf, BLKSIZE) 系统调用，要确定需要的（dev, blk）编号，然后查询 I/O 缓冲区高速缓存，以执行以下操作：

```
.get a buffer = (dev, blk);
.write data to the I/O buffer;
.mark buffer as dataValid and DIRTY (for delay-write to disk);
.release the buffer to buffer cache;
```

（10）设备 I/O：I/O 缓冲区上的物理 I/O 最终会仔细检查设备驱动程序，设备驱动程序

由上半部分的 start_io() 和下半部分的磁盘中断处理程序组成。

```
------------------- Upper-half of disk driver --------------------
start_io(bp): //bp=a locked buffer in dev_list, opcode=R|W(ASYNC)
{
   enter bp into dev's I/O_queue;
   if (bp is FIRST in I/O_queue)
      issue I/O command to device;
}

------------------ Lower-half of disk driver --------------------
Device_Interrupt_Handler:
{
   bp = dequeue(first buffer from dev.I/O_queue);
   if (bp was READ){
       mark bp data VALID;
       wakeup/unblock waiting process on bp;
   }
   else       // bp was for delay write
       release bp into buffer cache;

   if (dev.I/O_queue NOT empty)
      issue I/O command for first buffer in dev.I/O_queue;
}
```

7.3 低级别文件操作

7.3.1 分区

一个块存储设备，如硬盘、U 盘、SD 卡等，可以分为几个逻辑单元，称为分区。各分区均可以格式化为特定的文件系统，也可以安装在不同的操作系统上。大多数引导程序，如 GRUB、LILO 等，都可以配置为从不同的分区引导不同的操作系统。分区表位于第一个扇区的字节偏移 446（0xlBE）处，该扇区称为设备的**主引导记录**（MBR）。表有 4 个条目，每个条目由一个 16 字节的分区结构体定义，即：

```
stuct partition {
      u8  drive;          // 0x80 - active
      u8  head;           // starting head
      u8  sector;         // starting sector
      u8  cylinder;       // starting cylinder
      u8  sys_type;       // partition type
      u8  end_head;       // end head
      u8  end_sector;     // end sector
      u8  end_cylinder;   // end cylinder
     u32  start_sector;   // starting sector counting from 0
     u32  nr_sectors;     // number of sectors in partition
};
```

如果某分区是扩展类型（类型编号 =5），那么它可以划分为更多分区。假设分区 P4 是扩展类型，它被划分为扩展分区 P5、P6、P7。扩展分区在扩展分区区域内形成一个链表，如图 7.2 所示。

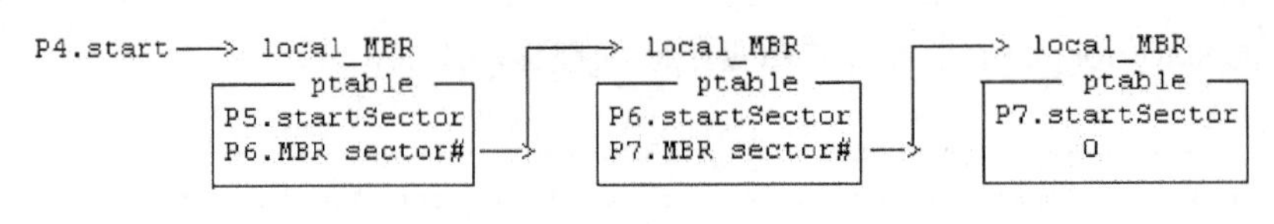

图 7.2　扩展分区链表

每个扩展分区的第一个扇区是一个**本地 MBR**。每个本地 MBR 在字节偏移量 0xlBE 处也有一个分区表，只包含两个条目。第一个条目定义了扩展分区的起始扇区和大小。第二个条目指向下一个本地 MBR。所有本地 MBR 的扇区编号都与 P4 的起始扇区有关。照例，链表以最后一个本地 MBR 中的 0 结尾。在分区表中，CHS 值仅对小于 8GB 的磁盘有效。对大于 8GB 但小于 4G 扇区的磁盘，只有最后两个条目 start_sector 和 nr_sector 有意义。接下来，我们将通过示例来演示 fdisk 和分区。由于使用计算机的真实磁盘进行操作会非常危险，所以我们要使用一个虚拟磁盘映像，它只是一个普通的文件，但是看起来像一个真实磁盘。

（1）在 Linux 下，例如 Ubuntu，创建一个名为 mydisk 的虚拟磁盘映像文件。

```
dd if=/dev/zero of=mydisk bs=1024 count=1440
```

dd 是一个将 1440（1KB）个 0 字节块写入目标文件 mydisk 的程序。我们选择 count=1440，因为它是旧软盘的 1KB 字节块的数量。必要时，读者可指定更大的库编号。

（2）在磁盘映像文件上运行 fdisk：

```
fdisk mydisk
```

fdisk 是一个交互程序。它有一个显示所有命令的帮助菜单。它收集用户的输入，在内存中创建一个分区表，该分区表仅在用户输入 w 命令时才被写入磁盘映像的 MBR。但是在内存中，它允许用户创建、检查和修改分区。将分区写入磁盘后，通过 q 命令退出 fdisk。图 7.3 显示了一个 fdisk 会话的结果，该会话将磁盘映像划分为 3 个主分区（P1 至 P3）和一个扩展分区 P4。扩展分区 P4 被进一步划分为更多的分区（P5 至 P7），所有这些分区都位于扩展分区的区域内。在开始创建时，分区类型都默认为 Linux。可以通过 t 命令将它们更改为不同的文件系统类型。每个文件系统类型都有一个用十六进制唯一值，例如 0x07 表示 HPFS/NTFS、0x83 表示 Linux 等，这些值可通过 t 命令显示出来（见图 7.3）。

```
Device     Boot Start  End Sectors   Size Id Type
mydisk1             1  719     719 359.5K  7 HPFS/NTFS/exFAT
mydisk2           720 1439     720   360K 83 Linux
mydisk3          1440 1799     360   180K 81 Minix / old Linux
mydisk4          1800 2879    1080   540K  5 Extended
mydisk5          1801 2159     359 179.5K  6 FAT16
mydisk6          2161 2519     359 179.5K 83 Linux
mydisk7          2521 2879     359 179.5K 82 Linux swap / Solaris

Command (m for help): q
```

图 7.3　分区表

练习 7.1：编写一个 C 程序，显示虚拟磁盘映像的分区和扩展分区（如有），其格式与 fdisk 相同。

帮助：对于缺乏编程经验的读者，以下内容可能会有所帮助。

（1）访问 MBR 中的分区表：

```
#include <stdio.h>
#include <fcntl.h>
char buf[512];

int fd = open("mydisk", O_RDONLY);  // open mydisk for READ
read(fd, buf, 512);                 // read MBR into buf[512]
struct partition *p = (struct partition *)&buf[0x1BE];
printf("p->start_sector = %d\n", p->start_sector); // access p->start_sector
p++;        // advance p to point to the next partition table in MBR, etc.
```

（2）假设 P4 是 start_sector = n 的扩展类型（类型 =5）。

```
lseek(fd, (long)(n*512), SEEK_SET);     // lseek to P4 begin
read(fd, char buf[ ], 512);             // read local MBR
p = (struct partition *)&buf[0x1BE]);   // access local ptable
```

（3）扩展分区形成一个“链表”，以一个 NULL“指针”结束。

相反，我们也可以编写一个简单的 mkfs 程序，通过以下步骤对虚拟磁盘进行分区。

（1）打开一个可用于 read/write(O_RDWR) 的虚拟磁盘。

（2）将 MBR 读取到一个 char buf[512] 中：在真实的磁盘中，MBR 可能包含引导程序的开头部分，一定不能打乱它。

（3）修改 buf[] 中的分区表条目：在每个分区表条目中，只有 start_sector 和 nr_sector 类型的字段是重要的。其他所有字段都不重要，因此它们可以保持不变，也可以设置为 0。

（4）将 buf[] 写回 MBR。

练习 7.2：编写一个 C 程序 myfdisk，将一个虚拟磁盘分成 4 个主分区。运行程序后，在虚拟磁盘上运行 Linux fdisk 来验证结果。

7.3.2 格式化分区

fdisk 只是将一个存储设备划分为多个分区。每个分区都有特定的文件系统类型，但是分区还不能使用。为了存储文件，必须先为特定的文件系统准备好分区。该操作习惯上称为**格式化**磁盘或磁盘分区。在 Linux 中，它被称为 mkfs，表示 Make 文件系统。Linux 支持多种不同类型的文件系统。每个文件系统都期望存储设备上有特定的格式。在 Linux 中，命令

```
mkfs -t TYPE [-b bsize] device nblocks
```

在一个 nblocks 设备上创建一个 TYPE 文件系统，每个块都是 bsize 字节。如果 bsize 未指定，则默认块大小为 1KB。具体来说，假设是 EXT2/3 文件系统，它是 Linux 的默认文件系统。因此，

```
mkfs -t ext2 vdisk 1440    或   mke2fs vdisk 1440
```

使用 1440（1KB）个块将 vdisk 格式化为 EXT2 文件系统。格式化后的磁盘应是只包含根目录的空文件系统。但是，Linux 的 mkfs 始终会在根目录下创建一个默认的 lost+found 目录。完成 mkfs 之后，设备就可以使用了。在 Linux 中，还不能访问新的文件系统。它必须挂载到根文件系统中的现有目录中。/mnt 目录通常用于挂载其他文件系统。由于虚拟文件系统不是真正的设备，它们必须作为循环设备挂载，如

```
sudo mount -o loop vdisk /mnt
```

将 vdisk 挂载到 /mnt 目录中。不带任何参数的 mount 命令会显示 Linux 系统的所有挂载设备。挂载完成后，挂载点 /mnt 改变，与挂载设备的根目录相同。用户可以将目录（cd）更改为 /mnt，像往常一样对设备进行平铺操作。挂载后的设备使用完成后，将 cd 从 /mnt 中取出，然后输入

```
sudo umount /mnt    或    sudo umount vdisk
```

以卸载设备，将其与根文件系统分离。设备上保存的文件应保留在该设备中。

要指出的是，目前大多数 Linux 系统，例如 Slackware 14.1 和 Ubuntu 15.10，都可以检测到包含文件系统的便携设备，并直接挂载它们。例如，当 Ubuntu 用户将 U 盘插入 USB 端口时，Ubuntu 可以检测设备，自动挂载设备，并在弹出窗口中显示文件，允许用户直接访问文件。用户可以输入 mount 命令来查看设备（通常是 /dev/sdbl）的挂载位置，以及挂载的文件系统类型。如果用户拔出 U 盘，Ubuntu 也会检测到，并自动**卸载**设备。但是，随意拔出便携设备可能会损坏设备上的数据。因为 Linux 内核通常使用延迟数据写入来写入设备。为了确保数据的一致性，用户应该先卸载设备，然后再断开连接。

Linux mount 命令可以挂载实际设备的分区或整个虚拟磁盘，但不能挂载虚拟磁盘的分区。如果某个虚拟磁盘包含多个分区，那么必须先将这些分区与**循环设备**关联起来。下面的例子演示了如何挂载虚拟磁盘的分区。

7.3.3　挂载分区

man 8 losetup：显示用于系统管理的 losetup 实用工具命令：

（1）用 dd 命令创建一个虚拟磁盘映像：

```
dd if=/dev/zero of=vdisk bs=1024 count=32768     #32K (1KB) blocks
```

（2）在 vdisk 上运行 fdisk 来创建一个分区 P1：

```
fdisk vdisk
```

输入 n（new）命令，使用默认的起始和最后扇区编号来创建一个分区 P1。然后，输入 w 命令将分区表写入 vdisk 并退出 fdisk。vdisk 应包含一个分区 P1 [start=2048, end=65535]。该分区的大小是 63488 个扇区。

（3）使用以下扇区数在 vdisk 的分区 1 上创建一个循环设备：

```
losetup -o $(expr 2048 \* 512) --sizelimit $(expr 65535 \* 512) /dev/loop1
vdisk
```

losetup 需要分区的开始字节（start_sector*512）和结束字节（end_sector*512）。读者可手动计算这些数值，并在 losetup 命令中使用它们。可用类似方法设置其他分区的循环设备。循环设备创建完成后，读进程可以使用命令

```
losetup -a
```

将所有循环设备显示为 /dev/loopN。

（4）格式化 /dev/loop1，它是一个 EXT2 文件系统：

```
mke2fs -b 4096 /dev/loop1 7936              # mke2fs with 7936 4KB blocks
```

该分区的大小是 63488 个扇区。4KB 块的扇区大小是 63488 / 8=7936。

（5）挂载循环设备：

```
mount /dev/loop1 /mnt                  # mount as loop device
```

（6）访问作为文件系统一部分的挂载设备：

```
(cd /mnt; mkdir bin boot dev etc user)  # populate with DIRs
```

（7）设备使用完毕后，将其卸载。

```
umount /mnt
```

（8）循环设备使用完毕后，通过以下命令将其断开：

```
losetup -d /dev/loop1                  # detach a loop device.
```

练习 7.3：将 vdisk 分为 4 个分区。创建 4 个用于分区的循环设备。将分区格式化为 EXT2 文件系统，挂载分区并将文件复制进去。然后，卸载并断开循环设备。如果读者知道如何编写 sh 脚本，则可以通过一个简单的 sh 脚本来完成上述所有步骤。

7.4 EXT2 文件系统简介

多年来，Linux 一直使用 EXT2（Card 等 1995; EXT2 2001）作为默认文件系统。EXT3（EXT3 2015）是 EXT2 的扩展。EXT3 中增加的主要内容是一个日志文件，它将文件系统的更改记录在日志中。日志可在文件系统崩溃时更快从错误中恢复。没有错误的 EXT3 文件系统与 EXT2 文件系统相同。EXT3 的最新扩展是 EXT4（Cao 等 2007）。EXT4 的主要变化是磁盘块的分配。在 EXT4 中，块编号是 48 位。EXT4 不是分配不连续的磁盘块，而是分配连续的磁盘块区，称为区段。

7.4.1 EXT2 文件系统数据结构

在 Linux 下，我们可以创建一个包含简单 EXT2 文件系统的虚拟磁盘，如下文所示。

```
(1). dd if=/dev/zero of=mydisk bs=1024 count=1440
(2). mke2fs -b 1024 mydisk 1440
```

得到的 EXT2 文件系统有 1440 个块，每个块大小为 1KB。我们之所以选择 1440 块，是因为它是（旧）软盘的块数。得到的磁盘映像可以直接作为虚拟（软）磁盘，在模拟基于 Intel x86 的大多数 PC 虚拟机上使用，例如 QUEM、VirtualBox 和 VMware 等。这种 EXT2 文件系统的布局如图 7.4 所示。

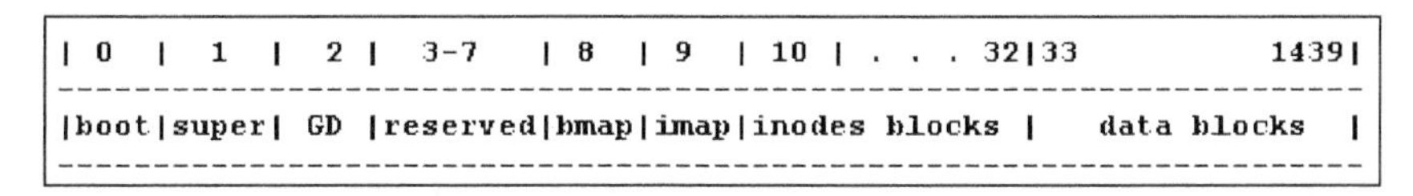

图 7.4 简单的 EXT2 文件系统布局

为了方便讨论，我们先假设使用这个基本文件系统布局。在适当的时候，我们会指出一些变化，包括硬盘上大型 EXT2/3 FS 中的变化。下面来简要解释一下磁盘块的内容。

Block#0：引导块 B0 是引导块，文件系统不会使用它。它用于容纳从磁盘引导操作系统的引导程序。

7.4.2 超级块

Block#1：**超级块**（在硬盘分区中字节偏移量为1024） B1是超级块，用于容纳关于整个文件系统的信息。下文说明了超级块结构中的一些重要字段。

```
struct ext2_super_block {
   u32  s_inodes_count;          /* Inodes count */
   u32  s_blocks_count;          /* Blocks count */
   u32  s_r_blocks_count;        /* Reserved blocks count */
   u32  s_free_blocks_count;     /* Free blocks count */
   u32  s_free_inodes_count;     /* Free inodes count */
   u32  s_first_data_block;      /* First Data Block */
   u32  s_log_block_size;        /* Block size */
   u32  s_log_cluster_size;      /* Allocation cluster size */
   u32  s_blocks_per_group;      /* # Blocks per group */
   u32  s_clusters_per_group;    /* # Fragments per group */
   u32  s_inodes_per_group;      /* # Inodes per group */
   u32  s_mtime;                 /* Mount time */
   u32  s_wtime;                 /* Write time */
   u16  s_mnt_count;             /* Mount count */
   s16  s_max_mnt_count;         /* Maximal mount count */
   u16  s_magic;                 /* Magic signature */
   // more non-essential fields
   u16  s_inode_size;            /* size of inode structure */
}
```

大多数超级块字段的含义都非常明显。只有少数几个字段需要详细解释。

s_first_data_block：0表示4KB块大小，1表示1KB块大小。它用于确定块组描述符的起始块，即s_first_data_block + 1。

s_log_block_size确定文件块大小，为1KB*(2**s_log_block_size)，例如0表示1KB块大小，1表示2KB块大小，2表示4KB块大小，等等。最常用的块大小是用于小文件系统的1KB和用于大文件系统的4KB。

s_mnt_count：已挂载文件系统的次数。当挂载计数达到max_mount_count时，fsck会话将被迫检查文件系统的一致性。

s_magic是标识文件系统类型的幻数。EXT2/3/4文件系统的幻数是0xEF53。

7.4.3 块组描述符

Block#2：**块组描述符块**（硬盘上的s_first_data_blocks-1） EXT2将磁盘块分成几个组。每个组有8192个块（硬盘上的大小为32K）。每组用一个块组描述符结构体描述。

```
struct ext2_group_desc {
   u32  bg_block_bitmap;         // Bmap block number
   u32  bg_inode_bitmap;         // Imap block number
   u32  bg_inode_table;          // Inodes begin block number
   u16  bg_free_blocks_count;    // THESE are OBVIOUS
   u16  bg_free_inodes_count;
   u16  bg_used_dirs_count;
   u16  bg_pad;                  // ignore these
   u32  bg_reserved[3];
};
```

由于一个软盘只有1440个块，B2只包含一个块组描述符。其余的都是0。在有大量块组的硬盘上，块组描述符可以跨越多个块。块组描述符中最重要的字段是bg_block_bitmap、bg_inode_bitmap和bg_inode_table，它们分别指向块组的块位图、索引节点位图和索引节点起始块。对于Linux格式的EXT2文件系统，保留了块3到块7。所以，bmap=8，imap=9，inode_table= 10。

7.4.4 位图

Block#8：块位图（Bmap）（bg_block_bitmap） 位图是用来表示某种项的位序列，例如磁盘块或索引节点。位图用于分配和回收项。在位图中，0位表示对应项处于FREE状态，1位表示对应项处于IN_USE状态。一个软盘有1440个块，但是Block#0未被文件系统使用。所以，位图只有1439个有效位。无效位视作IN_USE处理，设置为1。

Block#9：索引节点位图（Imap）（bg_inode_bitmap） 一个**索引节点**就是用来代表一个文件的数据结构。EXT2文件系统是使用有限数量的索引节点创建的。各索引节点的状态用B9中Imap中的一个位表示。在EXT2 FS中，前10个索引节点是预留的。所以，空EXT2 FS的Imap以10个1开头，然后是0。无效位再次设置为1。

7.4.5 索引节点

Block#10：索引（开始）节点块（bg_inode_table） 每个文件都用一个128字节（EXT4中的是256字节）的独特索引节点结构体表示。下面列出了主要索引节点字段。

```
struct ext2_inode {
  u16  i_mode;         // 16 bits = |tttt|ugs|rwx|rwx|rwx|
  u16  i_uid;          // owner uid
  u32  i_size;         // file size in bytes
  u32  i_atime;        // time fields in seconds
  u32  i_ctime;        // since 00:00:00,1-1-1970
  u32  i_mtime;
  u32  i_dtime;
  u16  i_gid;          // group ID
  u16  i_links_count;  // hard-link count
  u32  i_blocks;       // number of 512-byte sectors
  u32  i_flags;        // IGNORE
  u32  i_reserved1;    // IGNORE
  u32  i_block[15];    // See details below
  u32  i_pad[7];       // for inode size = 128 bytes
}
```

在索引节点结构体中，i_mode为u16或2字节无符号整数。

```
         | 4  | 3 |    9    |
i_mode = |tttt|ugs|rwxrwxrwx|
```

在i_mode字段中，前四位表示文件类型。例如，tttt= 1000表示REG文件，0100表示DIR文件等。接下来的3位ugs表示文件的特殊用法。最后9位是文件保护的rwx权限位。

i_size字段表示文件大小（以字节为单位）。各时间字段表示自1970年1月1日0时0分0秒以来经过的秒数。所以，每个时间字段都是一个非常大的无符号整数。可借助以下库函数将它们转换为日历形式：

```
char *ctime(&time_field)
```

将指针指向时间字段，然后返回一个日历形式的字符串。例如：

```
printf("%s", ctime(&inode.i_atime);     // note: pass & of time field
```

以日历形式打印 i_atime。

i_block[15] 数组包含指向文件磁盘块的指针，这些磁盘块有：

- **直接块**：i_block[0] 至 i-block[11]，指向直接磁盘块。
- **间接块**：i-block[12] 指向一个包含 256 个块编号（对于 1KB BLKSIZE）的磁盘块，每个块编号指向一个磁盘块。
- **双重间接块**：i_block[13] 指向一个指向 256 个块的块，每个块指向 256 个磁盘块。
- **三重间接块**：i_block[14] 是三重间接块。对于“小型”EXT2 文件系统，我们可以忽略它。

索引节点大小（128 或 256）用于平均分割块大小（1KB 或 4KB），所以，每个索引节点块都包含整数个索引节点。在简单的 EXT2 文件系统中，索引节点的数量是 184 个（Linux 默认值）。索引节点块数等于 184/8=23 个。因此，索引节点块为 B10 至 B32。每个索引节点都有一个独特的**索引节点编号**，即索引节点在索引节点块上的位置 +1。注意，索引节点位置从 0 开始计数，而索引节点编号从 1 开始计数。0 索引节点编号表示没有索引节点。根目录的索引节点编号为 2。同样，磁盘块编号也从 1 开始计数，因为文件系统从未使用块 0。块编号 0 表示没有磁盘块。

数据块：紧跟在索引节点块后面的是文件存储块。假设有 184 个索引节点，第一个实际数据块是 B33，它就是根目录 / 的 i_block[0]。

7.4.6　目录条目

EXT2 目录条目：目录包含 dir_entry 结构，即：

```
struct ext2_dir_entry_2{
   u32 inode;                // inode number; count from 1, NOT 0
   u16 rec_len;              // this entry's length in bytes
   u8  name_len;             // name length in bytes
   u8  file_type;            // not used
   char name[EXT2_NAME_LEN]; // name: 1-255 chars, no ending NULL
};
```

dir_entry 是一种可扩充结构。名称字段包含 1 到 255 个字符，不含终止 NULL 字节。所以 dir_entry 的 rec_len 也各不相同。

7.5　编程示例

在本节中，我们将演示如何通过示例程序访问和显示 EXT2 文件系统的内容。为了编译和运行这些程序，系统必须安装 ext2fs.h 头文件，它定义了 EXT2/3/4 文件系统的数据结构。Ubuntu Linux 用户可通过以下代码获取并安装 ext2fs 开发包：

```
sudo apt-get install ext2fs-dev
```

7.5.1　显示超级块

示例 7.1 中的 C 程序显示了 EXT2 文件系统的超级块。基本方法是将超级块（Block#1

或 1KB 的 1024 偏移量位置）读入 char buf[1024] 中。让 ext2_super_block *p 结构体指向 buf[]。然后，利用 p->field 访问超级块结构体的各个字段。该方法类似于访问 MBR 中的分区表。

示例 7.1：superblock.c 程序显示 EXT2 文件系统的超级块信息。

```
/*********** superblock.c program ************/
#include <stdio.h>
#include <stdlib.h>
#include <fcntl.h>
#include <ext2fs/ext2_fs.h>
// typedef u8, u16, u32 SUPER for convenience
typedef unsigned char   u8;
typedef unsigned short u16;
typedef unsigned int   u32;
typedef struct ext2_super_block SUPER;

SUPER *sp;
char buf[1024];
int fd, blksize, inodesize;

int print(char *s, u32 x)
{
  printf("%-30s = %8d\n", s, x);
}

int super(char *device)
{
  fd = open(device, O_RDONLY);
  if (fd < 0){
    printf("open %sfailed\n", device); exit(1);
  }
  lseek(fd, (long)1024*1, 0);  // block 1 or offset 1024
  read(fd, buf, 1024);
  sp = (SUPER *)buf;           // as a super block structure
  // check for EXT2 FS magic number:
  printf("%-30s = %8x ", "s_magic", sp->s_magic);
  if (sp->s_magic != 0xEF53){
     printf("NOT an EXT2 FS\n"); exit(2);
  }
  printf("EXT2 FS OK\n");
  print("s_inodes_count",      sp->s_inodes_count);
  print("s_blocks_count",      sp->s_blocks_count);
  print("s__r_blocks_count",   sp->s_r_blocks_count);
  print("s_free_inodes_count", sp->s_free_inodes_count);
  print("s_free_blocks_count", sp->s_free_blocks_count);
  print("s_first_data_blcok",  sp->s_first_data_block);
  print("s_log_block_size",    sp->s_log_block_size);
  print("s_blocks_per_group",  sp->s_blocks_per_group);
  print("s_inodes_per_group",  sp->s_inodes_per_group);
  print("s_mnt_count",         sp->s_mnt_count);
  print("s_max_mnt_count",     sp->s_max_mnt_count);
  printf("%-30s = %8x\n", "s_magic", sp->s_magic);
  printf("s_mtime = %s", ctime(&sp->s_mtime));
  printf("s_wtime = %s", ctime(&sp->s_wtime));
  blksize = 1024 * (1 << sp->s_log_block_size);
```

```
  printf("block size = %d\n", blksize);
  printf("inode size = %d\n",  sp->s_inode_size);
}

char *device = "mydisk";        // default device name
int main(int argc, char *argv[])
{
  if (argc>1)
     device = argv[1];
  super(device);
}
```

图 7.5 表示运行 super.c 示例程序的输出。

```
s_magic                     =    ef53  EXT2 FS OK
s_inodes_count              =     184
s_blocks_count              =    1440
s__r_blocks_count           =      72
s_free_inodes_count         =     173
s_free_blocks_count         =    1393
s_first_data_blcok          =       1
s_log_block_size            =       0
s_blocks_per_group          =    8192
s_inodes_per_group          =     184
s_mnt_count                 =       1
s_max_mnt_count             =      -1
s_magic                     =    ef53
s_mtime = Sun Oct  8 14:22:03 2017
s_wtime = Sun Oct  8 14:22:07 2017
block size = 1024
inode size = 128
```

图 7.5　EXT2 文件系统的超级块

练习 7.4：编写一个 C 程序来显示设备上 EXT2 文件系统的块组描述符。

7.5.2　显示位图

示例 7.2 中的 C 程序以十六进制形式显示索引节点位图（imap）。

示例 7.2：imap.c 程序显示 EXT2 文件系统的索引节点位图。

```
/*************** imap.c program **************/
#include <stdio.h>
#include <stdlib.h>
#include <fcntl.h>
#include <ext2fs/ext2_fs.h>
typedef struct ext2_super_block SUPER;
typedef struct ext2_group_desc  GD;

#define BLKSIZE 1024
SUPER *sp;
GD    *gp;
char buf[BLKSIZE];
int fd;

// get_block() reads a disk block into a buf[ ]
int get_block(int fd, int blk, char *buf)
{
```

```
  lseek(fd, (long)blk*BLKSIZE, 0);
  read(fd, buf, BLKSIZE);
}

int imap(char *device)
{
  int i, ninodes, blksize, imapblk;
  fd = open(dev, O_RDONLY);
  if (fd < 0){printf("open %s failed\n", device); exit(1);}
  get_block(fd, 1, buf);           // get superblock
  sp = (SUPER *)buf;
  ninodes = sp->s_inodes_count;    // get inodes_count
  printf("ninodes = %d\n", ninodes);
  get_block(fd, 2, buf);           // get group descriptor
  gp = (GD *)buf;
  imapblk = gp->bg_inode_bitmap;   // get imap block number
  printf("imapblk = %d\n", imapblk);
  get_block(fd, imapblk, buf);     // get imap block into buf[ ]
  for (i=0; i<=nidoes/8; i++){     // print each byte in HEX
      printf("%02x ", (unsigned char)buf[i]);
  }
  printf("\n");
}

char * dev="mydisk";               // default device
int main(int argc, char *argv[ ] )
{
  if (argc>1) dev = argv[1];
  imap(dev);
}
```

程序以十六进制数字形式打印每个字节的索引节点位图。输出如下所示。

```
|---------------- niodes = 184 bits (23 bytes)----------------------|
ff 07 00 00 00 00 00 00 00 00 00 00 00 00 00 00 00 00 00 00 00 00 00 ff
```

在 imap 中，位是从低位地址到高位地址线性存储的。前 16 位（从低到高）是 b’11111111 11100000’，但是被以十六进制形式打印成 ff 07，这并不能说明什么信息，因为位是按相反顺序打印的，即从高位地址到低位地址。

练习 7.5：修改 imap.c 程序，以 char map 的形式打印索引节点位图，即对于每个位而言，如果该位是 0，则打印“0”，如果该位是 1，则打印“1”。

提示：

（1）检查（1 <<j）的位模式，j=0 至 7。

（2）若 char c 是一个字节，则 C 语句

```
if ( c & (1<<j) )    (j=0 to 7)
```

测试 c 的第 j 位是 1 还是 0。

修改后的程序输出如图 7.6 所示，其中每个字符表示 imap 中的一位。

练习 7.6：编写一个 C 语言程序，同样以 char map 格式来显示 EXT2 文件系统的块位图。

```
ninodes = 184  imapblk = 9
11111111 11100000 00000000 00000000 00000000 00000000 00000000 00000000
00000000 00000000 00000000 00000000 00000000 00000000 00000000 00000000
00000000 00000000 00000000 00000000 00000000 00000000 00000000 11111111
```

图 7.6　EXT2 文件系统的索引节点位图

7.5.3　显示根索引节点

在 EXT2 文件系统中，第 2 个（从 1 开始计）索引节点是根目录 / 的索引节点。如果我们将根索引节点读入内存中，就能显示模式、uid、gid、文件大小、创建时间、硬链接数和数据块编号等多个字段。示例 7.3 中的程序显示了 EXT2 文件系统根目录的索引节点信息。

示例 7.3：inode.c 程序显示 EXT2 文件系统的根索引节点信息。

```
/*********** inode.c file **********/
#include <stdio.h>
#include <stdlib.h>
#include <fcntl.h>
#include <ext2fs/ext2_fs.h>

#define BLKSIZE 1024
typedef struct ext2_group_desc  GD;
typedef struct ext2_super_block SUPER;
typedef struct ext2_inode       INODE;
typedef struct ext2_dir_entry_2 DIR;
SUPER *sp;
GD    *gp;
INODE *ip;
DIR   *dp;
char buf[BLKSIZE];
int fd, firstdata, inodesize, blksize, iblock;
char *dev = "mydisk"; // default to mydisk

int get_block(int fd, int blk, char *buf)
{
  lseek(fd, blk*BLKSIZE, SEEK_SET);
  return read(fd, buf, BLKSIZE);
}
int inode(char *dev)
{
  int i;
  fd = open(dev, O_RDONLY) < 0);
  if (fd < 0){
    printf("open failed\n"); exit(1);
  }
  get_block(fd, 1, buf);              // get superblock
  sp = (SUPER *)buf;
  firstdata = sp->s_first_data_block;
  inodesize = sp->s_inode_size;
  blksize = 1024*(1<<sp->s_log_block_size);
  printf("first_data_block=%d block_size=%d inodesize=%d\n",
          firstdata, blksize, inodesize);
  get_block(fd, (firstdata+1), buf);   // get group descriptor
  gp = (GD *)buf;
  printf("bmap_block=%d imap_block=%d inodes_table=%d ",
```

```
        gp->bg_block_bitmap,
        gp->bg_inode_bitmap,
        gp->bg_inode_table,
        gp->bg_free_blocks_count);
  printf("free_blocks=%d free_inodes=%d\n",
        gp->bg_free_inodes_count,
        gp->bg_used_dirs_count);
  iblock = gp->bg_inode_table;
  printf("root inode information:\n", iblock);
  printf("-----------------------\n");
  get_block(fd, iblock, buf);
  ip = (INODE *)buf + 1;            // ip point at #2 INODE
  printf("mode=%4x ", ip->i_mode);
  printf("uid=%d  gid=%d\n", ip->i_uid, ip->i_gid);
  printf("size=%d\n", ip->i_size);
  printf("ctime=%s", ctime(&ip->i_ctime));
  printf("links=%d\n", ip->i_links_count);
  for (i=0; i<15; i++){             // print disk block numbers
    if (ip->i_block[i])             // print non-zero blocks only
      printf("i_block[%d]=%d\n", i, ip->i_block[i]);
  }
}
int main(int argc, char *argv[ ])
{
  if (argc>1) dev = argv[1];
  inode(dev);
}
```

图 7.7 为 EXT2 文件系统的根索引节点。

```
first_data_block=1 block_size=1024  inodesize=128
bmap_block=8 imap_block=9 inodes_table=10
free_blocks=1393 free_inodes=173 used_dirs=2
root inode information:
-----------------------
mode=41ed uid=0  gid=0
size=1024
links=3
ctime=Mon Oct  9 16:11:44 2017
i_block[0]=33
```

图 7.7　EXT2 文件系统的根索引节点

在图 7.7 中，i_mode =0x41ed 或 b'0100 0001 1110 1101'，以二进制形式表示。前 4 位 0100 表示文件类型（目录）。接着 3 位 000=ugs 都是 0，表示文件无特殊用法，例如它不是 setuid 程序。最后 9 位可分为三组，即 111、101、101，表示文件所有人、同组和其他人员的 rwx 权限位。对于普通文件，x 位 =1 表示文件可执行。对于目录，x 位 =1 表示可访问（即 cd into）该目录；0 x 位表示不可访问该目录。

7.5.4　显示目录条目

目录索引节点的各数据块均包含 dir_entries，分别是：

```
struct ext2_dir_entry_2 {
   u32 inode;                // inode number; count from 1, NOT 0
   u16 rec_len;              // this entry's length in bytes
```

```
    u8  name_len;            // name length in bytes
    u8  file_type;           // not used
    char name[EXT2_NAME_LEN]; // name: 1-255 chars, no ending NULL
};
```

因此，目录中每个数据块的内容都具有以下形式：

```
[inode rec_len name_len NAME] [inode rec_len name_len NAME] ......
```

其中NAME是一系列name_len字符（不含终止NULL）。下面的算法演示了如何单步遍历目录数据块中的dir_entries。

```
/**** Algorithm to step through entries in a DIR data block ****/
struct ext2_dir_entry_2 *dp;              // dir_entry pointer
char *cp;                                 // char pointer
int blk = a data block (number) of a DIR (e.g. i_block[0]);
char buf[BLKSIZE], temp[256];
get_block(fd, blk, buf);                  // get data block into buf[ ]

dp = (struct ext2_dir_entry_2 *)buf;      // as dir_entry
cp = buf;
while(cp < buf + BLKSIZE){
  strncpy(temp, dp->name, dp->name_len); // make name a string
  temp[dp->name_len] = 0;                // in temp[ ]
  printf("%d %d %d %s\n", dp->inode, dp->rec_len, dp->name_len, temp);
  cp += dp->rec_len;                      // advance cp by rec_len
  dp = (struct ext2_dir_entry_2 *)cp;     // pull dp to next entry
}
```

练习7.7：编写一个C语言程序来打印目录的dir_entries。为方便起见，我们可以假设目录索引节点最多有12个直接块，i_block[0]至i_block[11]。这一假设是合理的。单个磁盘块的块大小为1KB，平均文件名长度为16个字符，最多可以容纳1024/（8+16）=42个dir_entries。如果一个目录有12个磁盘块，则可容纳500多个条目。我们可以放心地假设，没有用户会在任何目录中放那么多文件。所以，我们最多只需要遍历12个直接块。在空EXT2文件系统中，程序输出如图7.8所示。

练习7.8：在Linux下挂载mydisk。创建新目录，并将文件复制到新目录中。卸载新目录。然后，在mydisk上再次运行dir.c程序，以查看输出，输出应该类似于图7.9。读者应该

```
check ext2 FS : OK
GD info: 8 9 10 1393 173 2
inodes begin block=10
******* root inode info ********
mode=41ed  uid=0  gid=0
size=1024
ctime=Sun Oct  8 14:04:52 2017
links=3
i_block[0]=33
*********************************
inode# rec_len name_len name
  2       12        1     .
  2       12        2     ..
 11     1000       10     lost+found
```

图7.8 目录条目

```
******* root inode info ********
mode=41ed  uid=0  gid=0
size=1024
ctime=Sun Oct  8 17:19:15 2017
links=7
i_block[0]=33
*********************************
inode# rec_len name_len name
  2       12        1     .
  2       12        2     ..
 11       20       10     lost+found
 12       12        1     a
 13       16        8     shortDir
 14       20       12     longNamedDir
 15       28       17     aVeryLongNamedDir
 16       16        7     super.c
 17       16        6     bmap.c
 18       16        6     imap.c
 19      856        5     dir.c
```

图7.9 目录的条目列表

注意到，每个条目的 name_len 对应 name 字段中的具体字符数量，每个 rec_len 是 4 的倍数（内存对齐），（8 + name_len）增大到 4 的下一个倍数，除了最后一个条目，它的 rec_len 覆盖了剩余块长度。

练习 7.9：假设 INODE 是 struct ext2_inode 类型。已知一个索引节点编号 ino，编写 C 语句，将指针返回至 INODE 结构体 。

提示：

（1）int fd = 打开 vidsk 进行读取。

（2）int offset = inodes_start_block * BLKSIZE +（ino-1）*inode_szie。

（3）根据偏移量 lseek fd。

（4）将 inode_size 字节读入 INODE 索引节点。

（5）返回（INODE *ip = &inode）。

练习 7.10：已知一个指向目录索引节点的 INODE 指针。编写：

```
int  search(INODE *dir, char *name)
```

函数，在目录中搜索名称字符串。如果有，则返回其索引节点编号；如果没有，则返回 0。

7.6 编程项目：将文件路径名转换为索引节点

在每个文件系统中，几乎每个操作都以文件名开始，例如：

```
cat filename
open filename for R|W
copy file1 to file2
etc.
```

各文件系统的根本问题是将文件名转换为文件系统中的文件数据结构表示形式。在 EXT2/3/4 文件系统中，相当于将路径名转换为文件的 INODE。该编程项目是将上述示例和练习中的知识和编程方法集中到一个程序中，在 EXT2 文件系统中查找文件并打印其信息。假设 vdisk 是一个虚拟磁盘，包含许多级别的目录和文件。已知某文件的路径名，例如 /a/b/c/d，编写一个 C 程序来查找文件并打印其信息，如文件类型、文件所有者 id、文件大小、创建日期及数据块编号（包括间接块和双重间接块）。

提示与帮助：

（1）将路径名标记为组件字符串。用 char *name[0]、*name[1]、……、*name[n-1] 表示各个字符串，其中 n 为组件字符串的数量。

（2）从 INODE *ip -> 根索引节点开始（ino=2）。

（3）
```
for (int i=0; i<n; i++) {
    int ino = search(ip, name[i]);
    if (!ino) exit(1);              // can't find name[i]
    ip ->inode of name[i];          // ip point at INODE of name[i]
    if (*ip is not a DIR) exit(2);  // name[i] is not a DIR
}
```

（4）如果到达这里：ip 必须指向文件索引节点。使用 ip-> 来访问索引节点字段。

（5）可用以下方法打印间接块编号：

```
read i_block[12] into a char buf[BLKSIZE];
u32 *up = (u32 *)buf;
*up is an indirect block number; up++ points to the next number, etc.
```

还可利用类似的方法来打印双重间接块编号。

7.7 习题

1. 修改练习 7.2 中的程序，使其支持扩展分区。
2. 大型 EXT2/3 文件系统包含超过 32K 万个块，块大小为 4KB。将项目程序扩展为大型 EXT2/3 文件系统。

参考文献

Card, R., Theodore Ts'o, T., Stephen Tweedie, S., "Design and Implementation of the Second Extended Filesystem", web.mit.edu/tytso/www/linux/ext2intro.html, 1995

Cao, M., Bhattacharya, S, Tso, T., "Ext4: The Next Generation of Ext2/3 File system", IBM Linux Technology Center, 2007

EXT2: www.nongnu.org/ext2-doc/ext2.html, 2001

EXT3: jamesthornton.com/hotlist/linux-filesystems/ext3-journal, 2015

使用系统调用进行文件操作

摘要

本章论述了如何使用系统调用进行文件操作；解释了系统调用的作用和 Linux 的在线手册页；展示了如何使用系统调用进行文件操作；列举并解释了文件操作中最常用的系统调用；阐明了硬链接和符号链接文件；具体解释了 stat 系统调用；基于 stat 信息，开发了一个类似于 ls 的程序来显示目录内容和文件信息；接着，讲解了 open-close-lseek 系统调用和文件描述符；然后，展示了如何使用读写系统调用来读写文件内容；在此基础上，说明了如何使用系统调用来显示和复制文件；还演示了如何开发选择性文件复制程序，其行为类似于一个简化的 Linux dd 实用程序。编程项目使用 Linux 系统调用来实现 C 程序，该程序将目录递归复制到目标中。该项目的目的是让读者练习程序的分层结构设计，并利用 stat()、open()、read()、write() 系统调用进行文件操作。

8.1 系统调用

在操作系统中，进程以两种不同的模式运行，即内核模式和用户模式，简称 Kmode 和 Umode。在 Umode 中，进程的权限非常有限。它不能执行任何需要特殊权限的操作。特殊权限的操作必须在 Kmode 下执行。系统调用（简称 syscall）是一种允许进程进入 Kmode 以执行 Umode 不允许操作的机制。复刻子进程、修改执行映像，甚至是终止等操作都必须在内核中执行。本章将讨论在 Unix/Linux 中使用系统调用进行文件操作。

8.2 系统调用手册页

在 Unix 以及大多数版本的 Linux 中，在线手册页保存在 /usr/man/ 目录中（Goldt 等 1995；Kerrisk 2010, 2017）。而在 Ubuntu Linux 中，则保存在 /usr/share/man 目录中。man2 子目录中列出了所有系统调用手册页。sh 命令 man 2 NAME 显示了系统调用名称的手册页。例如：

```
man 2 stat  : display man pages of stat(), fstat() and lstat() syscalls
man 2 open: display man pages of open() syscall
man 2 read: display man pages of read() syscall, etc.
```

许多系统调用需要特别包含头文件，手册页的 SYNOPSIS（概要）部分列出了这些文件。如果没有合适的头文件，C 编译器可能会因为 syscall 函数名称类型不匹配而发出许多警告。一些系统调用可能还需要特定的数据结构作为参数，必须在手册页中描述这些参数。

8.3 使用系统调用进行文件操作

系统调用必须由程序发出。它们的用法就像普通函数调用一样。每个系统调用都是一个库函数，它汇集系统调用参数，并最终向操作系统内核发出一个系统调用。

```
int syscall(int a, int b, int c, int d);
```

其中，第一个参数 a 是系统调用编号，b、c、d 是对应内核函数的参数。在基于 Intel x86 的 Linux 中，系统调用是由 INT 0x80 汇编指令实现的，可将 CPU 从用户模式切换到内核模式。内核的系统调用处理程序根据系统调用编号将调用路由到一个相应的内核函数。当进程结束执行内核函数时，会返回到用户模式，并得到所需的结果。返回值≥0 表示成功，-1 表示失败。如果失败，errno 变量（在 errno.h 中）会记录错误编号，它们会被映射到描述错误原因的字符串。下面的示例演示了如何使用一些简单的系统调用。

示例程序 C8.1：mkdir、chdir、getcwd 系统调用。

```
/************ C8.1.c file ************/
#include <stdio.h>
#include <errno.h>
int main()
{
    char buf[256], *s;
    int r;
    r = mkdir("newdir", 0766); // mkdir syscall
    if (r < 0)
       printf("errno=%d : %s\n", errno, strerror(errno));
    r = chdir("newdir");       // cd into newdir
    s = getcwd(buf, 256);      // get CWD string into buf[ ]
    printf("CWD = %s\n", s);
}
```

该程序发出一个 mkdir() 系统调用来创建新目录。mkdir() 系统调用需要一个路径名和一个权限（八进制的 0766）。如果没有新目录，则系统调用成功，返回值为 0。如果不止一次运行该程序，由于目录已经存在，则在第二次或后续任何运行时会失败，返回值为 -1。在这种情况下，该程序会打印消息：

```
errno=17 : File exists
```

除了 mkdir() 之外，该程序还演示了 chdir() 和 getcwd() 系统调用的用法。

练习：修改 C8.1 程序，在一次运行中创建多个目录，例如：

```
mymkdir dir1 dir2 dir3, ... dirn
```

提示：将 main() 写成 main(int argc, char *argv[])

简单的系统调用：下面列出了一些简单的文件操作系统调用。鼓励读者编写 C 程序来使用、测试它们。

access：检查对某个文件的权限

```
int access(char *pathname, int mode);
```

chdir：更改目录

```
int chdir(const char *path);
```

chmod：更改某个文件的权限

```
int chmod(char *path, mode_t mode);
```

chown：更改文件所有人

```
int chown(char *name, int uid, int gid);
```

chroot：将（逻辑）根目录更改为路径名

```
int chroot(char *pathname);
```

getcwd：获取 CWD 的绝对路径名

```
char *getcwd(char *buf,  int size);
```

mkdir：创建目录

```
int mkdir(char *pathname, mode_t mode);
```

rmdir：移除目录（必须为空）

```
int rmdir(char *pathname);
```

link：将新文件名硬链接到旧文件名

```
int link(char *oldpath, char *newpath);
```

unlink：减少文件的链接数；如果链接数达到 0，则删除文件

```
int unlink(char *pathname);
```

symlink：为文件创建一个符号链接

```
int symlink(char *oldpath, char *newpath);
```

rename：更改文件名称

```
int rename(char *oldpath, char *newpath);
```

utime：更改文件的访问和修改时间

```
int utime(char *pathname,  struct utimebuf *time)
```

以下系统调用需要超级用户权限。
mount：将文件系统添加到挂载点目录上

```
int  mount(char *specialfile, char *mountDir);
```

umount：分离挂载的文件系统

```
int umount(char *dir);
```

mknod：创建特殊文件

```
int mknod(char *path, int mode, int device);
```

8.4 常用的系统调用

本节，我们将讨论一些最常见的文件操作的系统调用。其中包括：
stat：获取文件状态信息

```
int stat(char *filename, struct stat *buf)
int fstat(int filedes, struct stat *buf)
int lstat(char *filename, struct stat *buf)
```

open：打开一个文件进行读、写、追加

```
int open(char *file, int flags, int mode)
```

close：关闭打开的文件描述符

```
int close(int fd)
```

read：读取打开的文件描述符

```
int read(int fd, char buf[ ], int count)
```

write：写入打开的文件描述符

```
int write(int fd, char buf[ ], int count)
```

lseek：重新定位文件描述符的读/写偏移量

```
int lseek(int fd, int offset, int whence)
```

dup：将文件描述符复制到可用的最小描述符编号中

```
int dup(int oldfd);
```

dup2：将 oldfd 复制到 newfd 中，如果 newfd 已打开，先将其关闭

```
int  dup2(int oldfd, int newfd)
```

link：将新文件硬链接到旧文件

```
int link(char *oldPath, char *newPath)
```

unlink：取消某个文件的链接；如果文件链接数为 0，则删除文件

```
int unlink(char *pathname);
```

symlink：创建一个符号链接

```
int symlink(char *target, char *newpath)
```

readlink：读取符号链接文件的内容

```
int readlink(char *path, char *buf, int bufsize)
```

umask：设置文件创建掩码；文件权限为（mask & ~umask）

```
int umask(int umask);
```

8.5 链接文件

在 Unix/Linux 中，每个文件都有一个路径名。但是，Unix/Linux 允许使用不同的路径名来表示同一个文件。这些文件叫作 LINK（链接）文件。有两种类型的链接，即硬链接和软链接或符号链接。

8.5.1 硬链接文件

硬链接：命令

```
ln  oldpath  newpath
```

创建从 newpath 到 oldpath 的硬链接。对应的系统调用为：

```
link(char *oldpath, char *newpath)
```

硬链接文件会共享文件系统中相同的文件表示数据结构（索引节点）。文件链接数会记录链接到同一索引节点的硬链接数量。硬链接仅适用于非目录文件。否则，它可能会在文件系统名称空间中创建循环，这是不允许的。相反，系统调用：

```
unlink(char *pathname)
```

会减少文件的链接数。如果链接数变为 0，文件会被完全删除。这就是 rm(file) 命令的作用。如果某个文件包含非常重要的信息，就最好创建多个链接到文件的硬链接，以防被意外删除。

8.5.2 符号链接文件

软链接：命令

```
ln -s  oldpath newpath     # ln command with the -s flag
```

创建从 newpath 到 oldpath 的软链接或**符号链接**。对应的系统调用是：

```
symlink(char *oldpath, char *newpath)
```

newpath 是 LNK 类型的普通文件，包含 oldpath 字符串。它可作为一个绕行标志，使访问指向链接好的目标文件。与硬链接不同，软链接适用于任何文件，包括目录。软链接在以下情况下非常有用。

（1）通过一个较短的名称来访问一个经常使用的较长路径名称，例如：

```
x -> aVeryLongPathnameFile
```

（2）将标准动态库名称链接到实际版本的动态库，例如：

```
libc.so.6 -> libc.2.7.so
```

当将实际动态库更改为不同版本时，库安装程序只需更改（软）链接以指向新安装的库。

软链接的一个缺点是目标文件可能不复存在了。如果是这样，绕行标志可能引导可怜的司机摔下悬崖。在 Linux 中，会通过 ls 命令以适当的深色 RED 显示此类危险，提醒用户链接已断开。此外，如果 foo -> /a/b/c 是软链接，open("foo", 0) 系统调用将打开被链接的文件 /a/b/c，而不是链接文件自身。所以 open()/read() 系统调用不能读取软链接文件，反而必须要用 readlink 系统调用来读取软链接文件的内容。

8.6 stat 系统调用

stat/lstat/fstat 系统调用可将一个文件的信息返回。命令 man 2 stat 会显示 stat 系统调用的手册页，如下文所述。

8.6.1 stat 文件状态

下面是关于 stat 系统调用的 Linux 开发者手册的内容。

名称

```
stat, fstat, lstat - get file status
```

概要

```
#include <sys/types.h>
#include <sys/stat.h>
#include <unistd.h>
int stat(const char *file_name, struct stat *buf);
int fstat(int filedes, struct stat *buf);
int lstat(const char *file_name, struct stat *buf);
```

描述

这些函数会返回指定文件的信息。不需要拥有文件的访问权限即可获取该信息，但是需要指向文件的路径中所有指定目录的搜索权限。

stat 按文件名统计指向文件，并在缓冲区中填写 stat 信息。

lstat 与 stat 相同，除非是符号链接，统计链接本身，而不是链接所引用文件。所以，stat 和 lstat 的区别是：stat 遵循链接，但 lstat 不是。

fstat 与 stat 相同，也只在文件名处说明 filedes（由 open（2）返回）所指向的打开文件。

8.6.2 stat 结构体

所有的 stat 系统调用都以 stat 结构体形式返回信息，其中包含以下字段：

```
struct stat{
   dev_t     st_dev;      /* device */
   ino_t     st_ino;      /* inode */
   mode_t    st_mode;     /* protection */
   nlink_t   st_nlink;    /* number of hard links */
   uid_t     st_uid;      /* user ID of owner */
   gid_t     st_gid;      /* group ID of owner */
   dev_t     st_rdev;     /* device type (if inode device) */
   off_t     st_size;     /* total size, in bytes */
   u32       st_blksize;  /* blocksize for filesystem I/O */
   u32       st_blocks;   /* number of blocks allocated */
   time_t    st_atime;    /* time of last access */
   time_t    st_mtime;    /* time of last modification */
   time_t    st_ctime;    /* time of last change */
};
```

st_size 是用字节表示的文件大小。符号链接的大小是指它所包含的路径名称长度，末尾没有 NULL。

st_blocks 值是用 512 字节块表示的文件大小。(可能小于 st_size /512，例如当文件有漏洞时。) st_blksize 值表示有效文件系统 I/O 的“首选”块大小。(以较小的块写入文件可能导致低效的读取－修改－重写。)

并非所有的 Linux 文件系统都能实现所有的时间字段。一些文件系统类型允许以这样一种方式挂载，即文件访问不会导致 st_atime 字段的更新。(见 mount（8）中的“noatime”。)

通过文件访问更改所含的 st_atime，例如 exec（2）、mknod（2）、pipe（2）、utime（2）和 read（2）(大于零字节)。其他例程，如 mmap（2），可能会，也可能不会更新 st_atime。

通过文件修改，如 mknod（2）、truncate（2）、utime（2）和 write（2）（大于零字节），

更改所包含的 st_mtime。此外，还可以通过创建或删除目录中的文件来更改目录的 st_mtime。所包含的 st_mtime 不会因为所有者、组、硬链接数或模式的变化而变化。

通过写入或设置索引节点信息（即所有者、组、链接数、模式等）更改所包含的 st_ctime。

定义以下 POSIX 宏来检查文件类型：

```
S_ISREG(m)  is it a regular file?
S_ISDIR(m)  directory?
S_ISCHR(m)  character device?
S_ISBLK(m)  block device?
S_ISFIFO(m) fifo?
S_ISLNK(m)  symbolic link? (Not in POSIX.1-1996.)
S_ISSOCK(m) socket? (Not in POSIX.1-1996.)
```

定义了 st_mode 字段的以下标志：

```
S_IFMT     0170000   bitmask for the file type bitfields
S_IFSOCK   0140000   socket
S_IFLNK    0120000   symbolic link
S_IFREG    0100000   regular file
S_IFBLK    0060000   block device
S_IFDIR    0040000   directory
S_IFCHR    0020000   character device
S_IFIFO    0010000   fifo

S_ISUID    0004000   set UID bit
S_ISGID    0002000   set GID bit (see below)
S_ISVTX    0001000   sticky bit (see below)

S_IRWXU    00700     mask for file owner permissions
S_IRUSR    00400     owner has read permission
S_IWUSR    00200     owner has write permission
S_IXUSR    00100     owner has execute permission
S_IRWXG    00070     mask for group permissions
S_IRGRP    00040     group has read permission
S_IWGRP    00020     group has write permission
S_IXGRP    00010     group has execute permission
S_IRWXO    00007     mask for permissions for others (not in group)
S_IROTH    00004     others have read permission
S_IWOTH    00002     others have write permission
S_IXOTH    00001     others have execute permission
```

返回值：如果成功，则返回零。如果错误，则返回 -1，并适当设置 errno。

另请参见 chmod（2）、chown（2）、readlink（2）、utime（2）。

8.6.3 stat 与文件索引节点

stat 与文件索引节点：首先，我们来阐明 stat 如何工作。每个文件都有一个独有的**索引节点**数据结构，包含文件的所有信息。下文给出了 Linux 中 EXT2 文件系统的索引节点结构体。

```
struct ext2_inode{
   u16  i_mode;
   u16  i_uid;
```

```
    u32  i_size;
    u32  i_atime;
    u32  i_ctime;
    u32  i_mtime;
    u32  i_dtime;
    u16  i_gid;
    u16  i_links_count;
    u32  i_blocks;
    u32  i_flags;
    u32  i_reserved1;
    u32  i_block[15];
    u32  pad[7];
}; // inode=128 bytes in ext2/3 FS; 256 bytes in ext4
```

每个索引节点在存储设备上都有唯一的索引节点编号（ino）。每个设备都由一对（主、次）设备号标识，例如 0x0302 表示 /dev/hda2，0x0803 表示 /dev/sda3 等。stat 系统调用只是查找文件的索引节点并将信息从索引节点复制到 stat 结构体中，但是 st_dev 和 st_ino 除外，它们分别是设备号和索引节点编号。在 Unix/Linux 中，所有时间字段都是自 1970 年 1 月 1 日 0 时 0 分 0 秒以来经过的秒数。它们可通过库函数 ctime(&time) 转换为日历形式。

8.6.4　文件类型和权限

在 stat 结构体中，大多数字段都无须解释。只有 st_mode 字段需要进行说明：

```
mode_t   st_mode;           /* copied from i_mode of INODE */
```

st_mode 的类型是一个 u16（16 位），这 16 位的含义如下：

```
|Type|   |permissions|
----------------------
|tttt|fff|uuu|ggg|ooo|
----------------------
```

前 4 位是文件类型，可以（以八进制形式）解释为：

```
S_IFMT      0170000   bitmask for the file type bitfields
S_IFSOCK    0140000   socket
S_IFLNK     0120000   symbolic link
S_IFREG     0100000   regular file
S_IFBLK     0060000   block device
S_IFDIR     0040000   directory
S_IFCHR     0020000   character device
S_IFIFO     0010000   fifo
```

目前，所有类 Unix 系统的手册页仍使用八进制数，可追溯至 20 世纪 70 年代的旧 PDP-11 时代。为方便起见，我们用十六进制重新定义它们，从而更加方便阅读，例如：

```
S_IFDIR     0x4000    directory
S_IFREG     0x8000    regular file
S_IFLNK     0xA000    symbolic link
```

st_mode 接下来的 3 位表示文件的特殊用法：

```
S_ISUID     0004000   set UID bit
S_ISGID     0002000   set GID bit
S_ISVTX     0001000   sticky bit
```

我们将在后文说明 setuid 程序的含义和用法。其余 9 位是文件保护权限位。可按进程的（有效）uid 和 gid 把这些位分为 3 类：

```
owner  group  other
 rwx    rwx    rwx
```

通过解释这些位，可将 st_mode 表示为：

```
-rwxr-xr-x       (REG file with r,x but w by owner only)
drwxr-xr-x       (DIR with r,x, but w by owner only)
lrw-r--r--       (LNK file with permissions)
```

其中第一个字母（-|d|l）表示文件类型，后面 9 个字符基于权限位。如果位是 1，则每个字符打印为 r|w|x；如果位是 0，则打印为 -。对于目录文件而言，x 位表示是否允许访问（cd 到）目录。

8.6.5 opendir-readdir 函数

目录也是一个文件。我们应该能像其他任何普通文件一样，打开一个 READ 目录，然后读取和显示它的内容。然而，根据文件系统的不同，目录文件的内容可能会有不同。因此，用户可能无法正确读取和解释目录的内容。鉴于此，POSIX 为目录文件指定了以下接口函数。

```
#include <dirent.h>
DIR *open(dirPath); // open a directory named dirPath for READ
struct dirent *readdir(DIR *dp); // return a dirent pointer
```

Linux 中的 dirent 结构体是：

```
struct dirent{
    u32 d_ino; // inode number
    u16 d_reclen;
    char d_name[ ]
}
```

在 dirent 结构体中，POSIX 只要求必须保留 d_name 字段。其他字段取决于具体的系统。opendir() 返回一个 DIR 指针 dirp。每个 readdir(dirp) 调用返回一个 dirent 指针，指向目录中下一个条目的 dirent 结构体。当目录中没有更多条目时，则返回一个 NULL 指针。我们用一个例子来说明它们的用法。下面的代码段可打印目录中的所有文件名称。

```
#include <dirent.h>
struct dirent *ep;
DIR *dp = opendir("dirname");
while (ep = readdir(dp)){
    printf("name=%s ", ep->d_name);
}
```

读者可参考 opendir 和 readdir 的 man 3 手册页，以了解更多详细信息。

8.6.6 readlink 函数

Linux 的 open() 系统调用遵循符号链接。因此，无法打开符号链接文件并读取其内容。

要想读取符号链接文件的内容，我们必须使用 readlink 系统调用，即：

```
int readlink(char *pathname, char buf[ ], int bufsize);
```

它将符号链接文件的内容复制到 bufsize 的 buf[] 中，并将实际复制的字节数返回。

8.6.7　ls 程序

下面给出了一个简单的 ls 程序，它的行为类似于 Linux 的 ls –l 命令。这里的目的并非要另外编写一个 ls 程序来重复操作。而是要说明如何使用各种系统调用来显示目录下的文件信息。通过学习示例程序代码，读者还应该能够理解如何实现 Linux 的 ls 命令。

```
/************* myls.c file **********/
#include <stdio.h>
#include <stdlib.h>
#include <string.h>
#include <sys/stat.h>
#include <time.h>
#include <sys/types.h>
#include <dirent.h>

struct stat mystat, *sp;
char *t1 = "xwrxwrxwr-------";
char *t2 = "----------------";

int ls_file(char *fname)
{
  struct stat fstat, *sp;
  int r, i;
  char ftime[64];
  sp = &fstat;
  if ( (r = lstat(fname, &fstat)) < 0){
     printf("can't stat %s\n", fname);
     exit(1);
  }
  if ((sp->st_mode & 0xF000) == 0x8000) // if (S_ISREG())
     printf("%c",'-');
  if ((sp->st_mode & 0xF000) == 0x4000) // if (S_ISDIR())
     printf("%c",'d');
  if ((sp->st_mode & 0xF000) == 0xA000) // if (S_ISLNK())
     printf("%c",'l');
  for (i=8; i >= 0; i--){
    if (sp->st_mode & (1 << i)) // print r|w|x
      printf("%c", t1[i]);
    else
      printf("%c", t2[i]);      // or print -
  }
  printf("%4d ",sp->st_nlink);  // link count
  printf("%4d ",sp->st_gid);    // gid
  printf("%4d ",sp->st_uid);    // uid
  printf("%8d ",sp->st_size);   // file size
  // print time
  strcpy(ftime, ctime(&sp->st_ctime)); // print time in calendar form
  ftime[strlen(ftime)-1] = 0;    // kill \n at end
```

```
  printf("%s  ",ftime);
  // print name
  printf("%s", basename(fname)); // print file basename
  // print -> linkname if symbolic file
  if ((sp->st_mode & 0xF000)== 0xA000){
     // use readlink() to read linkname
     printf(" -> %s", linkname); // print linked name
  }
  printf("\n");
}

int ls_dir(char *dname)
{
  // use opendir(), readdir(); then call ls_file(name)
}

int main(int argc, char *argv[])
{
  struct stat mystat, *sp = &mystat;
  int r;
  char *filename, path[1024], cwd[256];
  filename = "./";          // default to CWD
  if (argc > 1)
     filename = argv[1];  // if specified a filename
  if (r = lstat(filename, sp) < 0){
     printf("no such file %s\n", filename);
     exit(1);
  }
  strcpy(path, filename);
  if (path[0] != '/'){ // filename is relative : get CWD path
     getcwd(cwd, 256);
     strcpy(path, cwd); strcat(path, "/"); strcat(path,filename);
  }
  if (S_ISDIR(sp->st_mode))
     ls_dir(path);
  else
     ls_file(path);
}
```

练习 8.1：请填写上面示例程序中缺少的代码，即 ls_dir() 和 readlinko 函数，使它可以用于任何目录。

8.7 open-close-lseek 系统调用

open：打开一个文件进行读、写、追加

```
int open(char *file, int flags, int mode);
```

close：关闭打开的文件描述符

```
int close(int fd);
```

read：读取打开的文件描述符

```
int read(int fd, char buf[ ], int count);
```

write：写入打开的文件描述符

```
int write(int fd, char buf[ ], int count);
```

lseek：将文件描述符的字节偏移量重新定位为偏移量

```
int lseek(int fd, int offset, int whence);
```

umask：设置文件创建掩码；文件权限为（mask & ~umask）

8.7.1 打开文件和文件描述符

```
#include <sys/type.h>
#include <sys/stat.h>
#include <fcntl.h>
int open(char *pathname, int flags, mode_t mode)
```

open() 打开一个文件进行读、写或追加。它会返回一个进程可用的最小文件描述符，用于后续的 read()、write()、lseek() 和 close() 系统调用。标志字段必须包含下列一种访问模式，即 O_RDONLY、O_WRONLY 或者 O_RDWR。此外，这些标志可与其他标志（O_CREAT、O_APPEND、O_TRUNC、O_CLOEXEC）逐位进行 OR 组合。 所有这些符号常数都在 fcntl.h 头文件中定义。可选模式字段指定文件的权限（以八进制形式）。新创建文件或目录的权限是指定权限逐位与 ~umask 进行 AND 组合，其中 umask 在登录配置文件中设置为（八进制）022，这相当于删除非所有者的 w（写）权限位。umask 可用 umask() 系统调用进行更改。creat() 相当于 open()，标志相当于 O_CREAT|O_WRONLY|O_TRUNC，如果没有文件，则创建一个文件，打开该文件进行写操作，并将文件大小截断为零。

8.7.2 关闭文件描述符

```
#include <unistd.h>
int close(inf fd);
```

close() 关闭指定的文件描述符 fd，可重新用它来打开另一个文件。

8.7.3 lseek 文件描述符

```
#include <sys/type.h>
#include <unistd.h>
off_t lseek(int fd, off_t offset, int whence);
```

在 Linux 中，off_t 可定义为 u64。当打开某个文件进行读或写时，它的 RW- 指针被初始化为 0，这样就可以从文件的开头开始读 / 写。每次读 / 写 n 个字节后，RW- 指针就会前进 n 个字节进行下一次读 / 写。lssek() 将 RW- 指针重新定位到指定的偏移量，允许从指定的字节位置开始下一次读 / 写。whence 参数指定 SEEK_SET（从文件开头）、SEEK_CUR（当前 RW- 指针加上偏移量）、SEEK_EXD（文件大小加上偏移量）。

8.8 read() 系统调用

```
#include <unistd.h>
int read(int fd, void *buf, int nbytes);
```

read() 将 n 个字节从打开的文件描述符读入用户空间中的 buf[]。返回值是实际读取的字节数，如果 read() 失败，会返回 -1，例如当 fd 无效时。注意，buf[] 区必须有足够的空间来接收 n 个字节，并且返回值可能小于 n 个字节，例如文件小于 n 个字节，或者文件无更多需要读取的数据。还要注意，返回值是一个整数，而不是文件结束（EOF）符，因为文件中没有文件结束符。文件结束符是 I/O 库函数在文件流无更多数据时返回的一个特殊整数值（-1）。

8.9 write() 系统调用

```
#include <unistd.h>
int write(int fd, void *buf, int nbytes);
```

write() 将 n 个字节从用户空间中的 buf[ ] 写入文件描述符，必须打开该文件描述符进行写、读写或追加。返回值是实际写入的字节数，通常等于 n 个字节，如果 write() 失败，则为 -1，例如由于出现无效的 fd 或打开 fd 用于只读等。

以下代码段使用 open()、read()、lseek()、write() 和 close() 系统调用。它将文件的第一个 1KB 字节复制到 2048 字节。

```
char buf[1024];
int fd=open("file", O_RDWR);  // open file for READ-WRITE
read(fd, buf[ ], 1024);       // read first 1KB into buf[ ]
lseek(fd, 2048, SEEK_SET);    // lseek to byte 2048
write(fd, buf, 1024);         // write 1024 bytes
close(fd);                    // close fd
```

8.10 文件操作示例程序

系统调用适用于大数据块上的文件 I/O 操作，即不需要行、字符或结构化记录等的操作。下面几节内容展示了一些使用文件操作系统调用的示例程序。

8.10.1 显示文件内容

示例 8.2：显示文件内容。该程序的行为类似于 Linux cat 命令，将文件内容显示到 stdout。如果未指定文件名，则从默认 stdin 获取输入。

```
/********* C8.2 file ********/
#include <stdio.h>
#include <stdlib.h>
#include <fcntl.h>
#include <unistd.h>

#define BLKSIZE 4096
int main(int argc, char *argv[ ])
{
  int fd, i, m, n;
  char buf[BLKSIZE], dummy;
  fd = 0; // default to stdin
  if (argc > 1){
     fd = open(argv[1], O_RDONLY);
```

```
      if (fd < 0) exit(1);
   }
   while (n = read(fd, buf, BLKSIZE)){
     m = write(1, buf, n);
   }
}
```

当运行无文件名的程序时，它从 fd=0 收集输入，fd=0 是标准输入流 stdin。若要终止程序，可按下“Ctrl+D”组合键（0x04），这是 stdin 上的默认文件终止符。当运行有文件名的程序时，它会首先打开文件进行读取。然后，它会使用一个 while 循环来读取和显示文件内容，直到 read() 返回 0，表示文件无更多数据。在每个迭代中，它将达到 4KB 的字符读入 buf[]，并将 n 个字符写入文件描述符 1。在 Unix/ Linux 文件中，行被 LF=\n 字符终止。如果文件描述符引用一个终端特殊文件，则伪终端仿真程序自动为每个 \n 字符添加一个 \r，以产生正确的视觉效果。如果文件描述符引用一个普通文件，则不会向输出中添加额外的 \r 字符。

8.10.2 复制文件

示例 8.3：复制文件。该示例程序类似于 Linux cp src dest 命令，将一个 src 文件复制到 dest 文件中。

```
/******** c8.3.c file *******/
#include <stdio.h>
#include <stdlib.h>
#include <fcntl.h>
#include <unistd.h>

#define BLKSIZE 4096
int main(int argc, char *argv[ ])
{
  int fd, gd, n, total=0;
  char buf[BLKSIZE];
  if (argc < 3) exit(1);  // usage a.out src dest
  if ((fd = (open(argv[1], O_RDONLY)) < 0)
     exit(2);
  if ((gd = open(argv[2],O_WRONLY|O_CREAT)) < 0)
     exit(3);
  while (n = read(fd, buf, BLKSIZE)){
     write(gd, buf, n);
     total += n;
  }
  printf("total bytes copied=%d\n", total);
  close(fd); close(gd);
}
```

练习 8.2：示例程序 c8.3 有一个严重的缺陷，我们不应该将文件复制到该文件本身。除了浪费时间，读者还可以通过查看程序代码找出原因。如果 src 和 dest 文件是同一个文件，则不会进行复制，只会将文件大小截断为 0。修改程序，以确保 src 和 dest 不是同一个文件。注意，由于硬链接，不同的文件名可能引用同一个文件。

提示：给出两个路径名的信息，并比较它们的（st_dev, st_ino）。

8.10.3 选择性文件复制

示例 8.4：选择性文件复制。该示例程序是对示例 8.3 中简单文件复制程序的简化。它类似于 Linux dd 命令，复制文件的选定部分。读者可参考 Linux dd 的手册页来了解它的全部功能。由于我们在这里的目的是展示如何使用文件操作系统调用，所以示例程序是 dd 的简化版。该程序按以下方式运行。

```
a.out if=in of=out bs=size count=k [skip=m] [seek=n] [conv=notrunc]
```

其中 [ ] 表示可选条目。如果指定了这些条目，则 skip=m 表示跳过输入文件的 m 个块，seek=n 表示在写入之前将输出文件向前移动 n 个块，conv=notrunc 表示如果输出文件已退出，则不要截断它。skip 和 seek 的默认值为 0，即没有 skip 或 seek。为简单起见，我们假设命令行参数不包含空格，从而简化了命令行解析。该程序唯一的新特性是解析命令行参数来设置文件名、计算变量和标志。例如，如果指定了 conv=notrunc，则必须在不截断文件的情况下打开目标文件。其次，如果指定了 skip 和 seek，则打开的文件必须使用 lseek() 来设置相应的 RW 偏移量。

```
/************ c8.4.c file **********/
#include <stdio.h>
#include <stdlib.h>
#include <fcntl.h>
#include <unistd.h>
#include <string.h>

char in[128], out[128], buf[4096];
int bsize, count, skip, seek, trnc;
int records, bytes;
// parse command line parameters and set variables
int parse(char *s)
{
  char cmd[128], parm[128];
  char *p = index(s, '=');
  s[p-s] = 0; // tokenize cmd=parm by '='
  strcpy(cmd, s);
  strcpy(parm, p+1);
  if (!strcmp(cmd, "if"))
     strcpy(in, parm);
  if (!strcmp(cmd, "of"))
     strcpy(out, parm);
  if (!strcmp(cmd, "bs")
     bsize = atoi(parm);
  if (!strcmp(cmd, "count"))
     count = atoi(parm);
  if (!strcmp(cmd, "skip"))
     skip = atoi(parm);
  if (!strcmp(cmd, "seek"))
     seek = atoi(parm);
  if (!strcmp(cmd, "conv")){
     if (!strcmp(parm, "notrunc"))
        trnc = 0;
  }
}
```

```
int main(int argc, char *argv[])
{
  int fd, gd, n, i;
  if (argc < 3){
     printf("Usage: a.out if of ....\n"); exit(1);
  }
  in[0] = out[0] = 0;              // null file names
  bsize = count = skip = seek = 0; // all 0 to start
  trnc = 1;                        // default = trunc
  for (i=1; i<argc; i++){
      parse(argv[i]);
  }
  // error checkings
  if (in[0]==0) || out[0]==0{
     printf("need in/out files\n"); exit(2);
  }
  if (bsize==0 || count==0){
     printf("need bsize and count\n"); exit(3);
  }
  // ADD: exit if in and out are the same file
  if ((fd = open(in, O_RDONLY)) < 0){
     printf("open %s error\n", in); exit(4);
  }
  if (skip) lseek(fd, skip*bsize, SEEK_SET);
  if (trnc) // truncate out file
     gd = open(out, O_WRONLY|O_CREAT|O_TRUNC);
  else
     gd = open(out, O_WRONLY|O_CREAT); // no truncate
  if (gd < 0){
     printf("open %s error\n", out); exit(5);
  }
  if (seek) lseek(gd, seek*bsize, SEEK_SET);

  records = bytes = 0;
  while (n = read(fd, buf, bsize)){
    write(gd, buf, n);
    records++; bytes += n;
    count--;               // dec count by 1
    if (count==0) break;
  }
  printf("records=%d bytes=%d copied\n", records, bytes);
}
```

8.11　编程项目：使用系统调用递归复制文件

该编程项目将会使用 Linux 系统调用来编写一个 C 程序 mycp，将 src 递归复制到 dest 中。它的工作方式应与以下 Linux 命令完全相同：

```
cp -r src dest
```

将 src 递归复制到 dest 中。

8.11.1　提示和帮助

（1）分析允许复制的各种条件。

下面是一些示例案例，但它们并不完整。读者应在尝试开发任何代码之前完成案例分析。

1）src 必须存在，但是 dest 可能存在，也可能不存在。

2）如果 src 是一个文件 , dest 可能不存在，也可能是一个文件或目录。

3）如果 src 是一个目录，dest 一定是一个现有目录或不存在的目录。

4）如果 src 是一个目录且 dest 不存在，创建 dest 目录并将 src 复制到 dest 中。

5）如果 src 是一个目录且 dest 是一个现有目录：若 dest 是 src 的后代，不要复制它。否则，将 src 复制到 dest/ 中，即 dest/(basename(src))

6）永远不要将文件或目录复制到它本身中。

如有疑问，运行 Linux cp -r src dest 命令并比较结果。

（2）分 3 个层级来组织项目程序。

- 底层：cpf2f(file1, filef2)：将文件 1 复制到文件 2 中，处理 REG 和 LNK 文件。
- 中间层：cpf2d(file, dir)：将文件复制到现有目录中。
- 顶层：cpd2d(dirl,dir2)：将目录 1 递归复制到文件 2 中。

8.11.2 示例解决方案

编程项目的示例解决方案可从本书的网站上下载。可应作者要求提供源代码给讲师。

参考文献

Goldt, S. Van Der Meer, S., Burkett, S., Welsh, M., The Linux Programmer's Guide-The Linux Documentation Project, 1995

Kerrisk, M., The Linux Programming Interface, No Starch Press, Inc., 2010

Kerrisk, M. The Linux man-pages project, https://www.kernel.org/doc/man-pages/, 2017

I/O 库函数

摘要

本章讨论了I/O库函数；解释了I/O库函数的作用及其相对于系统调用的优势；使用示例程序来说明I/O库函数和系统调用之间的关系，并解释了它们之间的相似性和基本区别；详细介绍了I/O库函数的算法，包括fread、fwrite和fclose的算法，重点介绍了它们与read、write和close系统调用的交互；介绍了I/O库函数的不同模式，包括字符模式、行模式、结构化记录模式和格式化I/O操作；阐述了文件流缓冲方案，并通过示例程序说明了不同缓冲方案的效果；阐释了有不同参数的函数以及如何使用stdarg宏访问参数。

编程项目将本章的原理与编程技术相结合，以实现一个类printf函数，根据格式字符串格式化打印不同数基的字符、字符串和数字。类printf函数的基础是Linux的putchar()，但它的工作原理与库函数printf()完全相同。该项目的目的是让读者了解如何实现I/O库函数。

9.1 I/O 库函数

系统调用是文件操作的基础，但它们只支持数据块的读/写。实际上，用户程序可能希望以最适合应用程序的逻辑单元读/写文件，如行、字符、结构化记录等，而系统调用不支持这些逻辑单元。I/O库函数是一系列文件操作函数，既方便用户使用，又提高了整体效率（GNU I/O on streams 2017；GNU libc 2017；GNU Library Reference Manual 2017）。

9.2 I/O 库函数与系统调用

几乎每个支持C语言编程的操作系统都可提供文件I/O库函数。在Unix/Linux中，I/O库函数建立在系统调用的基础上。为了说明它们之间的密切关系，我们首先列举了其中几个进行比较。

- **系统调用函数**：open()、read()、write()、lseek()、close()；
- **I/O 库函数**：fopen()、fread()、fwrite()、fseek()、fclose()。

从它们的高度相似性中，读者大概可以猜出每个I/O库函数的根都在对应的系统调用函数中。的确如此，fopen()依赖于open()，fread()依赖于read()，等等。下面的C程序说明了它们的相同点和不同点。

示例 9.1：显示文件内容。

```
            系统调用                 |            I/O 库函数
-------------------------------------|-------------------------------
#include <fcntl.h>                   | #include <stdio.h>
int main(int argc, char *argv[ ])    | int main(int argc, char *argv[ ])
{                                    | {
1. int fd;                           |   FILE *fp;
   int i, n;                         |
   char buf[4096];                   |   int c;    // for EOF of stdin
   if (argc < 2) exit(1);            |   if (argc < 2) exit(1);
2. fd = open(argv[1], O_RDONLY);     |   fp = fopen(argv[1], "r");
```

```
   if (fd < 0) exit(2);                    |  if (fp==0) exit(2);
3. while (n = read(fd, buf, 4096)){       |  while ((c = fgetc(fp))!= EOF){
     for (i=0; i<n; i++){                  |
       write(1, &buf[i], 1);              |    putchar(c);
     }                                     |
   }                                       |  }
}                                          | }
-----------------------------------------------------------------------
```

左边显示的是一个使用系统调用的程序。右边显示的是一个使用I/O库函数的类似程序。这两个程序都可将文件内容打印到显示屏上。两个程序看起来很相似，但二者有根本的区别。

第1行：在系统调用程序中，文件描述符fd是一个整数。在库I/O程序中，fp是一个文件流指针。

第2行：系统调用open()打开一个文件进行读取，并返回一个整数文件描述符fd，如果open()失败，则返回-1。I/O库函数fopen()返回一个FILE结构体指针，如果fopen()失败，则返回NULL。

第3行：系统调用程序使用while循环读取/写入文件内容。在每个迭代中，它发出read()系统调用，将最多4KB的字符读入buf[]。然后，它将各字符从buf[]写到文件描述符1中，这是该进程的标准输出。正如前文所指出的，使用系统调用一次写入一个字节非常低效。相反，I/O库程序仅仅使用fgetc(fp)从文件流中获取字符，通过putchar()输出字符，直至文件结束符。

除了使用的语法和函数上的细微差别之外，这两个程序之间还有一些根本的区别，下面将进行更详细的说明。

第2行：fopen()发出open()系统调用以获取文件描述符fd。如果open()调用失败，将返回一个NULL指针。否则，它会在程序的堆区分配一个FILE结构体。FILE结构体包含一个内部缓冲区char fbuf[BLKSIZE]和一个整数fd字段。它记录open()在FILE结构体中返回的文件描述符，将fbuf[]初始化为空，并将FILE结构体的地址作为fp返回。

第3行：fgetc(c, fp)尝试从文件流fp中获取一个字符。如果FILE结构体中的fbuf[]为空，则发出read(fd, fbuf, BLKSIZE)系统调用，从文件中读取BLKSIZE字节，其中BLKSIZE与文件系统块大小匹配。然后它从fbuf[]返回一个char。随后，fgetc()从fbuf[]返回一个char，只要它仍然有数据。这样，库I/O read函数发出read()系统调用，仅用于重新填充fbuf[ ]，它们总是将BLKSZISE字节的数据从操作系统内核传输到用户空间。类似表述也适用于I/O write库函数。

练习9.1：在示例9.1的系统调用程序中，通过一个系统调用来写每个字符是非常低效的。用一个write()系统调用来替换for循环。

示例9.2：复制文件。同样，我们并列列出两个版本的程序，以说明它们的异同点。

```
             系统调用                      |            I/O 库函数
-----------------------------------------|-----------------------------
#include <fcntl.h>                       | #include <stdio.h>
#define BLKSIZE 4096                     | #define BLKSIZE 4096
int fd, gd;                              | FILE *fp, *gp;
char buf[4096];                          | char buf[4096];
                                         |
int main(int argc,char *argv[])          | int main(int argc,char *argv[])
```

```
{                                                 | {
  int n, total=0;                                 |  int n, total=0;
  if (argc < 3) exit(1);                          |  if (argc < 3) exit(1);
  // check for same file                          |  // check for same file
  fd = open(argv[1], O_RDONLY);                   |  fp = fopen(argv[1], "r");
  if (fd < 0) exit(2);                            |  if (fp == NULL) exit(2);
  gd = open(argv[2],O_WRONLY|O_CREAT);            |  gp = fopen(argv[2], "w");
  if (gd < 0) exit(3);                            |  if (gp == NULL) exit(3);
  while(n=read(fd, buf, BLKSIZE))                 |   while(n=fread(buf,1,BLKSIZE,fp))
  {                                               |  {
      write(gd, buf, n);                          |     fwrite(buf, 1, n, gp);
      total += n;                                 |     total += n;
  }                                               |  }
  printf("total=%d\n",total);                     |  printf("total = %d\n", total);
  close(fd); close(gd);                           |  fclose(fp); fclose(gp);
}                                                 | }
-------------------------------------------------------------------
```

这两个程序都会将 src 文件复制到 dest 文件中。由于第 6 章已经解释过系统调用程序，所以我们只讨论使用 I/O 库函数的程序。

（1）fopen() 使用字符串表示模式，其中“r”表示 READ，“w”表示 WRITE。它返回一个指向 FILE 结构体的指针。fopen() 首先发出 open() 系统调用来打开文件，以获取文件描述符编号 fd。如果 open() 系统调用失败，则 fopen() 会返回一个 NULL 指针。否则，它会在程序的堆区中分配一个 FILE 结构体。每个 FILE 结构体均包含一个内部缓冲区 fbuf [BLKSIZE]，其大小通常与文件系统的 BLKSIZE 相匹配。此外，它还包含用于操作 fbuf[] 的指针、计数器和状态变量，存储来自 open() 的文件描述符。它将 FILE 结构体初始化并返回指向 FILE 结构体的 fp。需要注意的是，FILE 结构体位于进程的用户模式映像中。这意味着对 I/O 库函数的调用是普通的函数调用，而不是系统调用。

（2）如有任何 fopen() 调用失败，程序将会终止。如前文所述，fopen() 在失败时会返回一个 NULL 指针，例如，文件不能在指定模式下打开时。

（3）然后，它使用一个 while 循环来复制文件内容。while 循环的每个迭代尝试从源文件读取 BLKSIZE 字节，并向目标文件写入 n 个字节，其中 n 是从 fread() 返回的值。fread() 和 fwrite() 的一般形式是：

```
int n = fread(buffer, size, nitems, FILEptr);
int n = fwrite(buffer,size, nitems, FILEptr);
```

其中 size 是记录大小（以字节为单位），nitems 是要读取或写入的记录数量，n 是实际读取或写入的记录数量。这些函数用于读 / 写结构化数据对象。例如，假设缓冲区包含以下结构化记录的数据对象：

```
struct record{.....}
```

我们可以使用：

```
n = fwrite(buffer, sizeof(struct record), nitem, FILEptr);
```

将 nitem 记录写入文件。同样，

```
n =   fread(buffer, sizeof(struct record), nitem, FILEptr);
```

从文件中读取 nitem 记录。

上面的程序尝试一次性读 / 写 BLKSIZE 字节。因此，它的 size = 1，nitems = BLKSIZE。事实上，size 和 nitems（size*nitems = BLKSIZE）的任意组合也适用。但是，使用 size > 1 可能会在最后一个 fread() 上出现问题，因为文件的剩余字节可能小于 size 字节。在这种情况下，返回的 n 是零，即使它已读取了一些数据。为了处理源文件的“尾”部，我们可以在 while 循环后面添加以下代码行：

```
fseek(fp, total, SEEK_SET);  // fseek to byte total
n = fread(buf, 1, size, fp); // read remaining bytes
    fwrite(buf,1, n, gp);    // write to dest file
total += n;
```

fseek() 的工作原理与 lseekf 完全相同。它将文件的 R|W 指针定位到总字节位置。这里，我们将文件读取为 1-byte 对象。这样可以读取所有剩余字节（如有），并将它们写入目标文件中。

（4）复制完成后，两个文件均由 fclose(FILE*p) 关闭。

9.3 I/O 库函数的算法

9.3.1 fread 算法

fread() 算法如下：

（1）在第一次调用 fread() 时，FILE 结构体的缓冲区是空的，fread() 使用保存的文件描述符 fd 发出一个

```
n = read(fd, fbuffer, BLKSIZE);
```

系统调用，用数据块填充内部的 fbuf[]。然后，它会初始化 fbuf[] 的指针、计数器和状态变量，以表明内部缓冲区中有一个数据块。接着，通过将数据复制到程序的缓冲区，尝试满足来自内部缓冲区的 fread() 调用。如果内部缓冲区没有足够的数据，则会再发出一个 read() 系统调用来填充内部缓冲区，将数据从内部缓冲区传输到程序缓冲区，直到满足所需的字节数（或者文件无更多数据）。将数据复制到程序的缓冲区之后，它会更新内部缓冲区的指针、计数器等，为下一个 fread() 请求做好准备。然后，它会返回实际读取的数据对象数量。

（2）在随后的每次 fread() 调用中，它都尝试满足来自 FILE 结构体内部缓冲区的调用。当缓冲区变为空时，它就会发出 read() 系统调用来重新填充内部缓冲区。因此，fread() 一方面接受来自用户程序的调用，另一方面向操作系统内核发出 read() 系统调用。除了 read() 系统调用之外，所有 fread() 处理都在用户模式映像中执行。它只在需要时才会进入操作系统内核，并且以一种最高效匹配文件的方式进入。它会提供自动缓冲机制，因此用户程序不必担心这些具体操作。

9.3.2 fwrite 算法

fwrite() 算法与 fread() 算法相似，只是数据传输方向不同。最开始，FILE 结构体的内部缓冲区是空的。在每次调用 fwrite() 时，它将数据写入内部缓冲区，并调整缓冲区的指针、计数器和状态变量，以跟踪缓冲区中的字节数。如果缓冲区已满，则发出 write() 系统

调用，将整个缓冲区写入操作系统内核。

9.3.3 fclose 算法

若文件以写的方式被打开，fclose() 会先关闭文件流的局部缓冲区。然后，它会发出一个 close(fd) 系统调用来关闭 FILE 结构体中的文件描述符。最后，它会释放 FILE 结构体，并将 FILE 指针重置为 NULL。

9.4 使用 I/O 库函数或系统调用

根据上面所讨论的内容，现在我们可以回答什么时候使用系统调用或库函数进行文件 I/O 的问题。fread() 依赖 read() 将数据从内核复制到内部缓冲区，然后从内部缓冲区将数据复制到程序的缓冲区。所以，它传输了两次数据。相反，read() 将数据从内核直接复制到程序的缓冲区，只复制了一次。因此，对于以 BLKSIZE 为单位的读 / 写数据来说，read() 本来就比 fread() 更高效，因为它只需要一个而不是两个复制操作。类似表述也适用于 write() 和 fwrite()。

需要注意，在 fread() 和 fwrite() 的一些实现中，例如在 GNU libc 库中，如果请求的大小以 BLKSIZE 为单位，它们可以使用系统调用将以 BLKSIZE 为单位的数据直接从内核传输到用户指定的缓冲区。即便如此，使用 I/O 库函数仍然需要其他的函数调用。因此，在上面的例子中，使用系统调用的程序实际上比使用 I/O 库函数的程序更高效。但是，如果不是以 BLKSIZE 为单位进行读 / 写，那么 fread() 和 fwrite() 可能更高效。例如，如果我们坚持一次读 / 写一个字节，fread() 和 fwrite() 会好得多，因为它们进入操作系统内核只是为了填充或清除内部缓冲区，并不是逐字节输入。这里，我们曾暗中假设，即进入内核模式比停留在用户模式的代价更高，事实的确如此。

9.5 I/O 库模式

fopen() 中的模式参数可以指定为："r"、"w"、"a"，分别代表读、写、追加。

每个模式字符串可包含一个 + 号，表示同时读写，或者在写入、追加情况下，如果文件不存在则创建文件。

"r+"：表示读 / 写，不会截断文件。

"w+"：表示读 / 写，但是会先截断文件；如果文件不存在，会创建文件。

"a+"：表示通过追加进行读 / 写；如果文件不存在，会创建文件。

9.5.1 字符模式 I/O

```
int fgetc(FILE *fp):             // get a char from fp, cast to int.
int ungetc(int c, FILE *fp);     // push a previously char got by fgetc()
                                    back to stream
int fputc(int c, FILE *fp);      // put a char to fp
```

注意，fgetc() 返回的是整数，而不是字符。这是因为它必须在文件结束时返回文件结束符。文件结束符通常是一个整数 -1，将它与文件流中的任何字符区分开。

对于 fp=stdin 或 stdout，可能会使用 c=getchar(); putchar(c) ; 来代替。对于运行时效来说，getchar() 和 putchar() 通常不是 getc() 和 putc() 的缩小版本。相反，可以将它们实现为

宏，以避免额外的函数调用。

示例 9.3：字符模式 I/O。

```
(1). /* file copy using getc(), putc() */
   #include <stdio.h>
   FILE *fp,*gp;
   int main()
   {
      int c;  /* for testing EOF */
      fp=fopen("source", "r");
      gp=fopen("target", "w");
      while ( (c=getc(fp)) != EOF )
          putc(c,gp);
      fclose(fp); fclose(gp);
    }
```

练习 9.2：编写一个 C 程序，将文本文件中的字母由小写转换为大写。

练习 9.3：编写一个 C 程序，计算文本文件的行数。

练习 9.4：编写一个 C 程序，计算文本文件的单词数。单词是由空格分开的一系列字符。

练习 9.5：Linux 的手册页是压缩的 gz 文件。使用 gunzip 解压手册页文件。未压缩手册页文件是一个文本文件，但是几乎不可读，因为它包含许多仅供手册页程序使用的特殊字符和字符序列。假设 ls.1 是 ls.1.gz 的未压缩手册页。分析未压缩的 ls.1 文件，找出有哪些特殊字符。然后编写一个 C 程序，删除所有特殊字符，使其成为纯文本文件。

9.5.2 行模式 I/O

char *fgets(char *buf, int size, FILE *fp)：从 fp 中读取最多为一行（以 \n 结尾）的字符。

int fputs(char *buf, FILE *fp)：将 buf 中的一行写入 fp 中。

示例 9.4：行模式 I/O。

```
#include <stdio.h>
FILE *fp,*gp;
char buf[256]; char *s="this is a string";
int main()
{
  fp = fopen("src",  "r");
  gp = fopen("dest", "w");
  fgets(buf, 256, fp); // read a line of up to 255 chars to buf
  fputs(buf, gp);      // write line to destination file
}
```

当 fp 是 stdin 或 stdout 时，也可以使用以下函数，但它们并非 fgets() 和 fputs() 的缩减版本。

```
gets(char *buf);      // input line from stdin but without checking length
puts(char *buf);      // write line to stdout
```

9.5.3 格式化 I/O

这些大概是最常用的 I/O 函数。

格式化输入：(FMT=格式字符串)

```
scanf(char *FMT, &items);          // from stdin
fscanf(fp, char *FMT, &items);     // from file stream
```

格式化输出：

```
printf(char *FMT, items);          // to stdout
fprintf(fp, char *FMT, items);     // to file stream
```

9.5.4 内存中的转换函数

```
sscanf(buf, FMT, &items);     // input from buf[ ] in memory
sprintf(buf,FMT, items);      // print to buf[ ] in memroy
```

注意，sscanf() 和 sprintf() 并非 I/O 函数，而是内存中的数据转换函数。例如，atoi() 是一个标准库函数，将一串 ASCII 数字转换成整数，但是大多数 Unix/Linux 系统没有 itoA() 函数，因为转换可由 sprintf() 完成，所以不需要它。

9.5.5 其他 I/O 库函数

- fseek()、ftell()、rewind()：更改文件流中的读 / 写字节位置。
- feof()、ferr()、fileno()：测试文件流状态。
- fdopen()：用文件描述符打开文件流。
- freopen()：以新名称重新打开现有的流。
- setbuf()、setvbuf()：设置缓冲方案。
- popen()：创建管道，复刻子进程来调用 sh。

9.5.6 限制混合 fread-fwrite

当某文件流同时用于读 / 写时，就会限制使用混合 fread() 和 fwrite() 调用。规范要求每对 fread() 和 fwrite() 之间至少有一个 fseek() 或 ftell()。

示例 9.5：混合 fread-fwrite。该程序在 HP-UX 和 Linux 下运行时会产生不同的结果。

```
#include <stdio.h>
FILE fp; char buf[1024];
int main()
{
  fp = fopen("file", "r+");    // for both R/W
  fread(buf, 1, 20, fp);       // read 20 bytes
  fwrite(buf,1, 20, fp);       // write to the same file
}
```

Linux 给出了正确结果，将字节从 20 修改为 39。HP-UX 向原始文件的末尾追加 40 个字节。区别在于两个系统对读 / 写指针的处理不一致。回顾前文的内容可以知道，fread()/fwrite() 会发出 read()/write() 系统调用来填充 / 清除内部缓冲区。当 read()/write() 使用文件 OFTE 中的读 / 写指针时，fread()/fwrite() 会使用 FILE 结构体中局部缓冲区的读 / 写指针。

如果没有 fseek() 来同步这两个指针，其结果就取决于它们在实现中的使用方式。为了避免出现任何不一致，请遵循手册页中的建议。对于示例 9.5 的程序，如果我们将下面一行：

```
fseek(fp, (long)20, 0);
```

插入 fread() 和 fwrite() 中间，结果会相同（而且正确）。

9.6 文件流缓冲

每个文件流都有一个 FILE 结构体，其中包含一个内部缓冲区。对文件流进行读写需要遍历 FILE 结构体的内部缓冲区。文件流可以使用三种缓冲方案中的一种。

- **无缓冲**：从非缓冲流中写入或读取的字符将尽快单独传输到文件或从文件中传输。例如，文件流 stderr 通常无缓冲。到 stderr 的所有输出都会立即发出。
- **行缓冲**：遇到换行符时，写入行缓冲流的字符以块的形式传输。例如，文件流 stdout 通常是行缓冲，逐行输出数据。
- **全缓冲**：写入全缓冲流或从中读取的字符以块大小传输到文件或从文件传输。这是文件流的正常缓冲方案。

通过 fopen() 创建文件流之后，在对其执行任何操作之前，用户均可发出一个

```
setvbuf(FILE *stream, char *buf, int node, int size)
```

调用来设置缓冲区（buf）、缓冲区大小（size）和缓冲方案（mode），它们必须是以下一个宏：

- _IONBUF：无缓冲。
- _IOLBUF：行缓冲。
- _IOFBUF：全缓冲。

此外，还有其他的 setbuf() 函数，是 setvbuf() 的变体。读者可参考 setvbuf 手册页，以了解更多详细信息。

对于行缓冲流或全缓冲流，可用 fflush(stream) 立即清除流的缓冲区。我们通过示例来说明不同的缓冲方案。

示例 9.6：文件流缓冲。思考下面的 C 程序。

```
#include <stdio.h>
int main()
{
  (1).  // setvbuf(stdout, NULL, _IONBF, 0);
        while(1){
  (2).     printf("hi ");  // not a line yet
  (3).     // fflush(stdout);
           sleep(1);       // sleep for 1 second
         }
}
```

在运行上面的程序时，尽管 print() 语句会每秒出现在第（2）行，但不会立即出现输出。因为 stdout 是行缓冲。只有当打印出来的字符与 stdout 的所有内部缓冲区相匹配时，输出才会出现，此时，所有以前打印的字符将同时出现。因此，如果我们不将行写入 stdout，它就会像是一个全缓冲流。如果取消注释第（3）行，就清除了 stdout，每个打印都会立即

出现，尽管它还不是一个行。如果取消注释第（1）行，即将 stdout 设置为无缓冲，则每一秒都会出现打印。

9.7　变参函数

在 I/O 库函数中，printf() 相当独特，因为多种不同类型的可变数量参数可以调用它。这是允许的，因为最初的 C 语言不是一种类型检查语言。目前，C 语言和 C++ 会强制执行类型检查，但是为了方便，这两种语言仍然允许参数数量可变的函数。这些函数必须至少使用一个参数进行声明，后跟 3 个点，如

```
int func(int m, int n . . .)          // n = last specified parameter
```

在函数内部，可以通过 C 语言库宏访问参数：

```
void va_start(va_list ap, last); // start param list from last parameter
type va_arg(va_list ap, type);   //  type = next parameter type
va_end(va_list ap);              // clear parameter list
```

我们用一个示例来说明这些宏的用法：

示例 9.7：使用 stdarg 宏访问参数列表。

```
/******* Example of accessing varying parameter list ********/
#include <stdio.h>
#include <stdarg>     // need this for va_list type

// assume: func() is called with m integers, followed by n strings
int func(int m, int n . . .) // n = last known parameter
{
     int i;
(1). va_list ap;          // define a va_list ap
(2). va_start(ap, n);     // start parameter list from last param n
     for (i=0; i<m; i++)
(3).   printf("%d ", va_arg(ap, int));    // extract int params
     for (i=0; i<n, i++)
(4).   printf("%s ", va_arg(ap, char *)) // extra char* params
(5).   va_end(ap);
}
int main()
{
   func(3, 2, 1, 2, 3, "test", "ok");
}
```

在示例程序中，我们假设函数 func() 会被两个已知参数 int m 和 int n 调用，后面是 m 个整数和 n 个字符串。

第（1）行定义了一个 va_list ap 变量。

第（2）行从最后一个已知参数（n）开始创建参数列表。

第（3）行使用 va_arg(ap, int) 将后面的 m 个参数提取为整数。

第（4）行使用 va_arg(ap, char *) 将后面的 n 个参数提取为字符串。

第（5）行通过将 ap 列表重置为 NULL 来结束参数列表。

读者可编译并运行示例程序。它可以打印：1 2 3 test OK。

9.8 编程项目：类 printf 函数

本编程项目会编写一个类 printf() 函数，用于格式化打印字符、字符串、无符号整数、十进制有符号整数和十六进制无符号整数。编程项目的目的是让读者了解 I/O 库函数是如何实现的。如果是可以打印不同类型可变数量项目的 printf()，基本操作是打印单个字符。

9.8.1 项目规范

在 Linux 中，putchar(char c) 可打印一个字符。只使用 putchar() 来实现函数

```
int myprintf(char *fmt, . . .)
```

用于格式化打印其他参数，其中 fmt 是格式字符串，包含：

```
%c : print char
%s : print string
%u : print unsigned integer
%d : print signed integer
%x : print unsigned integer in HEX
```

为简单起见，我们将忽略保留的宽度和精度。只按格式字符串中指定的内容打印参数（如有）即可。注意，要打印的项目数量和类型是由格式字符串中的 % 符号的数量隐式指定的。

9.8.2 项目基本代码

（1）读者应实现一个用于打印字符串的 prints(char *s) 函数。

（2）下文给出了一个 printu() 函数，可打印十进制无符号整数。

```
char *ctable = "0123456789ABCDEF";
int BASE = 10; // for printing numbers in decimal
int rpu(unsigned int x)
{
    char c;
    if (x){
       c = ctable[x % BASE];
       rpu(x / BASE);
       putchar(c);
    }
}
int printu(unsigned int x)
{
   (x==0)? putchar('0') : rpu(x);
   putchar(' ');
}
```

函数 rpu(x) 以 ASCII 递归生成 x % 10 数字，并在返回路径上打印它们。例如，如果 x=123，则按 '3'、'2'、'1' 的顺序生成数字，按应有的顺序打印为 '1'、'2'、'3'。

（3）借助 printu() 函数，读者应能够实现一个 printd() 函数，以打印有符号整数。

（4）实现一个 printx() 函数，以打印无符号整数十六进制（用于地址）。

（5）假设我们有 printc()、prints()、printd()、printu() 和 printx() 函数。然后通过以下算

法实现 myprintf(char *fmt，...)。

9.8.3　myprintf() 的算法

假设格式字符串 fmt = " char=%c string=%s integer=%d u32=%x\ n "。这意味着分别有 char、char *、int、unsigned int 和 type 的 4 个附加参数。myprint() 的算法如下：

（1）扫描格式字符串 fmt。打印任何不是 % 的字符。对于每个 '\n' 字符，打印一个额外的 '\r' 字符。

（2）当遇到 ' % ' 时，得到的下一个字符必须是 ' c '、' s '、' u '、' d ' 或 ' x ' 中的一个。使用 va_arg(ap, type) 来提取相应的参数。然后通过参数类型调用打印函数。

（3）当 fmt 字符串扫描结束时，算法结束。

练习 9.6：假设是 32 位 GCC。通过下面的算法实现 myprintf(char *fmt, . . .) 函数，并说明为什么该算法有效。

```
int myprintf(char *fmt, . . .)
{
   char *cp = fmt;
   int  *ip = (int *)&fmt + 1;
   // Use cp to scan the format string for %TYPE symbols;
   // Use ip to access and print each item by TYPE;
}
```

9.8.4　项目改进

（1）将制表键定义为 8 个空格。将 %t 添加到制表键的格式字符串中。

（2）修改 %u、%d 和 %x 以包含宽度，例如 %8d 打印一个 8 字符空间整数，并右对齐等。

9.8.5　项目演示和示例解决方案

通过驱动程序演示 myprintf() 函数。编程项目的示例解决方案可在本书的网站上下载。如果提出要求，可向讲师提供项目源代码。

9.9　习题

1. 给出下面几个 C 程序：

```
---------- System call ---------------- Library I/O -------------
#include <fcntl.h>                     | #include <stdio.h>
char buf[4096];                        | char buf[4096];
#define SIZE 1                         | #define SIZE 1
int main()                             | int main()
{                                      | {
  int i;                               |   int i;
  int fd = open("file", O_RDONLY);     |   FILE *fp = fopen("file","r")
  for (i=0; i<100; i++){               |    for (i=0; i<100; i++)
    read(fd, buf, SIZE);               |      fread(buf, SIZE, 1, fp);
  }                                    |    }
}                                      | }
---------------------------------------------------------------------
```

（1）哪个程序运行得更快？为什么？

（2）将 SIZE 更改为 4096，即 Linux 的文件块大小。哪个程序运行得更快？为什么？

2. 在复制文件时，我们绝不能把文件复制到文件自身。

（1）为什么？

（2）修改示例 9.2 的程序，避免将文件复制到自身。注意，比较文件名可能没有用，因为由于是硬链接，两个不同的文件名有可能是同一个文件。因此，问题是如何判断两个路径名是否表示同一个文件。提示：stat 系统调用。

3. 在复制目录时，绝不能将目录复制到目录本身中。假设 /a/b/c/ 是一个目录。允许将 /a/b/c 复制到 /a 中，但不允许将 /a/b 复制到 /a/b/c/ 中。

（1）为什么？

（2）如何确定一个目录位于另一个目录中？

（3）编写 C 代码来确定某个目录是否在其自己内部。

参考文献

The GNU C Library:I/O on Streams, www.gnu.org/s/libc/manual/html_node/I_002fO-Overview.html, 2017

The GNU C Library (glibc), www.gnu.org/software/libc/, 2017

The GNU C Library Reference Manual, www.gnu.org/software/libc/manual/pdf/libc.pdf, 2017

sh 编程

摘要

本章讨论了 sh 编程，阐述了 sh 脚本和不同版本的 sh；比较了 sh 脚本与 C 程序，并指出了解释语言和编译语言的区别；详细说明了如何编写 sh 脚本，包括 sh 变量、sh 语句、sh 内置命令、常规系统命令和命令替换；解释了 sh 控制语句，其中包括测试条件、for 循环、while 循环、do-until 循环、case 语句等，并示范了它们的用法；说明了如何编写 sh 函数以及使用参数调用 sh 函数；还举例说明了 sh 脚本的广泛应用，包括 Linux 系统的安装、初始化和管理。

读者可通过编程项目编写一个 sh 脚本，递归复制文件和目录。本项目由三个 sh 函数组成的层次结构组成；cpf2f() 将文件复制到文件，cpf2d() 将文件复制到目录，cpd2d() 递归复制目录。

10.1 sh 脚本

sh 脚本（Bourne 1982；Forouzan 和 Gilberg 2003）是一个包含 sh 语句的文本文件，命令解释程序 sh 要执行该语句。例如，我们可以创建一个文本文件 mysh，包含：

```
#! /bin/bash
# comment line
echo hello
```

使用 chmod +x mysh 使其可执行。然后运行 mysh。sh 脚本的第一行通常以 #! 组合开始，通常称为 shebang。当主 sh 见到 shebang 时，会读取脚本所针对的程序名并调用该程序。sh 有许多不同的版本，例如 Linux 的 bash、BSD Unix 的 csh 和 IBM AIX 的 ksh 等。所有 sh 程序基本上都执行相同的任务，但它们的脚本在语法上略有不同。shebang 允许主 sh 调用适当版本的 sh 来执行脚本。如果未指定 shebang，它将运行默认的 sh，即 Linux 中的 /bin/bash。当 bash 执行 mysh 脚本时，将会打印 hello。

10.2 sh 脚本与 C 程序

sh 脚本和 C 程序有一些相似之处，但它们在根本上是不同的。下面并列列出了一个 sh 脚本和一个 C 程序，以比较它们的语法形式和用法。

```
------------ sh ------------------------ C ------------------
INTERPRETER: read & execute |   COMPILE-LINKED to a.out
                            |
 mysh a  b  c  d            |   a.out a b c d
  $0  $1 $2 $3 $4           |   main(int argc, char *argv[ ])
-------------------------------------------------------------
```

首先，sh 是一个解释程序，逐行读取 sh 脚本文件并直接执行这些行。如果行是可执行命令且为内置命令，那么 sh 可直接执行。否则，它会复刻一个子进程来执行命令，并等待子进程终止后再继续，这与它执行单个命令行完全一样。相反，C 程序必须先编译链接到一个二进制可执行文件，然后通过主 sh 的子进程运行二进制可执行文件。其次，在 C 程序中，每个变量必须有一个类型，例如 char、int、float、派生类型（如 struct）等。相反，在 sh 脚本中，每个变量都是字符串。因此不需要类型，因为只有一种类型，即字符串。最后，每个 C 程序必须有一个 main() 函数，每个函数必须定义一个返回值类型和参数（如有）。相反，sh 脚本不需要 main 函数。在 sh 脚本中，第一个可执行语句是程序的入口点。

10.3 命令行参数

可使用与运行 sh 命令完全相同的参数调用 sh 脚本，如：

```
mysh one two three
```

在 sh 脚本中，可以通过位置参数 $0、$1、$2 等访问命令行参数。前 10 个命令行参数可以作为 $0~$9 被访问。其他参数必须称为 ${10}~${n}，其中 n>10。或者，可以通过稍后显示的 shift 命令查看它们。通常，$0 是程序名本身，$1 到 $n 是程序的参数。在 sh 中，可用内置变量 $# 和 $* 计数并显示命令行参数。

$# = 命令行参数 $1 到 $n 的数量

$* = 所有命令行参数，包括 $0

此外，sh 还有与命令执行相关的以下内置变量。

$S = 执行 sh 的进程 PID

$? = 最后一个命令执行的退出状态（如果成功，则为 0，否则为非 0）

示例 10.1：假设以下 mysh 脚本运行为

```
        mysh abc  D  E  F  G  H  I  J  K  L  M  N
           #  1   2  3  4  5  6  7  8  9  10 11 12
1. #! /bin/bash
2. echo \$# = $#             #  $# = 12
3. echo \$* = $*             #  $* = abc D E F G H I J K L M N
4. echo $1  $9  $10          #  abc K abc0  (note: $10 becomes abc0)
5. echo $1  $9  ${10}        #  abc K L        (note: ${10} is L)
6. shift                     # replace $1, $2 .. with $2, $3,...
7. echo $1  $9  ${10}        # D L M
```

在 sh 中，特殊字符 $ 表示替换。要按原样使用 $，它必须带有单引号或反引号 \，类似于 C 程序中的 \n、\r、\t 等。在第 2 行和第 3 行中，每个 \$ 按原样打印，没有替换。在第 4 行，$1 和 $9 打印正确，但是 $10 却被打印成了 abc0。这是因为 sh 将 $10 看作是 $1 与 0 连接。在回显之前，它会用 abc 替换 $1，从而将 $10 打印为 abc0。在第 5 行，${10} 被正确打印为 L。读者可尝试回显其他位置参数 ${11} 和 ${12} 等。第 6 行将位置参数向左移动一次，使得 $2=$1、$3=$2 等。移动后，$9 变成了 L，${10} 变成了 M。

10.4 sh 变量

sh 有许多内置变量，如 PATH、HOME、TERM 等。除了内置变量外，用户还可使用任何符号作为 sh 变量。不需要声明。所有的 sh 变量值都是字符串。未赋值的 sh 变量是 NULL 字符串。sh 变量可用以下方法设置或赋值：

```
variable=string      # NOTE: no white spaces allowed between tokens
```

如果 A 是一个变量，则 $A 是变量的值。

示例 10.2：

```
echo A      ==>     A
echo $A     ==>     (null if variable A is not set)
A="this is fun"     # set A value
echo $A     ==>     this is fun
B=A                 # assign "A" to B
echo $B     ==>     A (B was assigned the string "A")
B=$A                (B takes the VALUE of A)
echo $B     ==>     this is fun
```

10.5　sh 中的引号

sh 有许多特殊字符，如 $、/、*、>、< 等。要想把它们用作普通字符，可使用 \ 或单引号来引用它们。

示例 10.3：

```
A=xyz
echo \$A        ==> $A        # back quote $ as is
echo '$A'       ==> $A        # NO substitution within SINGLE quotes
echo "see $A" ==> see xyz   # substitute $A in DOUBLE quotes
```

通常，\ 用于引用单个字符。单引号用于引用长字符串。单引号内没有替换。双引号用于保留双引号字符串中的空格，但在双引号内会发生替换。

10.6　sh 语句

sh 语句包括所有 Unix/Linux 命令，以及可能的 I/O 重定向。

示例 10.4：

```
ls
ls > outfile
date
cp f1 f2
mkdir newdir
cat < filename
```

此外，sh 编程语言还支持控制 sh 程序执行的测试条件、循环、case 等语句。

10.7　sh 命令

10.7.1　内置命令

sh 有许多内置命令，这些命令由 sh 执行，不需要创建一个新进程。下面列出一些常用的内置 sh 命令。

- .file：读取并执行文件。
- break [n]：从最近的第 n 个嵌套循环中退出。
- cd [dirname]：更换目录。
- continue [n]：重启最近的第 n 个嵌套循环。

- eval [arg ...]：计算一次参数并让 sh 执行生成的命令。
- exec [arg ...]：通过这个 sh 执行命令，sh 将会退出。
- exit [n]：使 sh 退出，退出状态为 n。
- export [var ...]：将变量导出到随后执行的命令。
- read [var ...]：从 stdin 中读取一行并为变量赋值。
- set [arg ...]：在执行环境中设置变量。
- shift：将位置参数 $2 $3 ... 重命名为 $1 $2 ...。
- trap [arg] [n]：接收到信号 n 后执行参数。
- umask [ddd]：将掩码设置为八进制数 ddd 的。
- wait [pid]：等待进程 pid，如果没有给出 pid，则等待所有活动子进程。

read 命令：当 sh 执行 read 命令时，它会等待来自 stdin 的输入行。它将输入行划分为几个标记，分配给列出的变量。read 的一个常见用法是允许用户与正在执行的 sh 进行交互，如下面的示例所示。

```
echo -n "enter yes or no : "   # wait for user input line from stdin
read ANS                       # sh reads a line from stdin
echo $ANS                      # display the input string
```

在获得输入后，sh 可能会测试输入字符串，以决定下一步做什么。

10.7.2 Linux 命令

sh 可以执行所有的 Linux 命令。其中，有些命令几乎已经成为 sh 不可分割的一部分，因为它们广泛用于 sh 脚本中。下文列出并解释了其中一些命令。

echo 命令：echo 只是将参数字符串作为行回显到 stdout。它通常将相邻的多个空格压缩为一个空格，除非有引号。

示例 10.5：

```
echo This   is     a   line       # display This is a line
echo "This is     a   line"       # display This is     a    line
echo -n hi                        # display hi without NEWLINE
echo    there                     # display hithere
```

expr 命令：因为所有的 sh 变量都是字符串，所以我们不能直接把它们改为数值。例如：

```
I=123      # I assigned the string "123"
I=I + 1    # I assigned the string "I + 1"
```

最后一个语句并不是将 I 的数值增加 1，而仅仅是将 I 更改为字符串 “I+1”，这肯定不是我们所希望的结果（I=124）。可通过 expr 命令间接更改 sh 变量的值（数值）。expr 是一个程序，它的运行方式如下：

```
expr string1 OP string2      # OP = any binary operator on numbers
```

首先，它将两个参数字符串转换为数字，然后对数字执行（二进制）操作 OP，再将得到的数字转换回字符串。因此，

```
I=123
I=$(expr $I + 1)
```

将 I 从“123”更改为“124”。同样，expr 也可用于对其值为数字字符串的 sh 变量执行其他算术操作。

管道命令：在 sh 脚本中经常使用管道作为过滤器。

示例 10.6：

```
ps -ax | grep httpd
cat file | grep word
```

实用命令：除了上面的 Linux 命令之外，sh 还使用许多其他实用程序作为命令。其中包括：

- awk：数据处理程序。
- cmp：比较两个文件。
- comm：选择两个排序文件共有的行。
- grep：匹配一系列文件的模式。
- diff：找出两个文件的差异。
- join：通过使用相同的键来连接记录以比较两个文件。
- sed：流或行编辑命令。
- sort：排序或合并文件。
- tail：打印某个文件的最后 n 行。
- tr：一对一字符翻译。
- uniq：从文件中删除连续重复行。

10.8　命令替换

在 sh 中，$A 会被替换成 A 值。同样，当 sh 遇到 'cmd'（用引号括起来）或 $(cmd) 时，它会先执行 cmd，然后用执行的结果字符串替换 $(cmd)。

示例 10.7：

```
echo $(date)        # display the result string of date command
echo $(ls dir)      # display the result string of ls dir command
```

命令替换是一种非常强大的机制。我们将在本章的后半部分介绍它的用法。

10.9　sh 控制语句

sh 是一种编程语言，支持许多执行控制语句，类似于 C 语言中的语句。

10.9.1　if-else-fi 语句

if-else-fi 语句的语法是：

```
if [ condition ]    # NOTE: must have white space between tokens
   then
      statements
   else             # as usual, the else part is optional
      statements
fi                  # each if must end with a matching fi
```

每个语句必须在单独的一行上。但是，如果多个语句之间用分号分开，则 sh 允许多个语句在同一行，实际上，if-else-fi 语句通常写成：

```
if [ condition ]; then
    statements
 else
    statements
fi
```

示例 10.8：在默认情况下，sh 中的所有值都是字符串，因此可以通过以下语句将它们作为字符串进行比较：

```
if  [ s1 = s2 ]          # NOTE: white spaces needed between tokens
if  [ s1 != s2 ]
if  [ s1 \< s2 ]         # \< because < is a special char
if  [ s1 \> s2 ] etc.    # \> because > is a special char
```

在上述语句中，左括号符号 [实际上是一个测试程序，作为以下内容执行：

```
test string1 COMP string2  OR  [ string1 COMP string2 ]
```

它比较两个参数字符串，以确定条件是否为真。需要注意的是，在 sh 中，**0 为 TRUE**，而**非 0 为 FALSE**，这与在 C 程序中完全相反。这是因为当 sh 执行命令时，它会获得命令执行的退出状态，如果执行成功，则为 0，否则为非 0。由于 [是一个程序，所以如果执行成功，则退出状态为 0，即测试条件为真，如果测试条件为假，则退出状态为非 0。或者，用户也可以使用 sh 内置变量 $? 来测试最后一次命令执行的退出状态。

相反地，运算符 -eq、-ne、-lt、-gt 等将参数作为整数进行比较。

```
if [ "123" = "0123" ]      is false since they differ as strings
if [ "123" -eq "0123" ]    is true since they have the same numerical value
```

除了比较字符串或数值之外，测试程序还可以测试文件操作中经常需要的文件类型和文件属性。

```
if [ -e name ]      # test whether file name exists
if [ -f name ]      # test whether name is a (REG) file
if [ -d name ]      # test whether name is a DIR
if [ -r name ]      # test whether name is readable; similarly for -w, -x, etc.
if [ f1 -ef f2 ]    # test whether f1, f2 are the SAME file
```

练习：测试程序如何测试文件类型和文件属性，如可读、可写等？特别是，它如何确定 f1 -ef f2？

提示：stat 系统调用。

if-elif-else-fi 复合语句：这类似于 C 语言中的 if-else if-else，只是 sh 使用 elif 而不是 else if。if-elif-else-fi 复合语句的语法是：

```
if [ condition1 ]; then
     commands
  elif [ condition2 ]; then
     commands
  # additional elif [ condition3 ]; then etc.
  else
     commands
fi
```

复合条件：与在C语言中一样，sh也允许在复合条件中使用&&（AND）和||（OR），但是语法比C语言更加严格。条件必须用一对匹配的双括号[[和]] 括起来。

示例10.9：

```
if [[ condition1 && condition2 ]]; then
if [[ condition1 && condition2 || condition3 ]]; then
```

与在C语言中一样，复合条件也可以通过()进行分组，以执行计算命令。

```
if [[ expression1 && (expression2 || expression3) ]]; then
```

10.9.2　for语句

sh中的for语句作用类似于C语言中的for循环。

```
for VARIABLE in string1 string2 .... stringn
  do
    commands
  done
```

在每次迭代中，变量接受一个参数字符串值，并执行关键字do和done之间的命令。

示例10.10：

```
for FRUIT in apple orange banana cherry
  do
     echo $FRUIT          # print lines of apple orange banana cherry
  done

for NAME in $*
  do
     echo $NAME           # list all command-line parameter strings
     if [ -f $NAME ]; then
        echo $NAME is a file
     elif [ -d $NAME ]; then
          echo $NAME is a DIR
     fi
  done
```

10.9.3　while语句

sh的while语句类似于C语言中的while循环：

```
while [ condition ]
  do
    commands
  done
```

当条件为真时，sh将重复执行do-done关键字中的命令。预计条件会有变化，所以循环将在某个时间点退出。

示例10.11：下面的代码段创建dir0，dir1，…，dir10000目录：

```
I=0                    # set I to "0" (STRING)
while [ $I != 10000 ]  # compare as strings; OR while [ $I \< 1000 ] as numbers
  do
    echo $I            # echo current $I value
```

```
    mkdir dir$I          # make directories dir0, dir1, etc
    I=$(expr $I + 1)     # use expr to inc I (value) by 1
  done
```

10.9.4 until-do 语句

该语句类似于 C 语言中的 do-until 语句。

```
until [ $ANS = "give up" ]
  do
       echo -n "enter your answer : "
       read ANS
  done
```

10.9.5 case 语句

该语句也类似于 C 语言中的 case 语句，但在 sh 编程中很少使用。

```
case $variable in
     pattern1)   commands;;  # note the double semicolons ;;
     pattern2)   command;;
     patternN)   command;;
esac
```

10.9.6 continue 和 break 语句

与在 C 语言中一样，continue 重启最近循环的下一个迭代，break 退出最近循环。它们的工作原理与在 C 语言中完全相同。

10.10 I/O 重定向

当进入 sh 命令时，我们可以指示 sh 将 I/O 重定向到除默认 stdin、stdout 和 sterr 以外的文件。I/O 重定向有以下形式和含义：

> file　　stdout 转向文件，如果文件不存在，将会创建文件。

>> file　　stdout 追加到文件。

< file　　将文件用作 stdin；文件必须存在并具有 r 权限。

<< word　　从“here”文件中获取输入，直到只包含“word”的行。

10.11 嵌入文档

可以指示输出命令从 stdin 获取输入，将其回显到 stdout，直到遇到预先安排的关键字。

示例 10.12：

```
echo << END
   # keep enter and echo lines until a line with only
END
cat << DONE
   # keep enter and echo lines until
DONE
```

这些文档通常被称为**嵌入文档**。它们通常用在 sh 脚本中，以生成长块的描述性文本，不需要分别回显每一行。

10.12　sh 函数

sh 函数的定义为：

```
func()
{
  # function code
}
```

由于 sh 逐行执行命令，所以必须在任何可执行语句**之前**定义 sh 脚本中的所有函数。与 C 语言不同，在 sh 脚本中无法声明函数原型。sh 函数的调用方式与 sh 脚本文件的执行方式完全相同。sh 语句

```
func s1 s2 ... sn
```

调用 sh 函数，以参数（字符串）形式传递 s1~sn。在被调函数中，参数被引用为 $0、$1 到 $n。通常，$0 是函数名，$1 到 $n 是与命令行参数对应的位置参数。函数执行结束时，$? 表示其退出状态，如果成功，状态为 0，否则，状态为非 0。$? 值可用函数的显式返回值进行更改。但是，为了测试 $? 的最后一次执行，必须将其分配给一个变量，然后测试该变量。

示例 10.13：sh 函数。

```
#! /bin/bash
testFile()    # test whether $1 is a REG file
{
  if [ -f $1 ]; then
     echo $1 is a REG file
  else
     echo $1 is NOT a REG file
}
testDir()     # test whether $1 is a DIR
{
  if [ -d $1 ]; then
     echo $1 is a DIR
  else
     echo $1 is NOT a DIR
}
echo entry point here # entry point of the program
for A in $*         # for A in command-line parameter list
  do
   testFile $A     # call testFile on each command-line param
   testDir  $A     # call testDir  on each command-line param
  done
```

练习：在下面的 sh 程序中：

```
testFile()    # test whether $1 is a REG file; return 0 if yes, 1 if not
{
 if [ -f $1 ]; then
    return 0
 else
    return 1
}
for A in f1 D2        # assume f1 is a REG file, D2 is a DIRectory
```

```
do
   testFile $A      # testFile return $?=0 or 1
   if [ $? = 0 ]; then
      echo $A is a REG file
   else
      echo $A is not a REG file
   fi
done
```

即使 $A 是一个目录，得到的结果始终会是“$A 是一个 REG 文件”。请解释为什么。修改程序代码使其正常工作。

10.13 sh 中的通配符

星号通配符：sh 中最有用的通配符是 *，可扩展到当前目录中的所有文件。

示例 10.14：

- file *：列出当前目录中所有文件的信息。
- ls *.c：列出当前目录中所有以 .c 结尾的文件。

? 通配符：查询某文件名中的字符

示例 10.15：

- file ???：有 3 个字符的所有文件名。
- ls *.??：一个点号 . 后有 2 个字符的所有文件名。

[ ] 通配符：查询文件名中一对 [] 中的字符。

示例 10.16：

- file *[ab]*：包含字符 a 或 b 的所有文件名。
- ls *[xyz]*：列出所有包含 x、y 或 z 的文件名。
- ls *[a-m]*：列出包含 a 到 m 范围内字符的所有文件名。

10.14 命令分组

在 sh 脚本中，可以用 {} 或 () 对命令进行分组。

{ls;mkdir abc;ls;} ：通过当前 sh 执行 {} 中的命令列表。{} 命令分组的唯一用处是在相同环境下执行这些命令，例如，为分组中的所有命令重定向 I/O。

更有用的命令分组是 ()，由 subsh（进程）执行。

（cd newdir;ls;A=value;mkdir $A）：通过 subsh 进程执行 () 中的命令。subsh 进程可在不影响父 sh 的情况下更改其工作目录。此外，当 subsh 进程终止时，subsh 中的任何赋值变量都不起作用。

10.15 eval 语句

```
eval [arg1 arg1 .. argn]
```

eval 是 sh 的一个内置命令。它由 sh 自己执行，而不需要复刻新进程。它将输入参数字符串连接到一个字符串中，计算一次，即执行变量和命令替换，然后给出结果字符串供 sh 执行。

示例 10.17：

```
a="cat big.c | more"
```

$a # error： 因为 sh 会用 big.c、|、more 作为文件执行 cat 命令，所以 cat 命令遇到 | 时会失效，因为它不是文件。

eval $a # OK： eval 先用 "cat big.c | more" 替换 $a，然后让 sh 执行生成的命令行 cat big.c | more。

示例 10.18：假设

```
A='$B';  B='abc*'; C=newdir; CWD=/root; /root/newdir is a DIR
```

对于命令行：

```
cp $A `pwd`/$C   # grave quoted command pwd
```

sh 在执行命令行之前，按以下步骤计算该命令行。

（1）**参数替换**：扫描命令行，将任何 $x 替换为它的值，但只执行一次，即不能再次替换任何产生的 $ 符号。命令行变成：

```
cp $B `pwd`/newdir
```

（2）**命令替换**：用替换执行 'pwd'。sh 将会执行结果行

```
cp $B /root/newdir
```

如果当前目录没有任何名为 $B 的文件，则会导致一个错误。但是，如果我们将原来的命令行更改为

```
eval cp $A $(pwd)/$C
```

sh 将会在执行命令行之前先计算它。计算完成后，该命令行变成：

```
cp abc* /root/newdir
```

（3）**通配符扩展**：当 sh 执行该行时，它将 new* 展开为以 abc 开头的文件名。这将会把所有以 abc 开头的文件名复制到目标目录中。

需要注意的是，我们总是可以通过一些额外的语句来实现与手动计算相同的效果。使用 eval 可以省去一些替换语句，但也可能使代码难以理解。因此，应避免使用任何不必要的 eval。

10.16 调试 sh 脚本

sh 脚本可由带有 -x 选项的子 sh 运行，以进行调试，如：

```
bash -x mysh
```

子 sh 将在执行命令之前显示要执行的每个 sh 命令，包括变量和命令替换。它允许用户跟踪命令执行。如果出现错误，sh 将在错误行上停止并显示错误消息。

10.17 sh 脚本的应用

sh 脚本最常用于执行涉及冗长命令序列的常规作业。我们举了一些 sh 脚本在其中非常

有用的例子。

示例 10.19：大多数用户都有将 Linux 安装到计算机上的经验。Linux 安装包是用 sh 脚本编写的。在安装过程中，它可以与用户交互，查找用户计算机上可用的硬盘，分区并格式化磁盘分区，从安装媒体下载或解压文件并将文件安装到它们的目录中，直到安装过程完成。试图手动安装 Linux 会非常烦琐，几乎不可能实现。

示例 10.20：当用户登录 Linux 系统时，登录进程会执行一系列 sh 脚本，

```
.login, .profile, .bashrc等
```

以自动配置用户进程的执行环境。同样，手动执行所有这些步骤是不切实际的。

示例 10.21：与使用 makefile 不同，简单的编译链接任务可由包含编译和链接命令的 sh 脚本来执行。下面是一个 sh 脚本，在名为 VFD 的虚拟磁盘上生成 MTX 操作系统内核映像和用户模式命令程序（Wang 2015）。

```
#---------- mk script of the MTX OS -------------
VFD=mtximage
# generate MTX kernel image file
as86 -o ts.o ts.s            # assemble .s file
bcc  -c -ansi t.c            # compile .c files
ld86 -d -o mtx ts.o t.o OBJ/*.o mtxlib 2> /dev/null
# write MTX kernel image to /boot directory of VFD
mount -o loop $VFD /mnt
cp mtx /mnt/boot             # mtx kernel image in /boot directory
umount /mnt
# generate User mode command binaries in /bin of VFD
(cd USER; mkallu)            # command grouping
echo all done
#---------------- end
```

示例 10.22：在 Linux 机器上为 CS 类创建用户账户。在作者所在的机构中，CS360 系统编程课程每学期平均招收 70 名学生。在 Linux 机器上手动创建所有学生账户会非常烦琐。这项工作可以使用 sh 脚本通过以下步骤完成。

（1）Linux 命令

```
sudo useradd -m -k DIR -p PASSWORD -s /bin/bash LOGIN
```

创建一个新用户账户，登录名为 LOGIN 和密码为 PASSWORD。它创建一个用户主目录 /home/LOGIN，使用 -k DIR 目录中的文件填充。它在 /etc/passwd 中为新用户账户创建一个行，并在 /etc/shadow 中为加密密码添加一个行。但是，useradd 命令中使用的密码必须是加密密码，而不是原始密码字符串。可以通过 crypt 库函数将密码字符串转换为加密形式。下面的 C 程序 enc.c 也将密码字符串转换成了加密形式。

```
/******* enc.c file **********/
#define _XOPEN_SOURCE
#include <stdio.h>
#include <unistd.h>
int main(int argc, char *argv[ ])
{
  printf("%s\n", crypt(argv[1], "$1$"));
}
# gcc -o enc -lcrypt enc.c    # generate the enc file
```

（2）假设 roster.txt 是一个包含学生 ID 和姓名的行的班级花名册文件：

```
ID name                # name string in lowercase
```

（3）运行以下 sh 脚本 mkuser < roster.txt

```
# ---------------- mkuser sh script file ---------------------
#! /bin/bash
while read LINE        # read a line of roster file
do
  I=0
  for N in $LINE       # extract ID as NAME0, name as NAME1
  do
    eval NAME$I=$N     # set NAME0, NAME1, etc.
    I=$(expr $I + 1)   # inc I by 1
  done
  echo $NAME0 $NAME1   # echo ID and name
  useradd -m -k h -p $(./enc $NAME0) -s /bin/bash $NAME1
done
```

（4）用户主目录内容：每个用户主目录 /home/name 均可用一组默认的目录和文件来填充。对于 CS360 课程，每个用户主目录都用一个 public_html 目录填充，该目录包含一个 index.html 文件作为用户的初始网页，供用户以后练习 Web 编程。

10.18 编程项目：用 sh 脚本递归复制文件

读者可通过本编程项目使用大多数 sh 工具在 sh 脚本中编写一个有意义和有用的程序。要求是编写一个 sh 脚本：

```
myrcp f1 f. .... fn-1 fn
```

复制 f1、f2、fn-1 到 fn，其中每个 fi 可以是一个 REG/LNK 文件或者一个 DIR。为简单起见，可排除 I/O 设备等特殊文件。myrcp 程序的工作原理应与 Linux 的 cp -r 命令完全相同。下面是对各种程序需求情况的简要分析。

（1）n<2：显示用法并退出。

（2）n>2：fn 必须是现有目录。

（3）n=2：将文件复制到文件，将文件复制到目录，或者将目录 1 复制到目录 2。

（4）永远不要将文件或目录复制到它本身。此外，如果目录 2 是目录 1 的派生目录，不要进行复制。

上述案例分析并不详尽，读者必须在尝试编写任何代码之前，完成案例分析并制定算法。

提示和帮助：对于命令行 myrcp f1 f. ... .fn

```
(1). echo n = $#                              # n = value of n
(2). last = \S{$#}; eval echo last = $last    # last = fn
(3). for NAME in $*
     do
       echo $NAME                             # show f1 to fn
     done
(4). Implement the following functions
     cpf2f(): copy $1 to $2 as files, handle symbolic links
```

```
    cpf2d(): copy file $1 into an existing DIR $2 as $2/$(basename $1)
    cod2d(): recursively copy DIR $1 to DIR $2


# Sample solution of myrcp program

cpf2f()     # called as cpf2f  f1  f2
{
   if [ ! -e $1 ]; then
      echo no such source file $1
      return 1
   fi
   if [ $1 -ef  $2 ]; then
      echo "never copy a file to itself"
      return 1
   fi
  if [ -L $1 ]; then
     echo "copying symlink $1"
     link=$(readlink $1)
       ln -s $link $2
       return 0
   fi
   echo "copying $1 to $2"
   cp $1 $2  2> /dev/null
}
cpf2d()     # called as cpf2d file DIR
{
   newfile=$2/$(basename $1)
   cpf2f $1 $newfile
}
cpd2d()     # called as cpd2d dir1 dir2
{
   # reader FINISH cpd2d code
}
# **************** entry point of myrcp ****************
# case analysis;
# for each f1 to fn-1 call cpf2f() or cpf2d() or cpd2d()
```

参考文献

Bourne, S.R., The Unix System, Addison-Wesley, 1982

Forouzan, B.A., Gilberg, R.F., Unix and Shell Programming, Brooks/Cole, 2003

Wang, K.C., Design and Implementation of the MTX Operating System, Springer International Publishing AG, 2015

EXT2 文件系统

摘要

本章讨论 EXT2 文件系统。本章将引导读者实现一个完全与 Linux 兼容的完整 EXT2 文件系统。前提是，只要读者充分理解了一个文件系统，那么就可以轻松改编其他任何文件系统。本章首先描述了 EXT2 文件系统在 Linux 中的历史地位以及 EXT3/EXT4 文件系统的当前状况；用编程示例展示了各种 EXT2 数据结构以及如何遍历 EXT2 文件系统树；介绍了如何实现支持 Linux 内核中所有文件操作的 EXT2 文件系统；展示了如何通过虚拟磁盘的 mount_root 来构建基本文件系统；将文件系统的实现划分为 3 个级别，级别 1 扩展了基本文件系统，以实现文件系统树，级别 2 实现了文件内容的读 / 写操作，级别 3 实现了文件系统的挂载 / 装载和文件保护；描述了各个级别文件系统函数的算法，并通过编程示例演示了它们的实现过程；将所有级别融合到一个编程项目中；最后，将所有编程示例和练习整合到一个完全有效的文件系统中。

11.1 EXT2 文件系统

多年来，Linux 一直使用 EXT2（Card 等 1995）作为默认文件系统。EXT3（EXT3, 2014）是 EXT2 的扩展。EXT3 中增加的主要内容是一个日志文件，它将文件系统的变更记录在日志中。日志可在文件系统崩溃时更快地从错误中恢复。没有错误的 EXT3 文件系统与 EXT2 文件系统相同。EXT3 的最新扩展是 EXT4（Cao 等 2007）。EXT4 的主要变化是磁盘块的分配。在 EXT4 中，块编号为 48 位。EXT4 不是分配不连续的磁盘块，而是分配连续的磁盘块区，称为区段。除了这些细微的更改之外，文件系统结构和文件操作保持不变。本书的目的是讲授文件系统的原理。主要目标并非实现大的文件存储容量，而是重点论述文件系统设计和实现的原则，强调简单性以及与 Linux 的兼容性。为此，我们以 ETX2 作为文件系统。

11.2 EXT2 文件系统数据结构

11.2.1 通过 mkfs 创建虚拟磁盘

在 Linux 下，命令

```
mke2fs [-b blksize -N ninodes] device nblocks
```

在设备上创建一个带有 nblocks 个块（每个块大小为 blksize 字节）和 ninodes 个索引节点的 EXT2 文件系统。设备可以是真实设备，也可以是虚拟磁盘文件。如果未指定 blksize，则默认块大小为 1KB。如果未指定 ninoides，mke2fs 将根据 nblocks 计算一个默认的 ninodes 数。得到的 EXT2 文件系统可在 Linux 中使用。举个具体的例子，下面的命令

```
dd if=/dev/zero of=vdisk bs=1024 count=1440
mke2fs vdisk 1440
```

可在一个名为 vdisk 的虚拟磁盘文件上创建一个 EXT2 文件系统，有 1440 个大小为 1KB 的块。

11.2.2 虚拟磁盘布局

上述 EXT2 文件系统的布局如图 11.1 所示。

```
| 0  |  1  |  2 |  3-7    | 8  | 9  | 10 | . . . 32|33                1439|
--------------------------------------------------------------------------
|boot|super| GD |reserved|bmap|imap|inodes blocks |    data blocks     |
--------------------------------------------------------------------------
```

图 11.1 简单的 EXT2 文件系统布局

一开始，我们先使用这个基本文件系统布局进行假设。在适当的时候，我们会指出一些变化，包括硬盘上大型 EXT2/3 文件系统中的变化。下面来简要解释一下磁盘块的内容。

Block#0：引导块 B0 是引导块，文件系统不会使用它。它用来容纳一个引导程序，从磁盘引导操作系统。

11.2.3 超级块

Block#1：超级块（在硬盘分区中字节偏移量为 1024） B1 是超级块，用于容纳整个文件系统的信息。下文说明了超级块结构中的一些重要字段。

```
struct ext2_super_block {
   u32  s_inodes_count;          /* Inodes count */
   u32  s_blocks_count;          /* Blocks count */
   u32  s_r_blocks_count;        /* Reserved blocks count */
   u32  s_free_blocks_count;     /* Free blocks count */
   u32  s_free_inodes_count;     /* Free inodes count */
   u32  s_first_data_block;      /* First Data Block */
   u32  s_log_block_size;        /* Block size */
   u32  s_log_cluster_size;      /* Allocation cluster size */
   u32  s_blocks_per_group;      /* # Blocks per group */
   u32  s_clusters_per_group;    /* # Fragments per group */
   u32  s_inodes_per_group;      /* # Inodes per group */
   u32  s_mtime;                 /* Mount time */
   u32  s_wtime;                 /* Write time */
   u16  s_mnt_count;             /* Mount count */
   s16  s_max_mnt_count;         /* Maximal mount count */
   u16  s_magic;                 /* Magic signature */
   // more non-essential fields
   u16  s_inode_size;            /* size of inode structure */
}
```

大多数超级块字段的含义都非常明显。只有少数几个字段需要详细解释。

s_first_data_block：0 表示 4KB 块大小，1 表示 1KB 块大小。它用于确定块组描述符的起始块，即 s_first_data_block + 1。

s_log_block_size：确定文件块大小，为 1KB*(2**s_log_block_size)，例如：0 表示 1KB 块大小，1 表示 2KB 块大小，2 表示 4KB 块大小等。最常用的块大小是用于小文件系统的1KB和用于大文件系统的4KB。

s_mnt_count：已挂载文件系统的次数。当挂载计数达到max_mnt_count时，fsck会话将被迫检查文件系统的一致性。

s_magic：标识文件系统类型的幻数。EXT2/3/4文件系统的幻数是0xEF53。

11.2.4 块组描述符

Block#2：块组描述符块（硬盘上的s_first_data_block+1） EXT2将磁盘块分成几个组。每个组有8192个块（硬盘上的大小为32K）。每组用一个块组描述符结构体来描述。

```
struct ext2_group_desc {
   u32  bg_block_bitmap;        // Bmap block number
   u32  bg_inode_bitmap;        // Imap block number
   u32  bg_inode_table;         // Inodes begin block number
   u16  bg_free_blocks_count;   // THESE are OBVIOUS
   u16  bg_free_inodes_count;
   u16  bg_used_dirs_count;
   u16  bg_pad;                 // ignore these
   u32  bg_reserved[3];
};
```

由于一个虚拟软盘（FD）只有1440个块，B2就只包含一个块组描述符。其余的都是0。在有大量块组的硬盘上，块组描述符可以跨越多个块。块组描述符中最重要的字段是bg_block_bitmap、bg_inode_bitmap和bg_inode_table，它们分别指向块组的块位图、索引节点位图和索引节点起始块。对于Linux格式的EXT2文件系统，保留了块3到块7。所以，bmap=8，imap=9，inode_table= 10。

11.2.5 块和索引节点位图

Block#8：块位图（Bmap）(bg_block_bitmap） 位图是用来表示某种项的位序列，例如磁盘块或索引节点。位图用于分配和回收项。在位图中，0位表示对应项处于FREE状态，1位表示对应项处于IN_USE状态。一个软盘有1440个块，但是Block#0未被文件系统使用。所以，位图只有1439个有效位。无效位被视作IN_USE，设置为1。

Block#9：索引节点位图（Imap）（bg_inode_bitmap） 一个**索引节点**就是用来代表一个文件的数据结构。EXT2文件系统是使用有限数量的索引节点创建的。各索引节点的状态用B9的Imap中的一个位表示。在EXT2 FS中，前10个索引节点是预留的。所以，空EXT2 FS的Imap以10个1开头，然后是0。无效位再次设置为1。

11.2.6 索引节点

Block#10：索引（开始）节点（bg_inode_table） 每个文件都用一个128字节（EXT4中是256字节）的唯一索引节点结构体表示。下面列出了主要索引节点字段。

```
struct ext2_inode {
  u16  i_mode;          // 16 bits = |tttt|ugs|rwx|rwx|rwx|
  u16  i_uid;           // owner uid
  u32  i_size;          // file size in bytes
  u32  i_atime;         // time fields in seconds
  u32  i_ctime;         // since 00:00:00,1-1-1970
```

```
  u32  i_mtime;
  u32  i_dtime;
  u16  i_gid;          // group ID
  u16  i_links_count;  // hard-link count
  u32  i_blocks;       // number of 512-byte sectors
  u32  i_flags;        // IGNORE
  u32  i_reserved1;    // IGNORE
  u32  i_block[15];    // See details below
  u32  i_pad[7];       // for inode size = 128 bytes
}
```

在索引节点结构体中，i_mode 为 u16 或 2 字节无符号整数。

```
           | 4  | 3 |    9     |
i_mode = |tttt|ugs|rwxrwxrwx|
```

在 i_mode 字段中，前 4 位指定了文件类型，例如：tttt=1000 表示 REG 文件，0100 表示 DIR 文件等。接下来的 3 位 ugs 表示文件的特殊用法。最后 9 位是用于文件保护的 rwx 权限位。

i_size 字段表示文件大小（以字节为单位）。各时间字段表示自 1970 年 1 月 1 日 0 时 0 分 0 秒以来经过的秒数。所以，每个时间字段都是一个非常大的无符号整数。可借助以下库函数将它们转换为日历形式：

```
char *ctime(&time_field)
```

将指针指向时间字段，然后返回一个日历形式的字符串。例如：

```
printf("%s", ctime(&inode.i_atime); // note: pass & of time field prints i_atime
in calendar form.
```

i_block[15] 数组包含指向文件磁盘块的指针，这些磁盘块有：

直接块：i_block[0] 至 i_block[11]，指向直接磁盘块。

间接块：i_block[12] 指向一个包含 256 个块编号（对于 1KB BLKSIZE）的磁盘块，每个块编号指向一个磁盘块。

双重间接块：i_block[13] 指向一个指向 256 个块的块，每个块指向 256 个磁盘块。

三重间接块：i_block[14] 是三重间接块。对于“小型”EXT2 文件系统，可以忽略它。

索引节点大小（128 或 256）用于平均分割块大小（1KB 或 4KB），所以，每个索引节点块都包含整数个索引节点。在简单的 EXT2 文件系统中，索引节点的数量是 184 个（Linux 默认值）。索引节点块数等于 184/8=23 个。因此，索引节点块为 B10 至 B32。每个索引节点都有一个唯一的**索引节点编号**，即索引节点在索引节点块上的位置 +1。注意，索引节点位置从 0 开始计数，而索引节点编号从 1 开始计数。0 索引节点编号表示没有索引节点。根目录的索引节点编号为 2。同样，磁盘块编号也从 1 开始计数，因为文件系统从未使用块 0。块编号 0 表示没有磁盘块。

11.2.7 数据块

紧跟在索引节点块后面的是文件存储数据块。假设有 184 个索引节点，第一个实际数据块是 B33，它就是根目录 / 的 i_block[0]。

11.2.8　目录条目

目录包含 dir_entry 结构，即

```
struct ext2_dir_entry_2{
   u32 inode;                  // inode number; count from 1, NOT 0
   u16 rec_len;                // this entry's length in bytes
   u8  name_len;               // name length in bytes
   u8  file_type;              // not used
   char name[EXT2_NAME_LEN];  // name: 1-255 chars, no ending NULL
};
```

dir_entry 是一种可扩充结构。名称字段包含 1 到 255 个字符，不含终止 NULL。所以 dir_entry 的 rec_len 也各不相同。

11.3　邮差算法

在计算机系统中，经常出现下面这个问题。一个城市有 M 个街区，编号从 0 到 M−1。每个街区有 N 座房子，编号从 0 到 N−1。每座房子有一个唯一的街区地址，用（街区，房子）表示，其中 0≤街区 < M，0 ≤房子 < N。来自外太空的外星人可能不熟悉地球上的街区寻址方案，倾向于采用线性方法将这些房子地址编为 0，1，…，N−1，N，N+1 等。已知某个街区地址 BA=（街区，房子），怎么把它转换为线性地址 LA，反过来，已知线性地址，怎么把它转换为街区地址？如果都从 0 开始计数，转换就会非常简单。

```
Linear_address LA = N*block + house;
Block_address  BA = (LA / N, LA % N);
```

注意，只有都从 0 开始计数时，转换才有效。如果有些条目不是从 0 开始计数的，则不能直接在转换公式中使用。读者可以试着找出处理这种情况的一般方法。为方便表述，我们把这种转换方法称为邮差算法。下面给出了邮差算法的几种应用。

11.3.1　C 语言中的 Test-Set-Clear 位

在标准 C 语言程序中，最小的可寻址单元是一个字符或字节。在一系列位组成的位图中，通常需要对位进行操作。考虑字符 buf[1024]，它有 1024 个字节，用 buf[i] 表示，其中 i = 0，1，…，1023。它还有 8192 个位，编号为 0，1，2，…，8191。已知一个位号 BIT，例如 1234，那么哪个字节 i 包含这个位，以及哪个位 j 在该字节中呢？

```
i = BIT / 8;    j = BIT % 8;    // 8 = number of bits in a byte.
```

我们可在 C 语言中结合使用邮差算法和位屏蔽来进行下面的位操作。

```
.TST a bit for 1 or 0 :  if (buf[i] &   (1 << j))
.SET a bit to 1       :      buf[i] |=  (1 << j);
.CLR a bit to 0       :      buf[i] &= ~(1 << j);
```

注意，一些 C 语言编译器允许在结构体中指定位，如：

```
struct bits{
   unsigned int bit0      :  1;  // bit0 field is a single bit
   unsigned int bit123    :  3;  // bit123 field is a range of 3 bits
   unsigned int otherbits : 27;  // other bits field has 27 bits
   unsigned int bit31     :  1;  // bit31 is the highest bit
}var;
```

该结构体将 var. 定义为一个 32 位无符号整数，具有单独的位或位范围。那么，var.bit0=0；将 1 赋值给第 0 位，则有 var.bit123 = 5；将 101 赋值给第 1 位到第 3 位等。但是，生成的代码仍然依赖于邮差算法和位屏蔽来访问各个位。我们可以用邮差算法直接操作位图中的位，无须定义复杂的 C 语言结构体。

11.3.2 将索引节点号转换为磁盘上的索引节点

在 EXT2 文件系统中，每个文件都有一个唯一的索引节点结构。在文件系统磁盘上，索引节点从 inode_table 块开始。每个磁盘块包含

```
INODES_PER_BLOCK = BLOCK_SIZE/sizeof(INODE)
```

个索引节点。每个索引节点都有一个唯一的索引节点号，ino = 1，2，…，从 1 开始线性计数。已知一个 ino，如 1234，那么哪个磁盘块包含该索引节点，以及哪个索引节点在该块中呢？我们需要知道磁盘块号，因为需要通过块来读 / 写一个真正的磁盘。

```
block = (ino - 1)  /  INODES_PER_BLOCK + inode_table;
inode = (ino - 1) % INODES_PER_BLOCK;
```

同样，将 EXT2 文件系统中的双重和三重间接逻辑块号转换为物理块号也依赖于邮差算法。

将线性磁盘块号转换为 CHS =（柱面、磁头、扇区）格式：软盘和旧硬盘使用 CHS 寻址，但文件系统始终使用线性块寻址。在调用 BIOS INT13 时，可用该算法将磁盘块号转换为 CHS。

11.4 编程示例

在本节中，我们将演示如何通过示例程序访问和显示 EXT2 文件系统的内容。为了编译和运行这些程序，系统必须安装 ext2fs.h 头文件，它定义了 EXT2/3/4 文件系统的数据结构。Ubuntu Linux 用户可通过以下代码获取并安装 ext2fs 开发包：

```
sudo apt-get install ext2fs-dev
```

11.4.1 显示超级块

以下 C 程序显示了 EXT2 文件系统的超级块。基本方法如下。

（1）打开虚拟磁盘读取：int fd = open("vdisk", O_RDONLY)。

（2）将超级块（Block#1 或 1KB 的 1024 偏移量位置）读入 char buf[1024] 中。

```
char buf[1024];
lseek(fd, 1024, SEEK_SET); // seek to byte offset 1024
int n = read(fd, buf, 1024);
```

（3）让 ext2_super_block *sp 结构体指向 buf[]。然后，利用 sp->field 访问超级块结构体的各个字段。

```
struct ext2_super_block *sp = (struct ext2_super_block *)buf;
printf("s_magic = %x\n", sp->s_magic)                    // print s_magic
printf("s_inodes_count = %d\n", sp->s_inodes_count); // print s_inodes_
count
etc.
```

该方法也适用于真实磁盘或虚拟磁盘中的其他任何数据结构。

示例 11.1：superblock.c 程序显示 EXT2 文件系统的超级块信息。

```
/*********** superblock.c program ************/
#include <stdio.h>
#include <stdlib.h>
#include <fcntl.h>
#include <ext2fs/ext2_fs.h>

// typedef u8, u16, u32 SUPER for convenience
typedef unsigned char   u8;
typedef unsigned short u16;
typedef unsigned int   u32;
typedef struct ext2_super_block SUPER;

SUPER *sp;
char buf[1024];
int fd, blksize, inodesize;

int print(char *s, u32 x)
{
  printf("%-30s = %8d\n", s, x);
}

int super(char *device)
{
  fd = open(device, O_RDONLY);
  if (fd < 0){
    printf("open %sfailed\n", device); exit(1);
  }
  lseek(fd, (long)1024*1, 0);  // block 1 on FD, offset 1024 on HD
  read(fd, buf, 1024);
  sp = (SUPER *)buf;           // as a super block structure

  // check EXT2 FS magic number:
  printf("%-30s = %8x ", "s_magic", sp->s_magic);
  if (sp->s_magic != 0xEF53){
     printf("NOT an EXT2 FS\n");
     exit(2);
  }
  printf("EXT2 FS OK\n");
  print("s_inodes_count",      sp->s_inodes_count);
  print("s_blocks_count",      sp->s_blocks_count);
  print("s_r_blocks_count",   sp->s_r_blocks_count);
  print("s_free_inodes_count", sp->s_free_inodes_count);
  print("s_free_blocks_count", sp->s_free_blocks_count);
  print("s_first_data_block",  sp->s_first_data_block);
  print("s_log_block_size",    sp->s_log_block_size);
  print("s_blocks_per_group",  sp->s_blocks_per_group);
  print("s_inodes_per_group",  sp->s_inodes_per_group);
  print("s_mnt_count",         sp->s_mnt_count);
  print("s_max_mnt_count",     sp->s_max_mnt_count);
```

```
  printf("%-30s = %8x\n", "s_magic", sp->s_magic);
  printf("s_mtime = %s", ctime(&sp->s_mtime));
  printf("s_wtime = %s", ctime(&sp->s_wtime));
  blksize = 1024 * (1 << sp->s_log_block_size);
  printf("block size = %d\n", blksize);
  printf("inode size = %d\n",  sp->s_inode_size);
}

char *device = "mydisk";        // default device name
int main(int argc, char *argv[])
{
  if (argc>1)
     device = argv[1];
  super(device);
}
```

图 11.2 给出了运行 superblock.c 示例程序的输出。

```
s_magic                          =    ef53  EXT2 FS OK
s_inodes_count                   =     184
s_blocks_count                   =    1440
s__r_blocks_count                =      72
s_free_inodes_count              =     173
s_free_blocks_count              =    1393
s_first_data_blcok               =       1
s_log_block_size                 =       0
s_blocks_per_group               =    8192
s_inodes_per_group               =     184
s_mnt_count                      =       1
s_max_mnt_count                  =      -1
s_magic                          =    ef53
s_mtime = Sun Oct  8 14:22:03 2017
s_wtime = Sun Oct  8 14:22:07 2017
block size = 1024
inode size = 128
```

图 11.2 EXT2 文件系统的超级块

练习 11.1：编写一个 C 程序来显示设备上 EXT2 文件系统的块组描述符。（请参见习题 1。）

11.4.2 显示位图

示例 11.2 中的程序以十六进制形式显示索引节点位图（imap）。

示例 11.2：imap.c 程序显示 EXT2 文件系统的索引节点位图。

```
/*************** imap.c program **************/
#include <stdio.h>
#include <stdlib.h>
#include <fcntl.h>
#include <ext2fs/ext2_fs.h>
typedef struct ext2_super_block SUPER;
typedef struct ext2_group_desc  GD;
#define BLKSIZE 1024
SUPER *sp;
GD    *gp;
char buf[BLKSIZE];
int fd;

// get_block() reads a disk block into a buf[ ]
```

```
int get_block(int fd, int blk, char *buf)
{
  lseek(fd, (long)blk*BLKSIZE, SEEK_SET);
  return read(fd, buf, BLKSIZE);
}

int imap(char *device)
{
  int i, ninodes, blksize, imapblk;
  fd = open(dev, O_RDONLY);
  if (fd < 0){printf("open %s failed\n", device); exit(1);}
  get_block(fd, 1, buf);           // get superblock
  sp = (SUPER *)buf;
  // check magic number to ensure it's an EXT2 FS
  ninodes = sp->s_inodes_count;    // get inodes_count
  printf("ninodes = %d\n", ninodes);
  get_block(fd, 2, buf);           // get group descriptor
  gp = (GD *)buf;
  imapblk = gp->bg_inode_bitmap;   // get imap block number
  printf("imapblk = %d\n", imapblk);
  get_block(fd, imapblk, buf);     // get imap block into buf[ ]
  for (i=0; i<=nidoes/8; i++){     // print each byte in HEX
      printf("%02x ", (u8)buf[i]);
  }
  printf("\n");
}

char * dev="mydisk";               // default device
int main(int argc, char *argv[ ] )
{
  if (argc>1) dev = argv[1];
  imap(dev);
}
```

程序以十六进制数字形式打印每个字节的索引节点位图。输出如下所示。

```
|←-------------- niodes = 184 bits (23 bytes)---------------------|
ff 07 00 00 00 00 00 00 00 00 00 00 00 00 00 00 00 00 00 00 00 00 00 ff
```

在 imap 中，位是从低位地址到高位地址线性存储的。前 16 位（从低到高）是 b’11111111 11100000’，但以十六进制形式打印成 ff 07，这并不能提供什么信息，因为每个字节中的位是按相反顺序打印的，即从高位地址到低位地址。

练习 11.2： 修改 imap.c 程序，以 char map 的形式打印索引节点位图，即对于每个位而言，如果该位是 0，则打印“0”，如果该位是 1，则打印“1”。（请参见习题 2。）

修改后的程序输出如图 11.3 所示，其中每个 char 表示 imap 中的一个位。

```
ninodes = 184  imapblk = 9
11111111 11100000 00000000 00000000 00000000 00000000 00000000 00000000
00000000 00000000 00000000 00000000 00000000 00000000 00000000 00000000
00000000 00000000 00000000 00000000 00000000 00000000 00000000 11111111
```

图 11.3　EXT2 文件系统的索引节点位图

练习 11.3： 编写一个 C 语言程序，同样以 char map 格式来显示 EXT2 文件系统的块位图。

11.4.3 显示根索引节点

在 EXT2 文件系统中，第 2 个（从 1 开始计）索引节点是根目录“/”的索引节点。如果我们将根索引节点读入内存中，就能显示模式、uid、gid、文件大小、时间字段、硬链接数和数据块编号等多个字段。示例 11.3 中的程序显示了 EXT2 文件系统根目录的索引节点信息。

示例 11.3：inode.c 程序显示 EXT2 文件系统的根索引节点信息。

```
/*********** inode.c file **********/
#include <stdio.h>
#include <stdlib.h>
#include <fcntl.h>
#include <ext2fs/ext2_fs.h>
#define BLKSIZE 1024
typedef struct ext2_group_desc  GD;
typedef struct ext2_super_block SUPER;
typedef struct ext2_inode       INODE;
typedef struct ext2_dir_entry_2 DIR;
SUPER *sp;
GD    *gp;
INODE *ip;
DIR   *dp;
char buf[BLKSIZE];
int fd, firstdata, inodesize, blksize, iblock;
char *dev = "mydisk";

int get_block(int fd, int blk, char *buf)
{
  lseek(fd, blk*BLKSIZE, SEEK_SET);
  return read(fd, buf, BLKSIZE);
}

int inode(char *dev)
{
  int i;
  fd = open(dev, O_RDONLY);
  if (fd < 0){
    printf("open failed\n"); exit(1);
  }
  /*************************************
   same code as before to check EXT2 FS
  *************************************/
  get_block(fd, 2, buf);   // get group descriptor
  gp = (GD *)buf;
  printf("bmap_block=%d imap_block=%d inodes_table=%d ",
          gp->bg_block_bitmap,
          gp->bg_inode_bitmap,
          gp->bg_inode_table,
  iblock = gp->bg_inode_table;
  printf("---- root inode information ----\n");
  get_block(fd, iblock, buf);
  ip = (INODE *)buf;
  ip++;    // ip point at #2 INODE
  printf("mode = %4x ", ip->i_mode);
  printf("uid = %d  gid = %d\n", ip->i_uid, ip->i_gid);
```

```
  printf("size = %d\n", ip->i_size);
  printf("ctime = %s", ctime(&ip->i_ctime));
  printf("links = %d\n", ip->i_links_count);
  for (i=0; i<15; i++){          // print disk block numbers
    if (ip->i_block[i])          // print non-zero blocks only
       printf("i_block[%d] = %d\n", i, ip->i_block[i]);
  }
}

int main(int argc, char *argv[ ])
{
  if (argc>1) dev = argv[1];
  inode(dev);
}
```

图 11.4 为 EXT2 文件系统的根索引节点。

在图 11.4 中，i_mode =0x41ed 或 b'0100 0001 1110 1101'，以二进制形式表示。前 4 位 0100 表示文件类型（目录）。接着 3 位 000=ugs 都是 0，表示文件无特殊用法，例如它不是 setuid 程序。最后 9 位可分为三组，即 111、101、101，表示文件所有人、同一团队所有人和其他人员的 rwx 权限位。对于普通文件，x 位 =1 表示文件可执行。对于目录，x 位 =1 表示可访问（即 cd into）该目录；x 位 =0 表示不可访问该目录。

```
bmap = 8 imap = 9 iblock = 10
---- root inode information ----
mode = 41ed uid = 0  gid = 0
size = 1024
ctime = Mon Oct  9 16:11:44 2017
links = 3
i_block[0] = 33
```

图 11.4　EXT2 文件系统的根索引节点

11.4.4　显示目录条目

目录索引节点的各数据块均包含 dir_entries，分别是：

```
struct ext2_dir_entry_2 {
   u32 inode;                // inode number; count from 1, NOT 0
   u16 rec_len;              // this entry's length in bytes
   u8  name_len;             // name length in bytes
   u8  file_type;            // not used
   char name[EXT2_NAME_LEN]; // name: 1-255 chars, no ending NULL
};
```

目录中每个数据块的内容都具有以下形式：

[inode rec_len name_len NAME] [inode rec_len name_len NAME] ......

其中 NAME 是一系列 name_len 字符，不含终止 NULL 字节。每个 dir_entry 都有一个记录长度 rec_len。块中最后一个条目的 rec_len 覆盖了剩余的块长度，即从条目开始到块结束的距离。下面的算法演示了如何单步遍历目录数据块中的 dir_entries。

```
/******** Algorithm to step through entries in a DIR data block ********/
struct ext2_dir_entry_2 *dp;              // dir_entry pointer
char *cp;                                 // char pointer
int blk = a data block (number) of a DIR (e.g. i_block[0]);
char buf[BLKSIZE], temp[256];
get_block(fd, blk, buf);                  // get data block into buf[ ]

dp = (struct ext2_dir_entry_2 *)buf;      // as dir_entry
cp = buf;
```

```
    while(cp < buf + BLKSIZE){
      strncpy(temp, dp->name, dp->name_len); // make name a string
      temp[dp->name_len] = 0;                // ensure NULL at end
      printf("%d %d %d %s\n", dp->inode, dp->rec_len, dp->name_len, temp);
      cp += dp->rec_len;                     // advance cp by rec_len
      dp = (struct ext2_dir_entry_2 *)cp;    // pull dp to next entry
    }
```

练习 11.4：编写一个 C 语言程序来打印目录的 dir_entries。为方便起见，我们可以假设目录索引节点最多有 12 个直接块——i_block[0] 至 i_block[11]。这一假设是合理的。单个磁盘块的块大小为 1KB，平均文件名长度为 16 个字符，最多可以容纳 1024/（8+16）=42 个 dir_entries。如果一个目录有 12 个磁盘块，则可容纳 500 多个条目。我们可以放心地假设没有用户会在任何目录中放那么多文件。在空 EXT2 文件系统中，程序输出如图 11.5 所示。（请参见习题 4。）

练习 11.5：在 Linux 下挂载 mydisk。创建新的目录并将文件复制到挂载的文件系统中，然后卸载文件系统（请参见习题 5）。在 mydisk 上再次运行 dir.c 程序，以查看输出，输出如图 11.6 所示。读者可以验证，每个条目的 name_len 对应 name 字段中的具体字符数量，每个 rec_len 是 4 的倍数（对齐），（8 + name_len）增大到 4 的下一个倍数，除了最后一个条目，它的 rec_len 覆盖了剩余块长度。

```
check ext2 FS : OK
GD info: 8 9 10 1393 173 2
inodes begin block=10
******* root inode info ********
mode=41ed  uid=0  gid=0
size=1024
ctime=Sun Oct  8 14:04:52 2017
links=3
i_block[0]=33
********************************
inode# rec_len name_len name
  2        12        1     .
  2        12        2     ..
 11      1000       10     lost+found
```

图 11.5　目录条目

```
******* root inode info ********
mode=41ed  uid=0  gid=0
size=1024
ctime=Sun Oct  8 17:19:15 2017
links=7
i_block[0]=33
********************************
inode# rec_len name_len name
  2        12        1     .
  2        12        2     ..
 11        20       10     lost+found
 12        12        1     a
 13        16        8     shortDir
 14        20       12     longNamedDir
 15        28       17     aVeryLongNamedDir
 16        16        7     super.c
 17        16        6     bmap.c
 18        16        6     imap.c
 19       856        5     dir.c
```

图 11.6　目录条目列表

练习 11.6：已知一个指向目录索引节点的 INODE 指针。编写一个

```
int  search(INODE *dir, char *name)
```

函数，搜索具有给定名称的 dir_entry。如果找到，则返回其索引节点编号，否则返回 0。（请参见习题 6。）

11.5　遍历 EXT2 文件系统树

已知一个 EXT2 文件系统和一个文件的路径名，例如 /a/b/c，问题是如何找到这个文件。查找文件相当于查找其索引节点。

11.5.1 遍历算法

（1）读取超级块。检查幻数 s_magic（0xEF53），验证它确实是 EXT2 FS。

（2）读取块组描述符块（1 + s_first_data_block），以访问组 0 描述符。从块组描述符的 bg_inode_table 条目中找到索引节点的起始块编号，并将其称为 InodesBeginBlock。

（3）读取 InodeBeginBlock，获取 / 的索引节点，即 INODE #2。

（4）将路径名标记为组件字符串，假设组件数量为 n。例如，如果路径名 =/a/b/c，则组件字符串是 "a" "b" "c"，其中 n = 3。用 name[0]，name[1]，…，name[n−1] 来表示组件。

（5）从（3）中的根索引节点开始，在其数据块中搜索 name[0]。为简单起见，我们可以假设某个目录中的条目数量很少，因此一个目录索引节点只有 12 个直接数据块。有了这个假设，就可以在 12 个（非零）直接块中搜索 name[0]。目录索引节点的每个数据块都包含以下形式的 dir_entry 结构体：

```
[ino rec_len name_len NAME] [ino rec_len name_len NAME] ......
```

其中 NAME 是一系列 nlen 字符，不含终止 NULL。对于每个数据块，将该块读入内存并使用 dir_entry *dp 指向加载的数据块。然后使用 name_len 将 NAME 提取为字符串，并与 name[0] 进行比较。如果它们不匹配，则通过以下代码转到下一个 dir_entry：

```
dp = (dir_entry *)((char *)dp + dp->rec_len);
```

继续搜索。如果存在 name[0]，则可以找到它的 dir_entry，从而找到它的索引节点号。

（6）使用索引节点号 ino 来定位相应的索引节点。回想前面的内容，ino 从 1 开始计数。使用邮差算法计算包含索引节点的磁盘块及其在该块中的偏移量。

```
blk    = (ino - 1) /  INODES_PER_BLOCK  + InodesBeginBlock;
offset = (ino - 1) % INODES_PER_BLOCK;
```

然后在索引节点中读取 /a，从中确定它是否是一个目录（DIR）。如果 /a 不是目录，则不能有 /a/b，因此搜索失败。如果它是目录，并且有更多需要搜索的组件，那么继续搜索下一个组件 name[1]。现在的问题是：在索引节点中搜索 /a 的 name[1]，与第（5）步完全相同。

（7）由于（5）～（6）步将会重复 n 次，所以最好编写一个搜索函数：

```
u32 search(INODE *inodePtr, char *name)
{
   // search for name in the data blocks of current DIR inode
   // if found, return its ino; else return 0
}
```

然后我们只需调用 search() n 次，如下所示。

```
Assume: n,  name[0], ...., name[n-1] are globals
INODE *ip points at INODE of /
for (i=0; i<n; i++){
     ino = search(ip, name[i])
     if (!ino){ // can't find name[i], exit;}
     use ino to read in INODE and let ip point to INODE
}
```

如果搜索循环成功结束，ip 必须指向路径名的索引节点。遍历有多个组的大型 EXT2/3 文件系统也是类似操作。读者可参阅文献（Wang 2015）的第 3 章 以了解详细信息。

11.5.2　将路径名转换为索引节点

已知一个包含 EXT2 文件系统和路径名的设备，例如 /a/b/c/d，编写一个 C 函数。

```
INODE *path2inode(int fd, char *pathname)  // assume fd=file descriptor
```

返回一个指向文件索引节点的 INODE 指针；如果文件不可访问，则返回 0（请参见习题 7）。

稍后我们会了解到，path2inode() 函数是文件系统中最重要的函数。

11.5.3　显示索引节点磁盘块

编写一个 C 程序 showblock，可打印文件的所有磁盘块（编号）(请参见习题 8）。

11.6　EXT2 文件系统的实现

本节内容旨在说明如何实现一个完整的文件系统。首先，我们将展示文件系统的总体结构，及其实现的逻辑步骤。接下来，我们介绍一个基本文件系统来帮助读者入门。然后，我们将实现步骤组织成一系列编程项目，由读者完成。我们相信，这种经验对计算机科学的所有学生都是有益的。

11.6.1　文件系统的结构

图 11.7 显示了 EXT2 文件系统的内部结构。标签（1）至（5）对结构图进行了说明。

（1）是当前运行进程的 PROC 结构体。在实际系统中，每个文件操作都是由当前执行的进程决定的。每个进程都有一个 cwd，指向进程当前工作目录（CWD）的内存索引节点。它还有一个文件描述符数组 fd[]，指向打开的文件实例。

（2）是文件系统的根指针。它指向内存中的根索引节点。当系统启动时，选择其中一个设备作为根设备，它必须是有效的 EXT2 文件系统。根设备的根索引节点（inode #2）作为文件系统的根（/）加载到内存中。该操作称为“**挂载根文件系统**”。

（3）是一个 openTable 条目。当某个进程打开文件时，进程 fd 数组的某个条目会指向 openTable，openTable 指向打开文件的内存索引节点。

（4）是内存索引节点。当需要某个文件时，会把它的索引节点加载到 minode 槽中以供使用。因为索引节点是唯一的，所以在任何时候每个索引节点在内存中都只能有一个副本。在 minode 中，（dev, ino）会确定索引节点的来源，以便将修改后的索引节点写回磁盘。refCount 字段会记录使用 minode 的进程数。

dirty 字段表示索引节点是否已被修改。挂载标志表示索引节点是否已被挂载，如果已被挂载，mntabPtr 将指向挂载文件系统的挂载表条目。lock 字段用于确保内存索引节点一次只能由一个进程访问，例如在修改索引节点时，或者在读 / 写操作过程中。

（5）是已挂载的文件系统表。对于每个挂载的文件系统，挂载表中的条目用于记录挂载的文件系统信息，例如挂载的文件系统设备号。在挂载点的内存索引节点中，挂载标志打开，mntabPtr 指向挂载表条目。在挂载表条目中，mntPointPtr 指向挂载点的内存索引节点。

后面将会讲到，这些双链接指针允许我们在遍历文件系统树时跨越挂载点。此外，挂载表条目还可能包含挂载文件系统的其他信息，例如超级块、块组描述符、位图和索引节点启动块的值，以便快速访问。如果任何缓存项有修改，当卸载设备时，必须将它们写回设备。

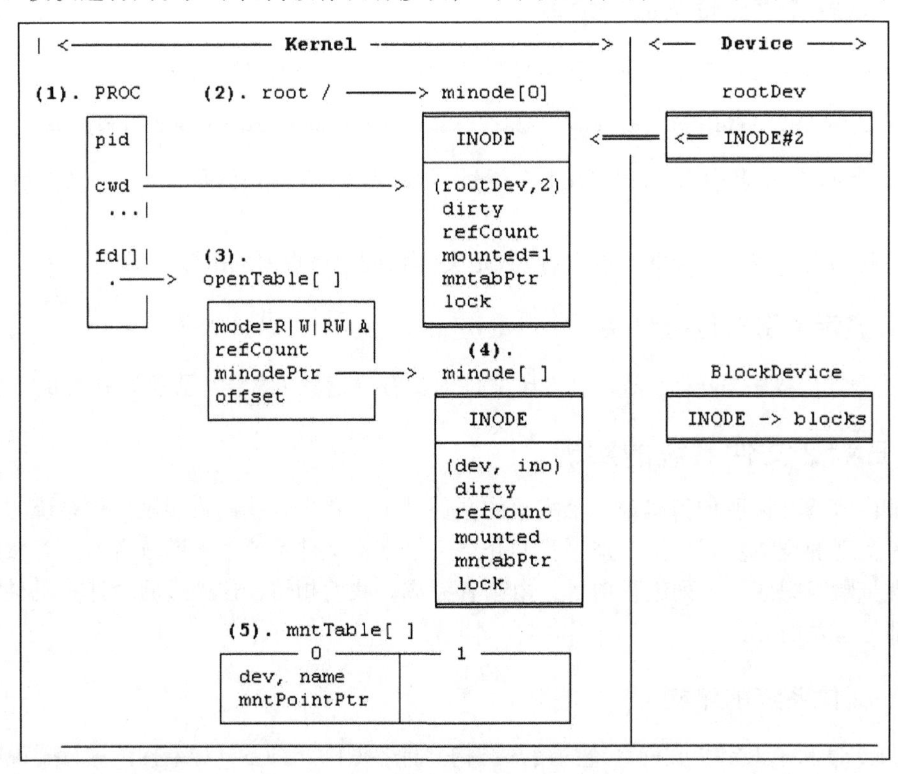

图 11.7　EXT2 文件系统数据结构

11.6.2　文件系统的级别

文件系统的实现分为三个级别。每个级别处理文件系统的不同部分。这使得实现过程模块化，更容易理解。在文件系统的实现过程中，FS 目录包含实现 EXT2 文件系统的文件。文件结构如下。

```
------------------ Common files of FS ------------------------------
  type.h  : EXT2 data structure types
  global.c: global variables of FS
  util.c    : common utility functions: getino(), iget(), iput(), search
  (), etc.
  allocate_deallocate.c : inodes/blocks management functions
```

第 1 级别实现了基本文件系统树。它包含以下文件，实现了指定函数。

```
------------------- Level-1 of FS -----------------------------
mkdir_creat.c        : make directory, create regular file
ls_cd_pwd.c          : list directory, change directory, get CWD path
rmdir.c              : remove directory
link_unlink.c        : hard link and unlink files
symlink_readlink.c   : symbolic link files
stat.c               : return file information
misc1.c              : access, chmod, chown, utime, etc.
-----------------------------------------------------------------
```

使用第 1 级别 FS 函数的用户命令程序有：

mkdir、creat、mknod、rmdir、link、unlink、symlink、rm、ls、cd 和 pwd 等。

第 2 级别实现了文件内容读 / 写函数。

```
---------------------- Level-2 of FS -------------------------------
open_close_lseek.c    : open file for READ|WRITE|APPEND, close file and lseek
read.c                : read from file descriptor of an opened regular file
write.c               : write to file descriptor of an opened regular file
opendir_readdir.c     : open and read directory
```

第 3 级别实现了文件系统的挂载、卸载和文件保护。

```
---------------------- Level-3 of FS ------------------------------
mount_umount.c       : mount/umount file systems
file protection      : access permission checking
file-locking         : lock/unlock files
-------------------------------------------------------------------
```

11.7 基本文件系统

11.7.1 type.h 文件

这类文件包含 EXT2 文件系统的数据结构类型，比如超块、组描述符、索引节点和目录条目结构。此外，它还包含打开文件表、挂载表、PROC 结构体和文件系统常数。

```
/***** type.h file for EXT2 FS *****/
#include <stdio.h>
#include <stdlib.h>
#include <fcntl.h>
#include <linux/ext2_fs.h>
#include <libgen.h>
#include <string.h>
#include <sys/stat.h>
// define shorter TYPES for convenience
typedef struct ext2_group_desc  GD;
typedef struct ext2_super_block SUPER;
typedef struct ext2_inode       INODE;
typedef struct ext2_dir_entry_2 DIR;
#define BLKSIZE        1024

// Block number of EXT2 FS on FD
#define SUPERBLOCK        1
#define GDBLOCK           2
#define ROOT_INODE        2

// Default dir and regular file modes
#define DIR_MODE          0x41ED
#define FILE_MODE         0x81AE
#define SUPER_MAGIC       0xEF53
#define SUPER_USER        0
// Proc status
#define FREE              0
#define BUSY              1
```

```
// file system table sizes
#define NMINODE          100
#define NMTABLE           10
#define NPROC              2
#define NFD               10
#define NOFT              40

// Open File Table
typedef struct oft{
  int   mode;
  int   refCount;
  struct minode *minodePtr;
  int   offset;
}OFT;

// PROC structure
typedef struct proc{
  struct Proc *next;
  int   pid;
  int   uid;
  int   gid;
  int   ppid;
  int   status;
  struct minode *cwd;
  OFT   *fd[NFD];
}PROC;

// In-memory inodes structure
typedef struct minode{
  INODE INODE;              // disk inode
  int   dev, ino;
  int   refCount;           // use count
  int   dirty;              // modified flag
  int   mounted;            // mounted flag
  struct mount *mntPtr;     // mount table pointer
  // int lock;              // ignored for simple FS
}MINODE;

// Open file Table          // opened file instance
typedef struct oft{
  int  mode;                // mode of opened file
  int  refCount;            // number of PROCs sharing this instance
  MINODE *minodePtr;        // pointer to minode of file
  int  offset;              // byte offset for R|W
}OFT;

// Mount Table structure
typedef struct mtable{
  int    dev;              // device number; 0 for FREE
  int    ninodes;          // from superblock
  int    nblocks;
  int    free_blocks       // from superblock and GD
  int    free_inodes
  int    bmap;             // from group descriptor
  int    imap;
```

```
  int    iblock;          // inodes start block
  MINODE *mntDirPtr;      // mount point DIR pointer
  char   devName[64];     //device name
  char   mntName[64];     // mount point DIR name
}MTABLE;
```

11.7.2 global.c 文件

这类文件包含文件系统的全局变量。全局变量的例子有：

```
MINODE minode[NMINODE];      // in memory INODEs
MTABLE mtable[NMTABLE];      // mount tables
OFT    oft[NOFT];            // Opened file instance
PROC   proc[NPROC]           // PROC structures
PROC   *running;             // current executing PROC
```

当文件系统启动时，我们初始化所有全局数据结构，并让运行点位于 PROC[0]，即超级用户的进程 P0（uid = 0）。在实际系统中，每个操作都是由当前运行的进程决定的。我们从超级用户进程开始，因为它不需要任何文件保护。通过权限检查以保护文件将在后面第 3 级别的 FS 实现中执行。

```
int fs_init()
{
  int i,j;
  for (i=0; i<NMINODE; i++)     // initialize all minodes as FREE
      minode[i].refCount = 0;
  for (i=0; i<NMTABLE; i++)     // initialize mtables as FREE
      mtable[i].dev = 0;
  for (i=0; i<NOFT; i++)        // initialize ofts as FREE
      oft[i].refCount = 0;
  for (i=0; i<NPROC; i++){      // initialize PROCs
      proc[i].status = READY;   // ready to run
      proc[i].pid = i;          // pid = 0 to NPROC-1
      proc[i].uid = i;          // P0 is a superuser process
      for (j=0; j<NFD; j++)
          proc[i].fd[j] = 0;    // all file descriptors are NULL
      proc[i].next = &proc[i+1];   // link list
  }
  proc[NPROC-1].next = &proc[0];   // circular list
  running = &proc[0];              // P0 runs first
}
```

文件系统操作过程中，全局数据结构被视为系统资源，可灵活使用和释放。每一组资源都由一对分配和释放函数管理。例如，mialloc() 分配一个空闲的 minode 供使用，而 midalloc() 则释放一个使用过的 minode。其他资源管理函数与此类似，后面在实际需要时会进行说明。

```
MINODE *mialloc()           // allocate a FREE minode for use
{
   int i;
   for (i=0; i<NMINODE; i++){
     MIONDE *mp = &minode[i];
     if (mp->refCount == 0){
```

```
      mp->refCount = 1;
      return mp;
    }
  }
  printf("FS panic: out of minodes\n");
  return 0;
}
int midalloc(MINODE *mip) // release a used minode
{
  mip->refCount = 0;
}
```

11.7.3　实用程序函数

util.c file：该文件包含文件系统常用的实用程序函数。最重要的实用程序函数是读/写磁盘块函数 iget()、iput() 和 getino()，下面将对这些函数进行更详细的说明。

（1）get_block/put_block 函数：我们假设某个块设备，例如真实磁盘或虚拟磁盘，只能以块大小为单位读写。对于真实磁盘，这是因为受到硬件的限制。对于虚拟磁盘，我们假设也是以块大小为单位读/写，这样就可以在需要时将代码移植到真实磁盘上。在虚拟磁盘上，我们先以读|写模式打开它，并使用文件描述符作为设备号。以下函数将虚拟磁盘块读/写到内存的缓冲区中。

```
int get_block(int dev, int blk, char *buf)
{
    lseek(dev, blk*BLKSIZE, SEEK_SET);
    int n = read(dev, buf, BLKSIZE);
    if (n<0) printf("get_block [%d %d] error\n", dev, blk);
}
int put_block(int dev, int blk, char *buf)
{
    lseek(dev, blk*BLKSIZE, SEEK_SET);
    int n = write(dev, buf, BLKSIZE);
    if (n != BLKSIZE)
       printf("put_block [%d %d] error\n", dev, blk);
}
```

（2）iget(dev, ino) 函数：该函数返回一个指针，指向包含 INODE(dev, ino) 的内存 minode。返回的 minode 是唯一的，即内存中只存在一个 INODE 副本。在实际文件系统中，返回的 minode 被锁定为独占使用，直到它被释放或解锁。为简单起见，我们假设 minode 锁定不是必要的，稍后将对此进行解释。

```
MINODE *iget(int dev, int ino)
{
  MINODE *mip;
  MTABLE *mp;
  INODE  *ip;
  int i, block, offset;
  char buf[BLKSIZE];

  // serach in-memory minodes first
  for (i=0; i<NMINODES; i++){
    MINODE *mip = &MINODE[i];
```

```
    if (mip->refCount && (mip->dev==dev) && (mip->ino==ino)){
       mip->refCount++;
       return mip;
    }
  }

  // needed INODE=(dev,ino) not in memory
  mip = mialloc();                    // allocate a FREE minode
  mip->dev = dev; mip->ino = ino;     // assign to (dev, ino)
  block = (ino-1)/8 + iblock;         // disk block containing this inode
  offset= (ino-1)%8;                  // which inode in this block
  get_block(dev, block, buf);
  ip = (INODE *)buf + offset;
  mip->INODE = *ip;                   // copy inode to minode.INODE
  // initialize minode
  mip->refCount = 1;
  mip->mounted = 0;
  mip->dirty = 0;
  mip->mountptr = 0;
  return mip;
}
```

（3）The iput(INODE *mip) 函数：该函数会释放一个 mip 指向的用完的 minode。每个 minode 都有一个 refCount，表示使用 minode. iput() 的用户数量为 refCount 减 1。如果 refCount 为非零，则表示 minode 仍有其他用户，那么调用者只是返回。如果调用者是 minode 的最后一个用户（refCount = 0），那么如果 INODE 被修改（dirty），它将被写回磁盘。

```
int iput(MINODE *mip)
{
  INODE  *ip;
  int i, block, offset;
  char buf[BLKSIZE];

  if (mip==0) return;
  mip->refCount--;                   // dec refCount by 1
  if (mip->refCount > 0) return;     // still has user
  if (mip->dirty == 0)   return;     // no need to write back

  // write INODE back to disk
  block  = (mip->ino - 1) / 8 + iblock;
  offset = (mip->ino - 1) % 8;

  // get block containing this inode
  get_block(mip->dev, block, buf);
  ip = (INODE *)buf + offset;        // ip points at INODE
  *ip = mip->INODE;                  // copy INODE to inode in block
  put_block(mip->dev, block, buf); // write back to disk
  midalloc(mip);                      // mip->refCount = 0;
}
```

（4）getino() 函数：getino() 函数可实现文件系统树遍历算法。它会返回指定路径名的 INODE 编号（ino）。首先，我们假设在 1 级文件系统的实现中，文件系统属于单个根设备，因此不存在挂载设备和挂载点交叉。稍后将在 3 级文件系统的实现中考虑挂载文件系统和挂载点交叉。因此，getino() 函数本质上返回的是路径名的（dev, ino）。首先，该函数使用 tokenize() 函数将路径名分解为组件字符串。我们假设标记化字符串位于全局数据区中，每个字符串由一个 name[i] 指针指向，标记字符串的数量为 nname。然后，它会调用 search() 函数来搜索连续目录中的标记字符串。下面给出了 tokenize() 和 search() 函数。

```
char *name[64];  // token string pointers
char gline[256]; // holds token strings, each pointed by a name[i]
int  nname;      // number of token strings

int tokenize(char *pathname)
{
  char *s;
  strcpy(gline, pathname);
  nname = 0;
  s = strtok(gline, "/");
  while(s){
    name[nname++] = s;
    s = strtok(0, "/");
  }
}

int search(MINODE *mip, char *name)
{
   int i;
   char *cp, temp[256], sbuf[BLKSIZE];
   DIR *dp;
   for (i=0; i<12; i++){ // search DIR direct blocks only
     if (mip->INODE.i_block[i] == 0)
        return 0;
     get_block(mip->dev, mip->INODE.i_block[i], sbuf);
     dp = (DIR *)sbuf;
     cp = sbuf;
     while (cp < sbuf + BLKSIZE){
        strncpy(temp, dp->name, dp->name_len);
        temp[dp->name_len] = 0;
        printf("%8d%8d%8u %s\n",
               dp->inode, dp->rec_len, dp->name_len, temp);
        if (strcmp(name, temp)==0){
           printf("found %s : inumber = %d\n", name, dp->inode);
           return dp->inode;
        }
        cp += dp->rec_len;
        dp = (DIR *)cp;
     }
   }
   return 0;
}

int getino(char *pathname)
{
  MINODE *mip;
  int i, ino;
  if (strcmp(pathname, "/")==0){
      return 2;              // return root ino=2
```

```
    }
    if (pathname[0] == '/')
        mip = root;           // if absolute pathname: start from root
    else
        mip = running->cwd;   // if relative pathname: start from CWD
    mip->refCount++;          // in order to iput(mip) later

    tokenize(pathname);       // assume: name[ ], nname are globals

    for (i=0; i<nname; i++){  // search for each component string
        if (!S_ISDIR(mip->INODE.i_mode)){  // check DIR type
           printf("%s is not a directory\n", name[i]);
           iput(mip);
           return 0;
        }
        ino = search(mip, name[i]);
        if (!ino){
           printf("no such component name %s\n", name[i]);
           iput(mip);
           return 0;
        }
        iput(mip);              // release current minode
        mip = iget(dev, ino);   // switch to new minode
    }
    iput(mip);
    return ino;
}
```

（5）getino()/iget()/iput() 的使用：在文件系统中，几乎每个操作都以一个路径名开头，例如 mkdir 路径名、cat 路径名等。只要指定了路径名，就必须将其索引节点加载到内存中备用。索引节点的一般使用方式是：

```
. ino  = getino(pathname);
. mip = iget(dev, ino);
.    use mip->INODE, which may modify the INODE;
. iput(mip);
```

这种使用方式只有少数例外情况。比如：

- 更改目录（chdir）：iget 新目录的 minode，但是 iput 旧目录的 minode。
- 打开（open）：iget 文件的 minode，当文件关闭时释放。
- 挂载（mount）：iget 挂载点的 minode，稍后通过卸载释放。

一般来说，iget 和 iput 要成对出现，就像一对配套的括号。我们可在实现代码过程中依赖这种使用方式，来确保每个 INODE 都能够正确加载和释放。

minode 锁定：在实际文件系统中，每个 minode 都有一个锁字段，确保一次只有一个进程可以访问 minode，例如在修改 INODE 时。Unix 内核使用忙碌标志和休眠 / 唤醒来同步访问同一 minode 的进程。在其他系统中，每个 minode 可能有一个互斥量或一个信号量锁。进程只有在持有 minode 锁的情况下才能访问 minode。minode 锁定的原因如下所示。

假设一个进程 Pi 需要内存中没有的索引节点（dev, ino）。Pi 必须将该索引节点加载到 minode 条目中。minode 必须标记为（dev, ino），以防止其他进程再次加载同一个索引节点。从磁盘加载索引节点时，Pi 可能会等待 I/O 完成，切换到另一个进程 Pj。如果 Pj 恰好需要相同的索引节点，它会发现所需的 minode 已经存在。如果没有锁，Pj 可能会在加载 minode 之前使用它。有了锁，Pj 就必须等待 minode 被 Pi 加载、使用和释放之后再使用。此外，当进程读 / 写某个打开的文件时，它必须锁定文件的 minode，以确保每个读 / 写操作都是原子操作。为简单起见，我们假设一次只运行一个进程，因此不需要锁。但是，读者要知道，在实际文件系统中，minode 锁定是必要的。

练习 11.7：设计并实现一个使用 minode[ ] 区域作为内存 INODE 缓存的方案。一旦 INODE 被加载到 minode 槽中，即使没有进程在主动使用它，也要尽可能长时间地将它留在内存中。

11.7.4 mount-root

mount_root.c 文件：该文件包含 mount_root() 函数，在系统初始化期间调用该函数来挂载根文件系统。它读取根设备的超级块，以验证该设备是否为有效的 EXT2 文件系统。然后，它将根设备的根 INODE（ino = 2）加载到 minode 中，并将根指针设置为根 minode。它还将所有进程的当前工作目录设置为根 minode。分配一个挂载表条目来记录挂载的根文件系统。根设备的一些关键信息，如 inode 和块的数量、位图的起始块和 inode 表，也记录在挂载表中，以便快速访问。

```
/************ FS1.1.c file ************/
#include "type.h"
#include "util.c"

// global variables
MINODE minode[NMINODES], *root;
MTABLE mtable[NMTABLE];
PROC proc[NPROC], *running;
int ninode, nblocks, bmap, imap, iblock;
int dev;
char gline[25], *name[16];  // tokenized component string strings
int  nname;                 // number of component strings
char *rootdev = "mydisk";   // default root_device

int fs_init()
{
  int i,j;
  for (i=0; i<NMINODES; i++)  // initialize all minodes as FREE
      minode[i].refCount = 0;
  for (i=0; i<NMOUNT; i++)    // initialize mtable entries as FREE
      mtable[i].dev = 0;
  for (i=0; i<NPROC; i++){    // initialize PROCs
      proc[i].status = READY; // reday to run
      proc[i].pid = i;        // pid = 0 to NPROC-1
      proc[i].uid = i;        // P0 is a superuser process
      for (j=0; j<NFD; j++)
          proc[i].fd[j] = 0;  // all file descriptors are NULL
      proc[i].next = &proc[i+1];
  }
  proc[NPROC-1].next = &proc[0];    // circular list
  running = &proc[0];               // P0 runs first
}
```

```
int mount_root(char *rootdev)  // mount root file system
{
  int i;
  MTABLE *mp;
  SUPER  *sp;
  GD     *gp;
  char buf[BLKSIZE];

  dev = open(rootdev, O_RDWR);
  if (dev < 0){
     printf("panic : can't open root device\n");
     exit(1);
  }
  /* get super block of rootdev */
  get_block(dev, 1, buf);
  sp = (SUPER *)buf;
  /* check magic number */
  if (sp->s_magic != SUPER_MAGIC){
     printf("super magic=%x : %s is not an EXT2 filesys\n",
            sp->s_magic, rootdev);
     exit(0);
  }
  // fill mount table mtable[0] with rootdev information
  mp = &mtable[0];      // use mtable[0]
  mp->dev = dev;
  // copy super block info into mtable[0]
  ninodes = mp->ninodes = sp->s_inodes_count;
  nblocks = mp->nblocks = sp->s_blocks_count;
  strcpy(mp->devName, rootdev);
  strcpy(mp->mntName, "/");
  get_block(dev, 2, buf);
  gp = (GD *)buf;
  bmap = mp->bmap = gp->bg_blocks_bitmap;
  imap = mp->imap = gp->bg_inodes_bitmap;
  iblock = mp->iblock = gp->bg_inode_table;
  printf("bmap=%d imap=%d iblock=%d\n", bmap, imap iblock);

  // call iget(), which inc minode's refCount
  root = iget(dev, 2);           // get root inode
  mp->mntDirPtr = root;          // double link
  root->mntPtr = mp;
  // set proc CWDs
  for (i=0; i<NPROC; i++)          // set proc's CWD
      proc[i].cwd = iget(dev, 2); // each inc refCount by 1
  printf("mount : %s  mounted on / \n", rootdev);
  return 0;
}

int main(int argc, char *argv[ ])
{
  char line[128], cmd[16], pathname[64];
  if (argc > 1)
     rootdev = argv[1];
  fs_init();
  mount_root(rootdev);
```

```
  while(1){
     printf("P%d running: ", running->pid);
     printf("input command : ");
     fgets(line, 128, stdin);
     line[strlen(line)-1] = 0;
     if (line[0]==0)
        continue;
     sscanf(line, "%s %s", cmd, pathname);
     if (!strcmp(cmd, "ls"))
        ls(pathname);
     if (!strcmp(cmd, "cd"))
        chdir(pathname);
     if (!strcmp(cmd, "pwd"))
        pwd(running->cwd);
     if (!strcmp(cmd, "quit"))
        quit();
  }
}

int quit() // write all modified minodes to disk
{
   int i;
   for (i=0; i<NMINODES; i++){
     MINODE *mip = &minode[i];
     if (mip->refCount && mip->dirty){
        mip->refCount = 1;
        iput(mip);
     }
   }
   exit(0);
}
```

如何执行 ls：ls [pathname] 列出了目录或文件的信息。在第 5 章（5.4 节）中，我们展示了一个 ls 程序，其工作原理如下文所示。

（1）ls_dir(dirname)：使用 opendir() 和 readdir() 获取目录中的文件名。对于每个文件名，调用 ls_file(filename)。

（2）ls_file(filename)：stat 文件名，以在 STAT 结构体中获取文件信息。然后，列出 STAT 信息。

由于 stat 系统调用实质上返回的是 minode 的相同信息，所以我们可以通过直接使用 minode 来修改原始的 ls 算法。下面是修改后的 ls 算法。

```
/************************ Algorithm of ls *******************************/
(1). From the minode of a directory, step through the dir_entries in the data
blocks of the minode.INODE. Each dir_entry contains the inode number, ino, and
name of a file. For each dir_entry, use iget() to get its minode, as in
        MINODE *mip = iget(dev, ino);
Then, call ls_file(mip, name).
(2). ls_file(MINODE *mip, char *name): use mip->INODE and name to list the file
information.
```

如何执行 chdir [pathname]：chdir 算法如下。

```
/*************** Algorithm of chdir ****************/
(1). int ino = getino(pathname);          // return error if ino=0
```

```
(2). MINODE *mip = iget(dev, ino);
(3). Verify mip->INODE is a DIR       // return error if not DIR
(4). iput(running->cwd);                       // release old cwd
(5). running->cwd = mip;                      // change cwd to mip
```

如何 pwd：下面展示了 pwd 算法，在目录 minode 上使用递归。

```
/*************** Algorithm of pwd ****************/
   rpwd(MINODE *wd){
    (1). if (wd==root) return;
    (2). from wd->INODE.i_block[0], get my_ino and parent_ino
    (3). pip = iget(dev, parent_ino);
    (4). from pip->INODE.i_block[ ]: get my_name string by my_ino as LOCAL
    (5). rpwd(pip);  // recursive call rpwd(pip) with parent minode
    (6). print "/%s", my_name;
   }

pwd(MINODE *wd){
   if (wd == root)    print "/"
   else                     rpwd(wd);
}
// pwd start:
pwd(running->cwd);
```

练习 11.8：将 pwd 算法的第（2）步作为实用程序函数来实现。

```
int get_myino(MINODE *mip, int *parent_ino)
```

会在 parent_ino 中返回 . 和 .. 的索引节点号。

练习 11.9：将 pwd 算法的第（4）步实现为一个实用程序函数。

```
int get_myname(MINODE *parent_minode, int my_ino, char *my_name)
```

会用 my_ino 标识在父目录中返回 dir_entry 的名称字符串。

11.7.5 基本文件系统的实现

本节的编程任务是将上面的 mount_root.c 程序实现为一个基本文件系统。然后用包含 EXT2 文件系统的虚拟磁盘进行运行。基本文件系统的输出如图 11.8 所示。图中只显示了 ls 命令的结果。它还应该支持 cd 和 pwd 命令。

```
checking EXT2 FS ....OK
bmp = 8 imap = 9 inodes_start = 10
init()
mount_root()
creating P0 as running process
mydisk mounted on / OK
input command : [ls|cd|pwd|quit] ls
drwxr-xr-x    5    0    0  Oct 21 09:02     1024    .
drwxr-xr-x    5    0    0  Oct 21 09:02     1024    ..
drwx------    2    0    0  Oct 21 09:01    12288    lost+found
drwxr-xr-x    4    0    0  Oct 21 09:03     1024    dir1
drwxr-xr-x    2    0    0  Oct 21 09:02     1024    dir2
-rw-r--r--    1    0    0  Oct 21 09:02        0    file1
-rw-r--r--    1    0    0  Oct 21 09:02        0    file2
input command : [ls|cd|pwd|quit]
```

图 11.8 mount_root 的样本输出

11.8　1 级文件系统函数

11.8.1　mkdir 算法

mkdir 命令

```
mkdir pathname
```

创建了一个带路径名的新目录。将新目录的权限位设置为默认值 0755（所有人可以访问和读写，其他人可以访问但只能读取）。

mkdir 使用包含默认 . 和 .. 条目的数据块创建一个空目录。mkdir 算法如下：

```
/********* Algorithm of mkdir pathname*********/
(1). divide pathname into dirname and basename, e.g. pathname=/a/b/c, then
        dirname=/a/b;   basename=c;
(2). // dirname must exist and is a DIR:
        pino = getino(dirname);
        pmip = iget(dev, pino);
        check pmip->INODE is a DIR
(3). // basename must not exist in parent DIR:
       search(pmip, basename) must return 0;
(4). call kmkdir(pmip, basename) to create a DIR;

        kmkdir() consists of 4 major steps:
        (4).1. Allocate an INODE and a disk block:
                  ino = ialloc(dev);
                  blk = balloc(dev);
        (4).2. mip = iget(dev, ino) // load INODE into a minode
                  initialize mip->INODE as a DIR INODE;
                  mip->INODE.i_block[0] = blk; other i_block[ ] = 0;
                  mark minode modified (dirty);
                  iput(mip);  // write INODE back to disk
        (4).3. make data block 0 of INODE to contain . and .. entries;
                 write to disk block blk.
        (4).4. enter_child(pmip, ino, basename); which enters
                  (ino, basename) as a dir_entry to the parent INODE;

(5). increment parent INODE's links_count by 1 and mark pmip dirty;
     iput(pmip);
```

mkdir 算法的大多数步骤都无须解释。只有第（4）步需要详细解释。我们通过示例代码来更详细地说明第（4）步。要创建一个目录，我们需要从索引节点位图中分配一个索引节点，并从块位图中分配一个磁盘块，分配操作依赖于测试，然后设置位图中的位。为了保持文件系统的一致性，分配一个索引节点后，必须将超级块和组块描述符中的空闲索引节点计数减 1。同样，分配一个磁盘块后，也必须将超级块和块组描述符中的空闲块数减 1。还要注意，位图中的位从 0 开始计数，而索引节点和块号从 1 开始计数。下面显示了（4）的详细步骤。

步骤（4）.1 分配索引节点和磁盘块：

```
// tst_bit, set_bit functions
int tst_bit(char *buf, int bit){
    return buf[bit/8] & (1 << (bit % 8));
}
int set_bit(char *buf, int bit){
    buf[bit/8] |= (1 << (bit % 8));
}
int decFreeInodes(int dev)
{
  // dec free inodes count in SUPER and GD
  get_block(dev, 1, buf);
  sp = (SUPER *)buf;
  sp->s_free_inodes_count--;
  put_block(dev, 1, buf);
  get_block(dev, 2, buf);
  gp = (GD *)buf;
  gp->bg_free_inodes_count--;
  put_block(dev, 2, buf);
}
int ialloc(int dev)
{
 int i;
 char buf[BLKSIZE];
 // use imap, ninodes in mount table of dev
 MTABLE *mp = (MTABLE *)get_mtable(dev);
 get_block(dev, mp->imap, buf);
 for (i=0; i<mp->ninodes; i++){
   if (tst_bit(buf, i)==0){
      set_bit(buf, i);
      put_block(dev, mp->imap, buf);
      // update free inode count in SUPER and GD
      decFreeInodes(dev);
      return (i+1);
   }
 }
 return 0; // out of FREE inodes
}
```

磁盘块的分配与此类似，只是它使用了块位图，并减少了超级块和组描述符中的空闲块数。这个问题留作练习。

练习 11.10：实现函数

```
int balloc(int dev)
```

从设备分配一个空闲磁盘块（编号）。

步骤（4）.2 创建索引节点：下面的代码段会在 minode 中创建 INODE=（dev, ino），并将索引节点写入磁盘。

```
MINODE *mip = iget(dev, ino);
INODE *ip = &mip->INODE;
ip->i_mode = 0x41ED;        // 040755: DIR type and permissions
ip->i_uid  = running->uid; // owner uid
ip->i_gid  = running->gid; // group Id
ip->i_size = BLKSIZE;       // size in bytes
```

```
ip->i_links_count = 2;     // links count=2 because of . and ..
ip->i_atime = ip->i_ctime = ip->i_mtime = time(0L);
ip->i_blocks = 2;          // LINUX: Blocks count in 512-byte chunks
ip->i_block[0] = bno;      // new DIR has one data block
ip->i_block[1] to ip->i_block[14] = 0;
mip->dirty = 1;            // mark minode dirty
iput(mip);                 // write INODE to disk
```

步骤（4）.3 为新目录创建包含 . 和 .. 条目的数据块。

```
char buf[BLKSIZE];
bzero(buf, BLKSIZE); // optional: clear buf[ ] to 0
DIR *dp = (DIR *)buf;
// make . entry
dp->inode = ino;
dp->rec_len = 12;
dp->name_len = 1;
dp->name[0] = '.';
// make .. entry: pino=parent DIR ino, blk=allocated block
dp = (char *)dp + 12;
dp->inode = pino;
dp->rec_len = BLKSIZE-12;    // rec_len spans block
dp->name_len = 2;
dp->name[0] = d->name[1] = '.';
put_block(dev, blk, buf);    // write to blk on diks
```

步骤（4）.4 在父目录中输入新的 dir_entry：函数

```
int enter_name(MINODE *pip,  int ino,  char *name)
```

将 [ino, name] 作为一个新的 dir_entry 输入父目录中。enter_name 算法包含以下几个步骤。

```
/****************** Algorithm of enter_name *******************/
 for each data block of parent DIR do  // assume: only 12 direct blocks
{
        if (i_block[i]==0) BREAK;     // to step (5) below
   (1). Get parent's data block into a buf[ ];
   (2). In a data block of the parent directory, each dir_entry has an ideal
   length

         ideal_length = 4*[ (8 + name_len + 3)/4 ]      // a multiple of 4

         All dir_entries rec_len = ideal_length, except the last entry. The
         rec_len of the LAST entry is to the end of the block, which may be
         larger than its ideal_length.
     (3). In order to enter a new entry of name with n_len, the needed length is

         need_length = 4*[ (8 + n_len + 3)/4 ]          // a multiple of 4

     (4). Step to the last entry in the data block:

         get_block(parent->dev, parent->INODE.i_block[i], buf);
         dp = (DIR *)buf;
         cp = buf;
```

```
      while (cp + dp->rec_len < buf + BLKSIZE){
         cp += dp->rec_len;
         dp = (DIR *)cp;
      }
      // dp NOW points at last entry in block
      remain = LAST entry's rec_len - its ideal_length;

      if (remain >= need_length){
         enter the new entry as the LAST entry and
         trim the previous entry rec_len to its ideal_length;
      }
      goto step (6);
   }
```

下图显示了将 [ino, name] 作为新条目输入前后的块内容。

```
Before                         |            LAST entry                 |
|-4---2----2--|----|----|------- rlen to end of block ----------|
|ino rlen nlen NAME|.....|ino rlen nlen|NAME                      |
-----------------------------------------------------------------

After                          |     Last entry     |      NEW entry      |
|-4---2----2--|----|----|----ideal_len-----|---- rlen=remain ----|
|ino rlen nlen NAME|....|ino rlen nlen|NAME|ino rlen nlen name    |
 -----------------------------------------------------------------
}
```

在步骤（5）中，如果现有数据块没有空间，则进行以下操作：

- 分配一个新的数据块，将父目录大小增加 BLKSIZE。
- 使用 rec_len=BLKSIZE 将新条目作为第一个条目输入新数据块中。

下图显示了只包含一个条目的新数据块。

```
|------------------- rlen = BLKSIZE ------------------------------
|ino rlen=BLKSIZE nlen name                                      |
-----------------------------------------------------------------
 (6).Write data block to disk;
}
```

11.8.2 creat 算法

creat 创建一个空的普通文件。creat 算法如下：

```
/************** Algorithm of creat  *************/
creat(char * pathname)
{
  This is similar to mkdir() except
  (1). the INODE.i_mode field is set to REG file type, permission bits set to
             0644 = rw-r--r--, and
  (2). no data block is allocated for it, so the file size is 0.
  (3). links_count = 1; Do not increment parent INODE's links_count
}
```

注意，上面的 creat 算法与 Unix/Linux 的不同之处在于，它不会以写模式打开文件并返回文件描述符。实际上，creat 很少作为独立的函数使用。它由 open() 函数进行内部使用，可能会创建一个文件，打开文件写入及返回一个文件描述符。打开操作稍后将在 2 级文件系统中实现。

11.8.3　mkdir-creat 的实现

本节的编程任务是实现 mkdir-creat 函数，并将相应的 mkdir 和 creat 命令添加到基本文件系统中。通过测试 mkdir 和 creat 函数来演示系统。图 11.9 所示为所得到文件系统的样本输出。它显示了执行命令 mkdir dir1 的详细步骤，如 mkdir 算法中所述。

```
fs_init()
mount_root
checking EXT2 FS : OK
bmp = 8 imap = 9 iblock = 10
mydisk mounted on / OK
root refCount = 3
input command : [ls|cd|pwd|mkdir|creat|quit] mkdir dir1
mkdir dir1
parent=.  child=dir1
getino: pathname=.
===========================================
getino: i=0 name[0]=.
search for . in MINODE = [3, 2]
. found . : ino = 2
search for dir1 in MINODE = [3, 2]
. .. lost+found : dir1 not yet exits
ialloc: ino=12 balloc: bno=47
making INODE
iput: dev=3 ino=12
making data block
writing data block 47 to disk
enter name: parent=(3 2) name=dir1
enter name: dir1 need_len=12
parent data blk[0] = 33
step to LAST entry in data block 33
. .. lost+found
found space: name=lost+found ideal=20 rlen=1000 remain=980
write parent data block[0]=33
-------------------------------------------
input command : [ls|cd|pwd|mkdir|creat|quit] ls
i_block[0] = 33
drwxr-xr-x    4   0    0  Oct 22 10:32     1024    .
drwxr-xr-x    4   0    0  Oct 22 10:32     1024    ..
drwx------    2   0    0  Oct 22 10:32    12288    lost+found
drwxr-xr-x    2   0    0  Oct 22 10:33     1024    dir1
i_block[1] = 0
input command : [ls|cd|pwd|mkdir|creat|quit] 
```

图 11.9　#2 项目的样本输出

rmdir 命令

rmdir dirname

可删除目录。在 Unix/Linux 中，要想删除目录，目录必须为空，原因如下。首先，删除非空目录意味着删除该目录中的所有文件和子目录。虽然可以实现 rmdir 递归操作，即删除整个目录树，但基本操作仍然是一次删除一个目录。其次，非空目录可能包含正在使用的文件，例如，打开进行读 / 写的文件等。显然，删除这类目录是不可接受的。虽然可以检查目录中是否有在用文件，但是这样做会产生过多的系统操作。最简单的方法是要求目录必须为空才能删除。rmdir 算法如下。

11.8.4 rmdir 算法

```
/************ Algorithm of rmdir *************/
  (1). get in-memory INODE of pathname:
         ino = getino(pathanme);
         mip = iget(dev, ino);

  (2). verify INODE is a DIR (by INODE.i_mode field);
         minode is not BUSY (refCount = 1);
         verify DIR is empty (traverse data blocks for number of entries = 2);

  (3). /* get parent's ino and inode */
         pino = findino();  //get pino from .. entry in INODE.i_block[0]
         pmip = iget(mip->dev, pino);

  (4). /* get name from parent DIR's data block
         findname(pmip, ino, name); //find name from parent DIR

  (5).  remove name from parent directory */
         rm_child(pmip, name);

  (6). dec parent links_count by 1; mark parent pimp dirty;
         iput(pmip);

  (7). /* deallocate its data blocks and inode */
         bdalloc(mip->dev, mip->INODE.i_blok[0];
         idalloc(mip->dev, mip->ino);
         iput(mip);
```

在 rmdir 算法中，大多数步骤都很简单而且不需要加以解释。我们只需要较为详细地解释以下步骤。

（1）如何测试目录为空：每个目录的 links_count 都从 2 开始（适用于 . 和 .. 条目）。每个子目录的 links_count 加 1，但是普通文件不会增加父目录的 links_count。因此，如果目录的 links_count 大于 2，目录肯定不为空。但是，如果 links_count = 2，目录可能仍然包含普通文件。在这种情况下，我们必须遍历目录的数据块来计算 dir_entries 的数量，该数量必须大于 2。这可以通过在 DIR INODE 的数据块中单步遍历 dir_entris 的相同方法实现。

（2）如何释放索引节点和数据块：当删除一个目录时，我们必须释放它的索引节点和数据块（编号）。函数

```
idalloc(dev, ino)
```

可释放一个索引节点（编号）。它将设备索引节点位图中的 ino 位清除为 0。然后，它将超级块和组描述符中的空闲 inode 计数增加 1。

```
int clr_bit(char *buf, int bit) // clear bit in char buf[BLKSIZE]
{   buf[bit/8] &= ~(1 << (bit%8)); }

int incFreeInodes(int dev)
{
  char buf[BLKSIZE];
  // inc free inodes count in SUPER and GD
  get_block(dev, 1, buf);
```

```
  sp = (SUPER *)buf;
  sp->s_free_inodes_count++;
  put_block(dev, 1, buf);
  get_block(dev, 2, buf);
  gp = (GD *)buf;
  gp->bg_free_inodes_count++;
  put_block(dev, 2, buf);
}
int idalloc(int dev, int ino)
{
  int i;
  char buf[BLKSIZE];
  MTABLE *mp = (MTABLE *)get_mtable(dev);
  if (ino > mp->ninodes){ // niodes global
     printf("inumber %d out of range\n", ino);
     return;
  }
  // get inode bitmap block
  get_block(dev, mp->imap, buf);
  clr_bit(buf, ino-1);
  // write buf back
  put_block(dev, mp->imap, buf);
  // update free inode count in SUPER and GD
  incFreeInodes(dev);
}
```

释放磁盘块号与此类似，只是它使用的是设备的块位图，并且增加了超级块和块组描述符中的空闲块数。这个问题留作练习。

练习 11.11：实现释放磁盘块（编号）bno 的 bdalloc(int dev, int bno) 函数。

（3）删除父目录中的 dir_entry：函数

```
rm_child(MINODE *pmip, char *name)
```

从 pmip 指向的父目录 minode 中删除 name 的 dir_entry。rm_child 算法如下。

```
/*************** Algorithm of rm_child ****************/
(1). Search parent INODE's data block(s) for the entry of name
(2). Delete name entry from parent directory by
(2).1. if (first and only entry in a data block){
       In this case, the data block looks like the following diagram.

       ----------------------------------------------
       |ino rlen=BLKSIZE nlen NAME                   |
       ----------------------------------------------

       deallocate the data block; reduce parent's file size by BLKSIZE;
       compact parent's i_block[ ] array to eliminate the deleted entry if it's
       between nonzero entries.
           }
(2).2. else if LAST entry in block{
       Absorb its rec_len to the predecessor entry, as shown in the following diagrams.

        Before                              |remove this entry    |
        -----------------------------------------------------------
```

```
        xxxxx|INO rlen nlen NAME |yyy  |zzz rec_len          |
        ----------------------------------------------------
        After
        ----------------------------------------------------
        xxxxx|INO rlen nlen NAME |yyy (add rec_len to yyy)  |
        ----------------------------------------------------
           }
(2).3. else: entry is first but not the only entry or in the middle of a block:
          {
            move all trailing entries LEFT to overlay the deleted entry;
            add deleted rec_len to the LAST entry; do not change parent's file
            size;

           The following diagrams illustrate the block contents before and after
           deleting such an entry.

           Before:  | delete this entry |<= move these LEFT  |
               -------------------------------------------------
               xxxxx|ino rlen nlen NAME |yyy|...|zzz          |
               -----|-------------------|------ size ---------
                    dp                  cp
           After:
               -----|---- after move LEFT -------------------
               xxxxx|yyy|...|zzz (rec_len += rlen)           |
               -------------------------------------------------

            How to move trailing entries LEFT? Hint: memcpy(dp, cp, size);
          }
```

11.8.5 rmdir 的实现

本节的编程任务是实现 rmdir 函数并将 rmdir 命令添加到文件系统中。编译并运行得到的程序来演示 rmdir 操作。图 11.10 为得到的文件系统的样本输出。该图显示了删除目录的详细步骤。

link 命令

```
link old_file  new_file
```

创建一个从 new_file 到 old_file 的硬链接。硬链接只能应用于普通文件，而不能应用于目录，因为链接到目录可能会在文件系统名称空间中创建循环。硬链接文件共享同一个索引节点。因此，它们必须在同一个设备上。link 算法如下。

11.8.6 link 算法

```
/******************** Algorithm of link *******************/
(1). // verify old_file exists and is not a DIR;
       oino = getino(old_file);
       omip = iget(dev, oino);
       check omip->INODE file type (must not be DIR).
(2). // new_file must not exist yet:
       getino(new_file) must return 0;
(3). creat new_file with the same inode number of old_file:
       parent = dirname(new_file);  child = basename(new_file);
```

```
        pino = getino(parent);
        pmip = iget(dev, pino);
        // creat entry in new parent DIR with same inode number of old_file
        enter_name(pmip, oino, child);
(4). omip->INODE.i_links_count++;  // inc INODE's links_count by 1
     omip->dirty = 1;              // for write back by iput(omip)
     iput(omip);
     iput(pmip);
```

```
fs_init()
mount_root
checking EXT2 FS : OK
bmp = 8 imap = 9 iblock = 10
mydisk mounted on / OK
root refCount = 3
input command : [ls|cd|pwd|mkdir|creat|rmdir|quit] ls
i_block[0] = 33
drwxr-xr-x   5   0   0  Oct 22 10:39    1024    .
drwxr-xr-x   5   0   0  Oct 22 10:39    1024    ..
drwx------   2   0   0  Oct 22 10:39   12288    lost+found
drwxr-xr-x   2   0   0  Oct 22 10:39    1024    dir1
drwxr-xr-x   2   0   0  Oct 22 10:39    1024    dir2
i_block[1] = 0
input command : [ls|cd|pwd|mkdir|creat|rmdir|quit] rmdir dir1
getino: pathname=dir1
tokenize dir1
dir1
===========================================
getino: i=0 name[0]=dir1
search for dir1 in MINODE = [3, 2]
search: i=0  i_block[0]=33
   i_number rec_len name_len    name
       2      12       1        .
       2      12       2        ..
      11      20      10        lost+found
      12      12       4        dir1
found dir1 : ino = 12
[3 12] refCount=1
last entry=[13 968] last entry=[13 980]
input command : [ls|cd|pwd|mkdir|creat|rmdir|quit] ls
i_block[0] = 33
drwxr-xr-x   4   0   0  Oct 22 10:39    1024    .
drwxr-xr-x   4   0   0  Oct 22 10:39    1024    ..
drwx------   2   0   0  Oct 22 10:39   12288    lost+found
drwxr-xr-x   2   0   0  Oct 22 10:39    1024    dir2
i_block[1] = 0
input command : [ls|cd|pwd|mkdir|creat|rmdir|quit] █
```

图 11.10　#3 项目的样本输出

我们通过一个例子来说明链接操作。假定：

link /a/b/c　/x/y/z

下图显示了 link 操作的结果。它将条目 z 添加到 /x/y 的数据块中。z 的索引节点号与 a/b/c 的索引节点号相同，因此它们共享同一个索引节点，其 links_count 递增 1。

```
        link    /a/b/c                          /x/y/z
                /a/b/   datablock               /x/y    datablock
-------------------------        ----------------------------
|....|ino rlen nlen c|...|        |....|ino rlen nlen z  .... |
-----|-|-------------------       -----|-|--------------------
       |                                |
       INODE <-- same INODE <-----------
       i_links_count=1 <=== increment i_links_count to 2
```

unlink：命令

unlink filename

取消文件链接。它将文件的 links_count 减 1 并从其父目录中删除文件名。当文件的 links_count 为 0 时，通过释放它的数据块和索引节点来真正删除文件。unlink 算法如下。

11.8.7 unlink 算法

```
/************** Algorithm of unlink *************/
(1). get filenmae's minode:
        ino = getino(filename);
        mip = iget(dev, ino);
        check it's a REG or symbolic LNK file; can not be a DIR
(2). // remove name entry from parent DIR's data block:
        parent = dirname(filename); child = basename(filename);
        pino = getino(parent);
        pimp = iget(dev, pino);
        rm_child(pmip, ino, child);
        pmip->dirty = 1;
        iput(pmip);
(3). // decrement INODE's link_count by 1
         mip->INODE.i_links_count--;
(4).     if (mip->INODE.i_links_count > 0)
             mip->dirty = 1;  // for write INODE back to disk
(5).     else{  // if links_count = 0: remove filename
             deallocate all data blocks in INODE;
             deallocate INODE;
         }
         iput(mip);            // release mip
```

symlink 命令

symlink old_file new_file

创建一个从 new_file 到 old_file 的符号链接。符号链接与硬链接不同，它可以链接到任何对象，包括目录，甚至不在同一个设备上的文件。symlink 算法如下。

11.8.8 symlink 算法

```
/********* Algorithm of symlink old_file new_file *********/
(1). check: old_file must exist and new_file not yet exist;
(2). creat new_file; change new_file to LNK type;
(3). // assume length of old_file name <= 60 chars
        store old_file name in newfile's INODE.i_block[ ] area.
        set file size to length of old_file name
        mark new_file's minode dirty;
        iput(new_file's minode);
(4). mark new_file parent minode dirty;
     iput(new_file's parent minode);
```

readlink 函数

int readlink(file, buffer)

读取符号文件的目标文件名并返回目标文件名的长度。readlink() 算法如下。

11.8.9　readlink 算法

```
/************* Algorithm of readlink (file, buffer) *************/
(1). get file's INODE in memory; verify it's a LNK file
(2). copy target filename from INODE.i_block[ ] into buffer;
(3). return file size;
```

11.8.10　其他 1 级函数

其他 1 级函数包括访问、chmod、chown、更改文件的时间字段等。所有这些函数的操作方式均相同。

```
(1). get the in-memory INODE of a file by
       ino = getino(pathname);
       mip = iget(dev,ino);
(2). get information from INODE or modify the INODE;
(3). if INODE is modified, set mip->dirty to zonzero for write back;
(4). iput(mip);
```

其他 1 级操作示例

（1）chmod oct 文件名：将文件名的权限位更改为八进制值。

（2）utime 文件名：将文件的访问时间更改为当前时间。

11.8.11　编程项目 1：1 级文件系统的实现

编程项目 1 可完成 1 级文件系统的实现并演示文件系统。图 11.11 显示了运行 1 级文件系统实现的样本输出。ls 命令表明 hi 是指向目录 dir1 的符号链接。

```
root@wang:~/abc/360/F17/TEST4# a.out
checking EXT2 FS ....OK
bmp=8 imap=9 inode_start = 10
init()
mount_root()
mydisk mounted on / OK
creating P0 as running process
input command: [ls|cd|pwd|mkdir|creat|rmdir|link|unlink|symlink|readlink|chmod|utime|quit] ls
cmd=ls path= param=
i_block[0] = 33
drwxr-xr-x   3   0   0  Apr 21 19:50    1024    .
drwxr-xr-x   3   0   0  Apr 21 19:50    1024    ..
drwx------   2   0   0  Apr 21 19:50   12288    lost+found
drwxr-xr-x   2   0   0  Apr 21 19:50    1024    dir2
drwxr-xr-x   2   0   0  Apr 26 16:03    1024    dir1
-rw-r--r--   1   0   0  Apr 26 16:03       0    file1
lrwxrwxrwx   1   0   0  Dec 23 07:18       4    hi -> dir1
i_block[1] = 0
input command: [ls|cd|pwd|mkdir|creat|rmdir|link|unlink|symlink|readlink|chmod|utime|quit] █
```

图 11.11　1 级文件系统的样本输出

11.9　2 级文件系统函数

2 级文件系统实现了文件内容的读 / 写操作。它由以下函数组成：open、close、lseek、read、write、opendir 和 readdir。

11.9.1 open 算法

在 Unix/Linux 中，系统调用

```
int open(char *filename, int flags);
```

打开一个文件进行读或写，标记是 O_RDONLY、O_WRONLY、O_RDWR 其中之一，可与 O_CREAT、O_APPEND、O_TRUNC 标记逐位进行 or 组合。这些符号常数在 fcntl.h 文件中定义。如果成功，open() 会返回一个文件描述符（编号），用于后续系统调用中，如 read()、write()、lseek() 和 close() 等。为简单起见，我们将假设参数标记 0|1|2|3 或 RD|WR|RW|AP 分别表示读 | 写 | 读写 | 追加。open() 算法如下。

```
/******************** Algorithm of open *********************/
(1). get file's minode:
       ino = getino(filename);
       if (ino==0){  // if file does not exist
           creat(filename);           // creat it first, then
           ino = getino(filename);    // get its ino
       }
       mip = iget(dev, ino);

(2). allocate an openTable entry OFT; initialize OFT entries:
       mode = 0(RD) or 1(WR) or 2(RW) or 3(APPEND)
       minodePtr = mip;               // point to file's minode
       refCount = 1;
       set offset = 0 for RD|WR|RW; set to file size for APPEND mode;

(3). Search for the first FREE fd[index] entry with the lowest index in PROC;
       fd[index] = &OFT;           // fd entry points to the openTable entry;

(4). return index as file descriptor;
```

图 11.12 所示为 open 创建的数据结构。图中，（1）是调用 open() 的进程 PROC 结构体。返回的文件描述符 fd 是 PROC 结构体中 fd[] 数组的索引。fd[fd] 的内容指向 OFT，OFT 指向文件的 minode。OFT 的 refCount 表示共享同一个打开文件实例的进程数量。当某个进程打开文件时，OFT 中的 refCount 会被设置为 1。当某个进程复刻子进程时，子进程继承父进程所有打开的文件描述符，将每个共享 OFT 的 refCount 增加 1。当某个进程关闭文件描述符时，它会将 OFT.refCount 减小 1。当 OFT.refCount 为 0 时，文件的 minode 会被释放，OFT 也会被释放。OFT 的偏移量是指向文件中当前读 / 写字节位置的概念指针。在读 | 写 | 读写模式下，会被初始化为 0；在追加模式下，会被初始化为文件大小。

11.9.2 lseek

在 Linux 中，系统调用

```
lseek(fd, position, whence); // whence=SEEK_SET or SEEK_CUR
```

将打开的文件描述符在 OFT 中的偏移量设置为从文件开头（SEEK_SET）或当前位置（SEEK_CUR）开始的字节位置。为简单起见，我们假设新位置始终从文件开头开始。设置完成后，下一个读 / 写将从当前偏移位置开始。lseek 算法很简单。它只需要检查请求的位置值是否在 [0, fileSize-1] 范围内。我们将 lseek 的实现留作练习。

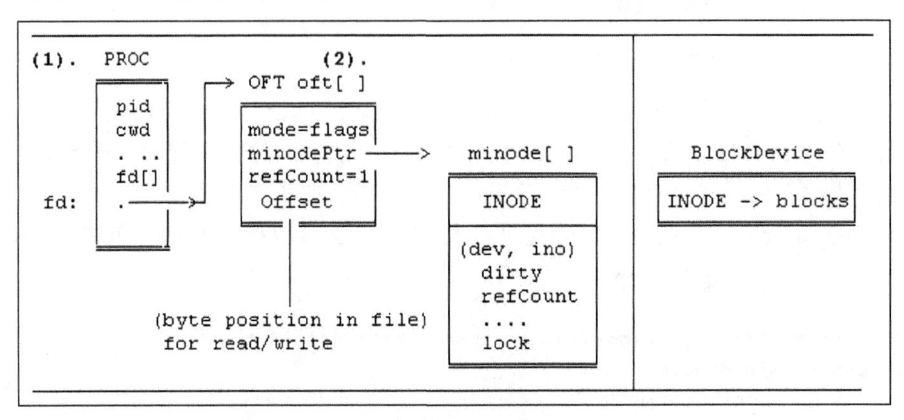

图 11.12　open 的数据结构

练习 11.12：编写 lseek 函数 int lseek(int fd, int position) 的 C 代码。

11.9.3　close 算法

close(int fd) 操作可关闭文件描述符。close 算法如下。

```
/***************** Algorithm of close ***************/
(1). check fd is a valid opened file descriptor;
(2). if (PROC's fd[fd] != 0){          // points to an OFT
  .      OFT.refCount--;                // dec OFT's refCount by 1
         if (refCount == 0)             // if last process using this OFT
             iput(OFT.minodePtr);       // release minode
     }
(4). PROC.fd[fd] = 0;                   // clear PROC's fd[fd] to 0
```

11.9.4　读取普通文件

在 Unix/Linux 中，系统调用

```
int read(int fd, char *buf, int nbytes);
```

将 n 个字节从打开的文件描述符读入用户空间中的缓冲区。read 系统调用被路由到操作系统内核中的 read 函数。普通文件的读取算法如下

```
/********* Algorithm of read(int fd, char *buf, int nbytes) *********/
(1). count = 0;              // number of bytes read
     offset = OFT.offset;    // byte offset in file to READ
     compute bytes available in file: avil = fileSize - offset;
(2). while (nbytes && avil){
         compute logical block: lbk   = offset  /  BLKSIZE;
         start byte in block:         start = offset % BLKSIZE;
(3).     convert logical block number, lbk, to physical block number, blk,
         through INODE.i_block[ ] array;
(4).     get_block(dev, blk, kbuf);   // read blk into char kbuf[BLKSIZE];
         char *cp = kbuf + start;
```

```
        remain = BLKSIZE - start;
(5).    while (remain){   // copy bytes from kbuf[ ] to buf[ ]
           *buf++ = *cp++;
           offset++; count++;          // inc offset, count;
           remain--; avil--; nbytes--;  // dec remain, avail, nbytes;
           if (nbytes==0 || avil==0)
               break;
        } // end of while(remain)
    }     //  end of while(nbytes && avil)
(6). return count;
```

图 11.13 对 read() 算法进行了最详细的解释。假设 fd 已打开可进行读取。OFT 中的偏移量指向我们想从文件中读取 n 个字节的当前字节位置。对于内核中的文件系统而言，文件只是一系列编号为 0 到 fileSize−1 的连续字节。如图 11.13 所示，当前字节位置（偏移量）位于逻辑块中。

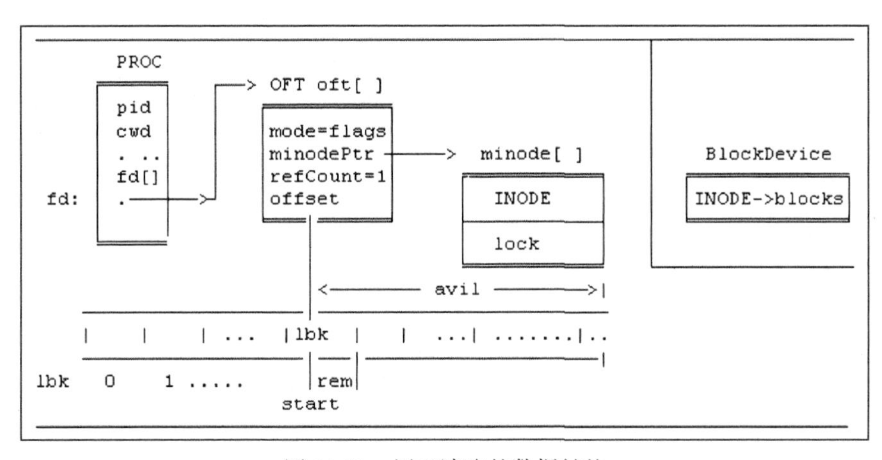

图 11.13　用于读取的数据结构

```
lbk = offset / BLKSIZE,
```

开始读取的字节是：

```
start = offset % BLKSIZE
```

逻辑块中剩余的字节数为：

```
remain = BLKSIZE - start.
```

此时，文件有

```
avil = fileSize - offset
```

个字节可以读取。该字节数可用于读取算法。

对于小的 EXT2 文件系统，块大小为 1KB，并且文件最多有两个间接块。在读取时，将逻辑块转换成物理块的算法如下：

```
/* Algorithm of Converting Logical Blocks to Physical Blocks */
 int map(INODE, lbk){                  // convert lbk to blk via INODE
      if (lbk < 12)                    // direct blocks
         blk = INODE.i_block[lbk];
```

```
        else if (12 <= lbk < 12+256){      // indirect blocks
            read INODE.i_block[12] into int ibuf[256];
            blk = ibuf[lbk-12];
        }
        else{   // doube indirect blocks; see Exercise 11.13 below.
         }
         return blk;
  }
```

练习 11.13：完成将双重间接块转换成物理块的算法。提示：邮差算法。

练习 11.14：为简单清晰起见，read 算法的第（5）步一次传输一个字节，在每次字节传输时都要更新控制变量，这样不是很有效率。通过一次传输最大的数据块来优化代码。

11.9.5　写普通文件

在 Unix/Linux 中，系统调用

```
int write(int fd, char buf[ ], int nbytes);
```

将 n 个字节从用户空间缓冲区写入打开的文件描述符，并返回写入的实际字节数。write 系统调用被路由到操作系统内核中的 write 函数。普通文件的写入算法如下：

```
/******* Algorithm of write(int fd, char *buf, int nbytes) *******/
(1). count = 0;          // number of bytes written
(2). while (nbytes){
       compute logical block: lbk = oftp->offset  / BLOCK_SIZE;
       compute start byte:      start = oftp->offset % BLOCK_SIZE;
(3).   convert lbk to physical block number, blk, through the i_block[ ] array;
(4).   read_block(dev, blk, kbuf); // read blk into kbuf[BLKSIZE];
       char *cp = kbuf + start;  remain = BLKSIZE - start;
(5)    while (remain){                  // copy bytes from buf[ ] to kbuf[ ]
          *cp++ = *buf++;
          offset++;  count++;        // inc offset, count;
          remain --; nbytes--;       // dec remain, nbytes;
          if (offset > fileSize)  fileSize++; // inc file size
          if (nbytes <= 0) break;
        } // end while(remain)
(6).    write_block(dev, blk, kbuf);
     } // end while(nbytes)
(7). set minode dirty = 1;  // mark minode dirty for iput() when fd is closed
     return count;
```

图 11.14 对写入算法进行了最详细的解释。图中，OFT 中的偏移量是指要写入文件的当前字节位置，位于逻辑块上。它使用偏移量来计算逻辑块号、lbk、起始字节位置和逻辑块中剩余的字节数。然后，通过文件的 INODE.i_block 数组将逻辑块转换为物理块。然后，它将物理块读入 kbuf，并将数据从 buf 传输到 kbuf，如果偏移量超过当前文件大小，则会增加文件大小。将数据传输到 kbuf 后，它会将缓冲区写回磁盘。当全部 n 个字节都写入完成时，写入算法停止。除非写入失败，否则返回值是 n 个字节。

将逻辑块转换为物理块进行写操作的算法与读操作类似，但有以下不同之处。在写入期间，预期数据块可能不存在。如果不存在一个直接块，则必须在 INODE 中分配并记录。如果间接块 i_block[12] 不存在，则必须进行分配并初始化为 0。如果间接数据块不存在，则必须在间接数据块中分配和记录。同样，如果双重间接块 i_block[13] 不存在，则必须进行分配并初始化为 0。如果双重间接块中的条目不存在，则必须进行分配并初始化为 0。如果双重间接块不存在，则必须进行分配并记录在一个双间重接块条目中，等等。

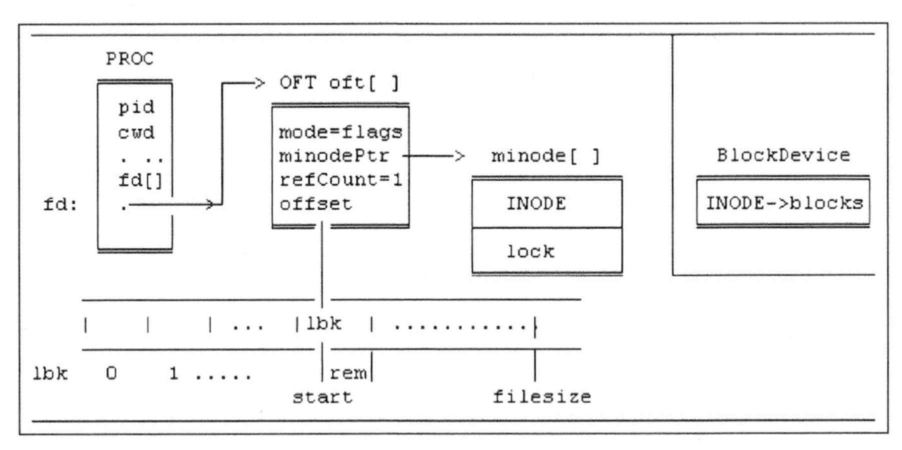

图 11.14 用于写入的数据结构

练习 11.15：在写入算法中，以写入模式打开的文件可能根本没有数据块。将逻辑块号转换为物理块的算法完成，包括直接块、间接块和双重间接块。

练习 11.16：为简单清晰起见，write 算法的第（5）步一次传输一个字节，在每次字节传输时都要更新控制变量。通过一次传输最大的数据块来优化代码。

练习 11.17：使用 open() 和 read()，实现 cat filename 命令，显示文件内容。

练习 11.18：使用 open()、read() 和 write()，实现 cp src dest 命令，将文件 src 复制到 dest 文件中。

练习 11.19：实现命令函数 mv file1 file2。

11.9.6 opendir-readdir

Unix 把所有目录都看作一个文件。因此，我们应该能够像打开普通文件一样打开一个目录进行读取。从技术角度来看，不需要一组单独的 opendir() 和 readdir() 函数。但是，不同的 Unix 系统可能有不同的文件系统。用户可能很难解释目录文件的内容。由于这个原因，POSIX 指定了 opendir 和 readdir 操作，它们独立于文件系统。对 opendir 的支持并不重要，它是同一个 open 系统调用，但是 readdir() 的形式如下：

```
struct dirent *ep = readdir(DIR *dp);
```

在每次调用时返回一个指向 dirent 结构体的指针。这可以在用户空间中作为 I/O 库函数实现。相反，我们将直接实现 opendir() 和 readir()。

```
int opendir(pathaname)
{   return open(pathname, RD|O_DIR); }
```

其中 O_DIR 是将文件作为目录打开的一个位模式。在打开的文件表中，mode 字段包含 O_DIR 位。

```
int readdir(int fd, struct udir *udirp) // struct udir{DIR udir;};
{
```

```
    // same as read() except:
    use the current byte offset in OFT to read the next dir_entry;
    copy the dir_entry into *udirp;
    advance offset by dir_entry's rec_len for reading next dir_entry;
}
```

11.9.7 编程项目 2：2 级文件系统的实现

编程项目 2 可完成 2 级文件系统的实现并演示文件系统中的读、写、cat、cp 和 mv 操作。读者可参考前文的 cat 和 cp 操作章节。关于 mv 操作，请参见习题 12。图 11.15 所示为运行完整 2 级文件系统的样本输出。

```
root@wang:~/abc/360/F17/TEST5# a.out
checking EXT2 FS ....OK
bmp=8 imap=9 inode_start = 10
init()
mount_root()
root refCount = 1
creating P0 as running process
root refCount = 2
input command: [ls|cd|pwd|mkdir|creat|rmdir|link|symlink|unlink
               |open|close|lseek|read|write|cat|cp|mv|quit] : ls
cmd=ls path= param=
i_block[0] = 33
drwxr-xr-x   3   0   0  Nov  9 12:54    1024    .
drwxr-xr-x   3   0   0  Nov  9 12:54    1024    ..
drwx------   2   0   0  Nov  3 09:07   12288    lost+found
-rw-r--r--   1   0   0  Nov  9 12:54      42    f1
drwxr-xr-x   2   0   0  Apr 21 18:49    1024    dir2
-rw-r--r--   1   0   0  Apr 10 08:48       0    file2
i_block[1] = 0
input command: [ls|cd|pwd|mkdir|creat|rmdir|link|symlink|unlink
               |open|close|lseek|read|write|cat|cp|mv|quit] : 
```

图 11.15　2 级文件系统

11.10 3 级文件系统

3 级文件系统支持文件系统的挂载、卸载和文件保护。

11.10.1 挂载算法

挂载操作命令

```
mount filesys mount_point
```

可将某个文件系统挂载到 mount_point 目录上。它允许文件系统包含其他文件系统作为现有文件系统的一部分。挂载中使用的数据结构是挂载表和 mount_point 目录的内存 minode。挂载算法如下。

```
/********* Algorithm of mount **********/
 1. If no parameter, display current mounted file systems;
 2. Check whether filesys is already mounted:
    The MOUNT table entries contain mounted file system (device) names
    and their mounting points. Reject if the device is already mounted.
    If not, allocate a free MOUNT table entry.
 3. Open the filesys virtual disk (under Linux) for RW; use (Linux) file
    descriptor as new dev. Read filesys superblock to verify it is an EXT2 FS.
```

```
4. Find the ino, and then the minode of mount_point:
      ino = getino(pathname);     // get ino:
      mip = iget(dev, ino);       // load its inode into memory;
5. Check mount_point is a DIR and not busy, e.g. not someone's CWD.
6. Record new dev and filesys name in the MOUNT table entry, store its
   ninodes, nblocks, bmap, imap and inodes start block, etc. for quick
   access.
7. Mark mount_point minode as mounted on (mounted flag = 1) and let it point
   at the MOUNT table entry, which points back to the mount_point minode.
```

11.10.2 卸载算法

卸载文件系统操作可卸载已挂载的文件系统。它将挂载的文件系统与挂载点分开，其中文件系统可以是虚拟的 diak 名称或挂载点目录名称。卸载算法如下。

```
/******************* Algorithm of umount ***********************/
  1. Search the MOUNT table to check filesys is indeed mounted.
  2. Check whether any file is active in the mounted filesys; If so, reject;
  3. Find the mount_point in-memory inode, which should be in memory while
     it's mounted on. Reset the minode's mounted flag to 0; then iput() the
     minode.
```

若要检查挂载的文件系统是否包含正在使用的文件，可搜索内存 minode 中是否有任何条目的设备与挂载文件系统的设备号匹配。

11.10.3 交叉挂载点

虽然可以轻松实现挂载和卸载，但是会有一定的影响。对于挂载，我们必须修改 getino(pathname) 函数，来支持交叉挂载点。假设某文件系统 newfs 已被挂载到目录 /a/b/c/ 上。当遍历一个路径名时，两个方向的挂载点可能会出现交叉。

（1）**向下遍历**：当遍历路径名 /a/b/c/x 时，一旦到达 /a/b/c 的 minode，我们就能看到 minode 已经被挂载（挂载标志 =1）。我们不是在 /a/b/c 的索引节点中搜索 x，而是必须要：

- 跟随 minode 的 mntPtr 指针来定位挂载表条目。
- 从挂载表的设备号中，将其根索引节点（ino=2）放入内存。
- 然后，继续在挂载设备的根索引节点下搜索 x。

（2）**向上遍历**：假设我们在目录 /a/b/c/x/ 上，然后向上遍历，例如 cd ../../，将会与挂载点 /a/b/c 交叉。当到达挂载文件系统的根索引节点时，我们就能看到它是一个根目录（ino=2），但是它的设备号与实根的设备号不同，因此它现在还不是实根。我们可以使用它的设备号，找到它的挂载表条目，它指向 /a/b/c/ 的 minode。然后，切换到 /a/b/c/ 的 minode，继续向上遍历。因此，交叉挂载点就像一只猴子或松鼠从一棵树跳到另一棵树上，然后又跳了回来。

由于交叉挂载点会更改设备号，因此一个全局设备号是不够的。我们必须要将 getino() 函数修改为

```
int getino(char *pathname, int *dev)
```

并将 getino() 调用修改为

```
int dev;                    // local in function that calls getino()
if (pathnme[0]=='/')
   dev = root->dev;         // absolute pathname
```

```
else
   dev = running->cwd->dev; // relative pathname

int ino = getino(pathname, &dev); // pass &dev as an extra parameter
```

在修改后的 getino() 函数中，当路径名与挂载点交叉时，要修改设备号。因此，修改后的 getino() 函数实际上返回的是路径名的（dev, ino），其中 dev 表示最终设备号。

11.10.4 文件保护

在 Unix/Linux 中，可通过文件索引节点中的权限位实现文件保护。每个文件的索引节点都有一个 i_mode 字段，其中下面的 9 位是权限。9 个权限位为：

```
owner    group    other
-------  -------  -------
r w x    r w x    r w x
```

前 3 位适用于文件所有人，中间 3 位适用于与所有人同一组的用户，最后 3 位适用于其他所有用户。对于目录，x 位表示某进程是否可进入目录。每个进程都有一个 uid 和 gid。当某进程试图访问某个文件时，文件系统会根据文件的权限位检查进程 uid 和 gid，以确定它是否能以目标操作模式访问文件。如果该进程没有适当的权限，访问会被拒绝。为简单起见，我们可忽略进程 gid，只使用进程 uid 来检查访问权限。

11.10.5 实际 uid 和有效 uid

在 Unix/Linux 中，每个进程都有一个**实际** uid 和一个**有效** uid。文件系统通过进程的有效 uid 检查进程的访问权限。在正常情况下，进程的有效 uid 和实际 uid 是相同的。当某进程执行 setuid 程序时，该程序打开文件 i_mode 字段中的 setuid 位，该进程的有效 uid 就变成了该程序的 uid。在执行 setuid 程序时，进程实际上成了程序文件的所有者。例如，当进程执行邮件程序（超级用户所有的 setuid 程序），它会写入另一个用户的邮件文件中。当进程执行完 setuid 程序时，它会返回到实际 uid。为简单起见，我们仍将忽略有效 uid。

11.10.6 文件锁定

文件锁定机制允许进程对一个文件或文件的某些部分设置文件锁，以防止在更新文件时出现竞态条件。文件锁可共享（允许同步读取），也可独占（执行独占写入）。文件锁既可以是强制性的，也可以是建议性的。例如，Linux 既支持共享文件锁，也支持独占文件锁，但文件锁定只是建议性的。在 Linux 中，文件锁可通过 fcntl() 系统调用设置，也可通过 flock() 系统调用操作。为简单起见，我们假设一种非常简单的文件锁定。当一个进程试图打开一个文件时，将会检查目标操作模式的兼容性。唯一兼容模式是读模式。如果已经为更新模式打开了一个文件，即写 | 读写 | 追加，则该文件无法再次打开。但是，这并不会阻止相关进程（例如父进程和子进程）修改父进程打开的同一文件，而且在 Unix/Linux 中同样如此。在这种情况下，文件系统只能保证每个写操作是原子操作，但不能保证进程的写入顺序，写入顺序取决于进程调度。

11.10.7 编程项目 3：整个文件系统的实现

编程项目 3 通过将函数并入 3 级文件系统中来完成文件系统的实现。该文件系统的样本解决方案可从本书的网站下载。如果提出要求，可向讲师提供整个文件系统的源代码。

11.11 文件系统项目的扩展

简单的 EXT2 文件系统使用 1KB 块大小，只有一个磁盘块组。它可以轻松进行以下扩展。

（1）多个组：组描述符的大小为 32 字节。对于 1KB 大小的块，一个块可能包含 1024/32 = 32 组描述符。32 个组的文件系统大小可以扩展为 32*8 = 256MB。

（2）4KB 大小的块：对于 4KB 大小的块和一个组，文件系统大小应为 4*8 = 32MB。对于一个组描述符块，文件系统可能有 128 个组，可将文件系统大小扩展到 128*32 = 4GB。对于 2 个组描述符块，文件系统大小为 8GB 等。大多数扩展都很简单，适合用于编程项目。

（3）管道文件：管道可实现为普通文件，这些文件遵循管道的读 / 写协议。此方案的优点是：它统一了管道和文件索引节点，并允许可被不相关进程使用的命名管道。为支持快速读 / 写操作，管道内容应在内存中，比如在 RAMdisk 中。必要时，读者可将命名管道实现为 FIFO 文件。

（4）I/O 缓冲：在编程项目中，每个磁盘块都是直接读写的。这会产生过多的物理磁盘 I/O 操作。为提高效率，实际文件系统通常使用一系列 I/O 缓冲区作为磁盘块的缓存内存。文件系统的 I/O 缓冲将会在第 12 章中讨论，但是可以把它合并到文件系统项目中。

11.12 习题

1. 编写一个 C 程序来显示设备上 EXT2 文件系统的块组描述符。
2. 修改 imap.c 程序，以 char map 的形式打印索引节点位图，即对于每个位而言，如果该位是 0，则打印“0”，如果该位是 1，则打印“1”。
3. 编写一个 C 语言程序，同样以 char map 格式来显示 EXT2 文件系统的块位图。
4. 编写一个 C 语言程序来打印目录的 dir_entries。
5. 虚拟磁盘可作为循环设备挂载在 Linux 下，如：

```
sudo mount -o loop mydisk /mnt      # mount mydisk to /mnt
mkdir /mnt/newdir                   # create new dirs and copy files to /mnt
sudo umount /mnt                    # umount when finished
```

   在 mydisk 上再次运行 dir.c 程序，验证新创建的目录和文件是否仍然存在。
6. 已知一个指向目录索引节点的 INODE 指针，编写

```
int  search(INODE *dir, char *name)
```

   函数，搜索具有给定名称的 dir_entry。如果找到，则返回其索引节点编号，否则返回 0。
7. 已知一个包含 EXT2 文件系统和路径名（例如 /a/b/c/d）的设备，编写一个 C 函数：

```
INODE *path2inode(int fd, char *pathname)  // assume fd=file descriptor
```

   可返回一个指向文件索引节点的 INODE 指针；如果文件不可访问，则返回 0。
8. 假设：一个小的 EXT2 文件系统的块大小为 1KB，只有一个块组。
   （1）证明一个文件可能只有一些双重间接块，但是没有三重间接块。
   （2）已知一个文件的路径名，例如 /a/b/c/d，编写一个 C 程序 showblock，它可打印文件的所有磁盘

块号。提示：如果有双重间接块，可以使用邮差算法。

9. 在编程项目 #1 中，当它的 refCount 达到 0 时，内存 minode 将被释放为 FREE。设计并实现一个使用 minode[] 区域作为内存 INODE 缓存的方案。一旦 INODE 被加载到 minode 槽中，即使没有进程在主动使用它，也要尽可能长时间地将它留在内存中。当进程在内存中发现一个需要的索引节点时，会把它算作缓存命中。否则，当作缓存未命中。修改 iget(dev, ino) 函数来计算缓存命中的次数。
10. 将 pwd 算法的第（2）步实现为一个实用程序函数：

```
int get_myino(MINODE *mip, int *parent_ino)
```

会在 parent_ino 中返回 . 和 .. 的索引节点号。

11. 将 pwd 算法的第（4）步实现为一个实用程序函数：

```
int get_myname(MINODE *parent_minode, int my_ino, char *my_name)
```

会用 my_ino 标识在父目录中返回 dir_entry 的名称字符串。

12. 实现 mv file1 file2 操作，该操作将 file 1 移动到 file 2 中。对于同一设备上的文件，mv 应该将 file 1 重命名为 file 2，而不复制文件内容。提示：链接 - 取消链接。如果文件在不同的设备上，如何实现 mv？

参考文献

Card, R., Theodore Ts'o, T., Stephen Tweedie, S., "Design and Implementation of the Second Extended Filesystem", web.mit.edu/tytso/www/linux/ext2intro.html, 1995

Cao, M., Bhattacharya, S, Tso, T., "Ext4: The Next Generation of Ext2/3 File system", IBM Linux Technology Center, 2007

EXT2: http://www.nongnu.org/ext2-doc/ext2.html, 2001

EXT3: http://jamesthornton.com/hotlist/linux-filesystems/ext3-journal, 2015

块设备 I/O 和缓冲区管理

摘要

本章讨论了块设备 I/O 和缓冲区管理；解释了块设备 I/O 的原理和 I/O 缓冲的优点；论述了 Unix 的缓冲区管理算法，并指出了其不足之处；还利用信号量设计了新的缓冲区管理算法，以提高 I/O 缓冲区的缓存效率和性能；表明了简单的 PV 算法易于实现，缓存效果好，不存在死锁和饥饿问题；还提出了一个比较 Unix 缓冲区管理算法和 PV 算法性能的编程方案。编程项目还可以帮助读者更好地理解文件系统中的 I/O 操作。

12.1 块设备 I/O 缓冲区

在第 11 章中，我们展示了读写普通文件的算法。这些算法依赖于两个关键操作，即 get_block 和 put_block，这两个操作将磁盘块读写到内存缓冲区中。由于与内存访问相比，磁盘 I/O 速度较慢，所以不希望在每次执行读写文件操作时都执行磁盘 I/O。因此，大多数文件系统使用 I/O 缓冲来减少进出存储设备的物理 I/O 数量。合理设计的 I/O 缓冲方案可显著提高文件 I/O 效率并增加系统吞吐量。

I/O 缓冲的基本原理非常简单。文件系统使用一系列 I/O 缓冲区作为块设备的缓存内存。当进程试图读取（dev, blk）标识的磁盘块时，它首先在缓冲区缓存中搜索分配给磁盘块的缓冲区。如果该缓冲区存在并且包含有效数据，那么它只需从缓冲区中读取数据，而无须再次从磁盘中读取数据块。如果该缓冲区不存在，它会为磁盘块分配一个缓冲区，将数据从磁盘读入缓冲区，然后从缓冲区读取数据。当某个块被读入时，该缓冲区将被保存在缓冲区缓存中，以供任意进程对同一个块的下一次读 / 写请求使用。同样，当进程写入磁盘块时，它首先会获取一个分配给该块的缓冲区。然后，它将数据写入缓冲区，将缓冲区标记为脏，以延迟写入，并将其释放到缓冲区缓存中。由于脏缓冲区包含有效的数据，因此可以使用它来满足对同一块的后续读 / 写请求，而不会引起实际磁盘 I/O。脏缓冲区只有在被重新分配到不同的块时才会写入磁盘。

在讨论缓冲区管理算法之前，我们先来介绍以下术语。在 read_file/write_file 中，我们假设它们从内存中的一个专用缓冲区进行读 / 写。对于 I/O 缓冲，将从缓冲区缓存中动态分配缓冲区。假设 BUFFER 是缓冲区（见下文定义）的结构类型，而且 getblk(dev, blk) 从缓冲区缓存中分配一个指定给（dev, blk）的缓冲区。定义一个 bread(dev, blk) 函数，它会返回一个包含有效数据的缓冲区（指针）。

```
BUFFER *bread(dev,blk) // return a buffer containing valid data
{
   BUFFER *bp = getblk(dev,blk); // get a buffer for (dev,blk)
   if (bp data valid)
      return bp;
   bp->opcode = READ;            // issue READ operation
   start_io(bp);                 // start I/O on device
   wait for I/O completion;
```

```
   return bp;
}
```

从缓冲区读取数据后，进程通过 brelse(bp) 将缓冲区释放回缓冲区缓存。同理，定义一个 write_block(dev, blk, data) 函数，如：

```
write_block(dev, blk, data)      // write data from U space
{
   BUFFER *bp = bread(dev,blk);  // read in the disk block first
   write data to bp;
   (synchronous write)? bwrite(bp) : dwrite(bp);
}
```

其中 bwrite(bp) 表示同步写入，dwrite(bp) 表示延迟写入，如下文所示。

```
---------------------------------------------------------------------
 bwrite(BUFFER *bp){              | dwrite(BUFFER *bp){
    bp->opcode = WRITE;           |   mark bp dirty for delay_write;
    start_io(bp);                 |   brelse(bp); // release bp
    wait for I/O completion;      | }
    brelse(bp); // release bp     |
 }                                |
---------------------------------------------------------------------
```

同步写入操作等待写操作完成。它用于顺序块或可移动块设备，如 USB 驱动器。对于随机访问设备，例如硬盘，所有的写操作都是延迟写操作。在延迟写操作中，dwrite(bp) 将缓冲区标记为脏，并将其释放到缓冲区缓存中。由于脏缓冲区包含有效数据，因此可用来满足同一个块的后续读 / 写请求。这不仅减少了物理磁盘 I/O 的数量，而且提高了缓冲区缓存的效果。脏缓冲区只有在被重新分配到不同的磁盘块时才会被写入磁盘，此时缓冲区将被以下代码写入：

```
awrite(BUFFER *bp)
{
    bp->opcode = ASYNC;      // for ASYNC write;
    start_io(bp);
 }
```

awrite() 会调用 start_io() 在缓冲区开始 I/O 操作，但是不会等待操作完成。当异步（ASYNC）写操作完成后，磁盘中断处理程序将释放缓冲区。

物理块设备 I/O：每个设备都有一个 I/O 队列，其中包含等待 I/O 操作的缓冲区。缓冲区上的 start_io() 操作如下：

```
start_io(BUFFER *bp)
{
    enter bp into device I/O queue;
    if (bp is first buffer in I/O queue)
       issue I/O command for bp to device;
 }
```

当 I/O 操作完成后，设备中断处理程序会完成当前缓冲区上的 I/O 操作，并启动 I/O 队列中的下一个缓冲区的 I/O（如果队列不为空）。设备中断处理程序的算法如下：

```
InterruptHandler()
{
   bp = dequeue(device I/O queue); // bp = remove head of I/O queue
  (bp->opcode == ASYNC)? brelse(bp) : unblock process on bp;
   if (!empty(device I/O queue))
      issue I/O command for first bp in I/O queue;
}
```

12.2 Unix I/O 缓冲区管理算法

Unix I/O 缓冲区管理算法最早出现在第 6 版 Unix 中（Ritchie 和 Thompson 1978；Lion 1996）。Bach 著作的第 3 章对该算法进行了详细论述（Bach 1990）。Unix 缓冲区管理子系统由以下几部分组成。

（1）I/O 缓冲区：内核中的一系列 NBUF 缓冲区用作缓冲区缓存。每个缓冲区用一个结构体表示。

```
typdef struct buf{
  struct buf *next_free;      // freelist pointer
  struct buf *next_dev;       // dev_list pointer
  int dev,blk;                // assigned disk block;
  int opcode;                 // READ|WRITE
  int dirty;                  // buffer data modified
  int async;                  // ASYNC write flag
  int valid;                  // buffer data valid
  int busy;                   // buffer is in use
  int wanted;                 // some process needs this buffer
  struct semaphore lock=1;    // buffer locking semaphore; value=1
  struct semaphore iodone=0;  // for process to wait for I/O completion;
  char buf[BLKSIZE];          // block data area
} BUFFER;
BUFFER buf[NBUF], *freelist; // NBUF buffers and free buffer list
```

缓冲区结构体由两部分组成：用于缓冲区管理的缓冲头部分和用于数据块的数据部分。为了保护内核内存，状态字段可以定义为一个位向量，其中每个位表示一个唯一的状态条件。为了便于讨论，这里将它们定义为 int。

（2）设备表：每个块设备用一个设备表结构表示。

```
struct devtab{
  u16  dev;             // major device number
  BUFFER *dev_list;     // device buffer list
  BUFFER *io_queue;     // device I/O queue
} devtab[NDEV];
```

每个设备表都有一个 dev_list，包含当前分配给该设备的 I/O 缓冲区，还有一个 io_queue，包含设备上等待 I/O 操作的缓冲区。I/O 队列的组织方式应确保最佳 I/O 操作。例如，它可以实现各种磁盘调度算法，如电梯算法或线性扫描算法等。为了简单起见，Unix 使用 FIFO I/O 队列。

（3）缓冲区初始化：当系统启动时，所有 I/O 缓冲区都在空闲列表中，所有设备列表和 I/O 队列均为空。

（4）缓冲区列表：当缓冲区分配给（dev, blk）时，它会被插入设备表的 dev_list 中。如果缓冲区当前正在使用，则会将其标记为 BUSY（繁忙）并从空闲列表中删除。繁忙缓冲区也可能会在设备表的I/O队列中。由于一个缓冲区不能同时处于空闲状态和繁忙状态，所以可通过使用相同的next_free指针来维护设备I/O队列。当缓冲区不再繁忙时，它会被释放回空闲列表，但仍保留在dev_list中，以便可能重用。只有在重新分配时，缓冲区才可能从一个dev_list更改到另一个dev_list中。如前文所述，读/写磁盘块可以表示为bread、bwrite和dwrite，它们都要依赖于getblk和brelse。因此，getblk和brelse构成了Unix缓冲区管理方案的核心。getblk和brelse算法如下。

（5）Unix getblk/brelse算法（(Lion 1996；(Bach 1990）第3章）。

```
/* getblk: return a buffer=(dev,blk) for exclusive use */
BUFFER *getblk(dev,blk){
   while(1){
     (1). search dev_list for a bp=(dev, blk);
     (2). if (bp in dev_lst){
             if (bp BUSY){
                set bp WANTED flag;
                sleep(bp);      // wait for bp to be released
                continue;       // retry the algorithm
             }
             /* bp not BUSY */
             take bp out of freelist;
             mark bp BUSY;
             return bp;
           }
     (3). /* bp not in cache; try to get a free buf from freelist */
          if (freelist empty){
              set freelist WANTED flag;
              sleep(freelist); // wait for any free buffer
              continue;        // retry the algorithm
          }
     (4). /* freelist not empty */
          bp = first bp taken out of freelist;
          mark bp BUSY;
          if (bp DIRTY){        // bp is for delayed write
              awrite(bp);       // write bp out ASYNC;
              continue;         // from (1) but not retry
          }
     (5). reassign bp to (dev,blk); // set bp data invalid, etc.
          return bp;
}

/** brelse: releases a buffer as FREE to freelist **/
brelse(BUFFER *bp){
  if (bp WANTED)
     wakeup(bp);        // wakeup ALL proc's sleeping on bp;
  if (freelist WANTED)
     wakeup(freelist); // wakeup ALL proc's sleeping on freelist;
  clear bp and freelist WANTED flags;
  insert bp to (tail of) freelist;
}
```

注意，在文献（Bach 1990）中，缓冲区存放在散列队列中。当缓冲区的数量很大时，散列可以减少搜索时间。当缓冲区的数量很少时，由于额外系统开销，散列实际上可能会增加执行时间。此外，研究（Wang 2002）表明，散列对缓冲区缓存性能几乎没有影响。实际上，我们可以通过简单的散列函数 hash(dev, blk) = dev 将设备列表视为散列队列。这样，将设备列表用作散列队列不会损失通用性。Unix 算法非常简单，易于理解。也许是因为它太过简单，大多数人第一眼看到它时并没有产生深刻的印象。由于不断的重试循环，有些人甚至认为它很不成熟。但是，你越研究它，就会发现它越有意义。这一异常简单而有效的算法证明了 Unix 最初设计者的独创性。下面是关于 Unix 算法的一些具体说明。

（1）数据一致性：为了确保数据一致性，getblk 一定不能给同一个（dev, blk）分配多个缓冲区。这可以通过让进程从休眠状态唤醒后再次执行“重试循环”来实现。读者可以验证分配的每个缓冲区都是唯一的。其次，脏缓冲区在重新分配之前被写出来，这保证了数据的一致性。

（2）缓存效果：缓存效果可通过以下方法实现。释放的缓冲区保留在设备列表中，以便可能重用。标记为延迟写入的缓冲区不会立即产生 I/O，并且可以重用。缓冲区会被释放到空闲列表的末尾，但分配是从空闲列表的前面开始的。这是基于 LRU（最近最少使用）原则，它有助于延长所分配缓冲区的使用期，从而提高它们的缓存效果。

（3）临界区：设备中断处理程序可操作缓冲区列表，例如从设备表的 I/O 队列中删除 bp，更改其状态并调用 brelse(bp)。所以，在 getblk 和 brelse 中，设备中断在这些临界区中会被屏蔽。这些都是隐含的，但没有在算法中表现出来。

Unix 算法的缺点

虽然 Unix 算法非常简单和简洁，但它也有以下缺点。

（1）效率低下：该算法依赖于重试循环。例如，释放缓冲区可能会唤醒两组进程：需要释放的缓冲区的进程，以及只需要空闲缓冲区的进程。由于只有一个进程可以获取释放的缓冲区，所以，其他所有被唤醒的进程必须重新进入休眠状态。从休眠状态唤醒后，每个被唤醒的进程必须从头开始重新执行算法，因为所需的缓冲区可能已经存在。这会导致过多的进程切换。

（2）缓存效果不可预知：在 Unix 算法中，每个释放的缓冲区都可被获取。如果缓冲区由需要空闲缓冲区的进程获取，那么将会重新分配缓冲区，即使有些进程仍然需要当前的缓冲区。

（3）可能会出现饥饿：Unix 算法基于“自由经济”原则，即每个进程都有尝试的机会，但不能保证成功。因此，可能会出现进程饥饿。

（4）该算法使用只适用于单处理器系统的休眠 / 唤醒操作。

12.3　新的 I/O 缓冲区管理算法

在本节中，我们将展示一种用于 I/O 缓冲区管理的新算法。我们将在信号量上使用 P/V 来实现进程同步，而不是使用休眠 / 唤醒。与休眠 / 唤醒相比，信号量的主要优点是：

（1）计数信号量可用来表示可用资源的数量，例如：空闲缓冲区的数量。

（2）当多个进程等待一个资源时，信号量上的 V 操作只会释放一个等待进程，该进程不必重试，因为它保证拥有资源。

这些信号量属性可用于设计更有效的缓冲区管理算法。我们正式对这个问题做以下详细说明。

使用信号量的缓冲区管理算法

假设有一个单处理器内核（一次运行一个进程）。使用计数信号量上的P/V来设计满足以下要求的新的缓冲区管理算法：

（1）保证数据一致性。

（2）良好的缓存效果。

（3）高效率：没有重试循环，没有不必要的进程“唤醒”。

（4）无死锁和饥饿。

注意，仅通过信号量上的P/V来替换Unix算法中的休眠/唤醒并不可取，因为这样会保留所有的重试循环。我们必须重新设计算法来满足所有上述要求，并证明新算法的确优于Unix算法。首先，我们定义以下信号量。

```
BUFFER buf[NBUF];          // NBUF I/O buffers
SEMAPHORE free = NBUF;  // counting semaphore for FREE buffers
SEMAPHORE buf[i].sem = 1; // each buffer has a lock sem=1;
```

为了简化符号，我们将用缓冲区本身来表示每个缓冲区的信号量。与Unix算法一样，最开始，所有缓冲区都在空闲列表中，所有设备列表和I/O队列均为空。下面展示一个使用信号量的简单缓冲区管理算法。

12.4 PV算法

```
BUFFER *getblk(dev, blk)
{
    while(1){
 (1).   P(free);              // get a free buffer first
 (2).   if (bp in dev_list){
 (3).       if (bp not BUSY){
                remove bp from freelist;
                P(bp);      // lock bp but does not wait
                return bp;
            }
            // bp in cache but BUSY
            V(free);          // give up the free buffer
 (4).       P(bp);            // wait in bp queue
            return bp;
        }
        // bp not in cache, try to create a bp=(dev, blk)
 (5).   bp = frist buffer taken out of freelist;
        P(bp);                // lock bp, no wait
 (6).   if (bp dirty){
            awrite(bp);       // write bp out ASYNC, no wait
            continue;         // continue from (1)
        }
 (7).   reassign bp to (dev,blk); // mark bp data invalid, not dirty
        return bp;
    }                         // end of while(1)
}

brelse(BUFFER *bp)
{
```

```
 (8). if (bp queue has waiter){ V(bp); return; }
 (9). if (bp dirty && free queue has waiter){ awrite(bp); return; }
(10). enter bp into (tail of) freelist; V(bp); V(free);
}
```

接下来，我们要证明PV算法是正确的，并且满足要求。

（1）**缓冲区唯一性**：在getblk()中，如果有空闲缓冲区，则进程不会在（1）处等待，而是会搜索dev_list。如果所需的缓冲区已经存在，则进程不会重新创建同一个缓冲区。如果所需的缓冲区不存在，则进程会使用一个空闲缓冲区来创建所需的缓冲区，而这个空闲缓冲区保证是存在的。如果没有空闲缓冲区，则需要同一个缓冲区的几个进程可能在（1）处阻塞。当在（10）处释放出一个空闲缓冲区时，它仅释放一个进程来创建所需的缓冲区。一旦创建了缓冲区，它就会存在于dev_list中，这将防止其他进程再次创建同一个缓冲区。因此，分配的每个缓冲区都是唯一的。

（2）**无重试循环**：进程重新执行while(1)循环的唯一位置是在（6）处，但这不是重试，因为进程正在不断地执行。

（3）**无不必要唤醒**：在getblk()中，进程可以在（1）处等待空闲缓冲区，也可以在（4）处等待所需的缓冲区。在任意一种情况下，在有缓冲区之前，都不会唤醒进程重新运行。此外，当在（9）处有一个脏缓冲区即将被释放并且在（1）处有多个进程等待空闲缓冲区时，该缓冲区不会被释放而是直接被写入。这样可以避免不必要的进程唤醒。

（4）**缓存效果**：在Unix算法中，每个释放的缓冲区都可被获取。而在新的算法中，始终保留含等待程序的缓冲区以供重用。只有缓冲区不含等待程序时，才会被释放为空闲。这样可以提高缓冲区的缓存效果。

（5）**无死锁和饥饿**：在getblk()中，信号量锁定顺序始终是单向的，即P(free)，然后是P(bp)，但决不会反过来，因此不会发生死锁。如果没有空闲缓冲区，所有请求进程都将在（1）处阻塞。这意味着，虽然有进程在等待空闲缓冲区，但所有正在使用的缓冲区都不能接纳任何新用户。这保证了繁忙缓冲区最终将被释放为空闲缓冲区。因此，不会发生空闲缓冲区饥饿的情况。

虽然我们已经证明了新算法是正确的，但是它是否能比Unix算法表现得更好仍是一个有待验证的问题，只能通过真实的定量数据来回答。因此，我们设计了一个编程项目来比较缓冲区管理算法的性能。编程项目还可以帮助读者更好地理解文件系统中的I/O操作。

12.5 编程项目：I/O缓冲区管理算法比较

该编程项目将会实现一个模拟系统，以比较Unix I/O缓冲区管理算法和使用信号量的PV算法的性能。虽然当前的目标是I/O缓冲区性能，但真正的目标是让读者理解文件系统的I/O操作。下面介绍该项目。

12.5.1 系统组织

图12.1所示为模拟系统的组织结构。

Box#1：用户界面 这是模拟系统的用户界面部分。它会提示输入命令、显示命令执行、显示系统状态和执行结果等。在开发过程中，读者可以手动输入命令来执行任务。在最后测试过程中，任务应该有自己的输入命令序列。例如，各任务可以读取包含命令的输入文件。完成任务切换之后，将从另一个文件读取任务输入。

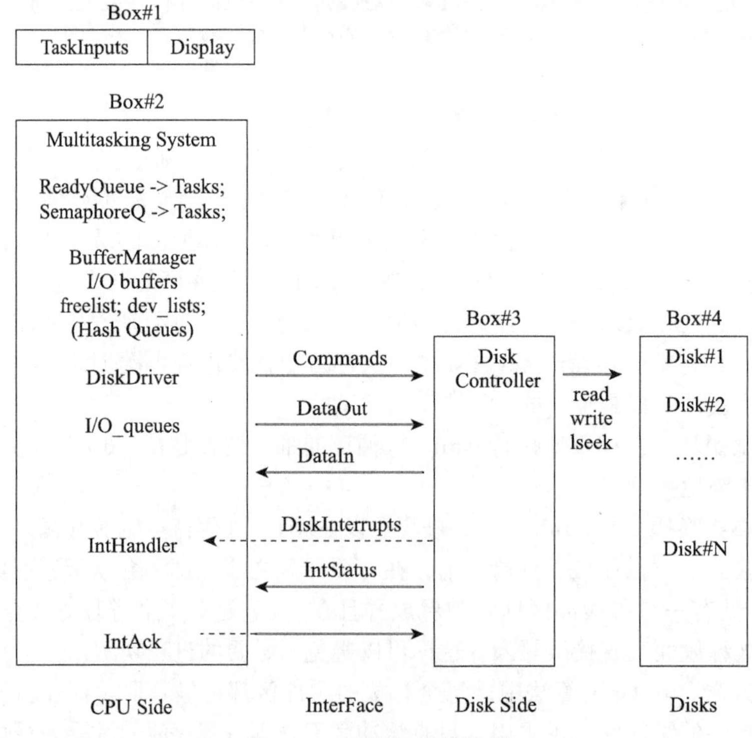

图 12.1　系统组织结构图

12.5.2 多任务处理系统

Box#2　这是多任务处理系统的 CPU 端，模拟单处理器（单 CPU）文件系统的内核模式。实际上，它与第 4 章中所述的用于用户级线程的多任务系统相同，只是以下修改除外。当系统启动时，它会创建并运行一个优先级最低的主任务，但它会创建 ntask 工作任务，所有任务的优先级都是 1，并将它们输入 readyQueue。然后，主任务执行以下代码，该代码将任务切换为从 readyQueue 运行工作任务。

```
/******* Main Task Code *******/
// after system initialization
while(1){
  while(task && readyQ == 0);   // loop if no task runnable
  if (readyQ)                   // if readyQueue nonempty
     kswitch();                 // switch to run a working task
  else
     end_task();                // all working tasks have ended
}
```

由于主任务的优先级最低，所以如果没有可运行的任务或所有任务都已结束，它将再次运行。在后一种情况下，主任务执行 end_task()，在其中收集并显示模拟结果，然后终止，从而结束模拟运行。

所有工作任务都执行同一个 body() 函数，其中每个任务从输入文件中读取命令来执行读或写磁盘块操作，直到命令文件结束。

```
#define CMDLEN 10
int cmdfile[NTASK]; // opened command file descriptors
int task = ntask;   // number of active tasks
int body()
{
   int dev, blk;
   char opcode, cmd[CMDLEN];
   while(1){
      if (read(cmdfile[running->pid], cmd, CMDLEN)==0){
          running->status = DEAD; // task ends
          task--;                 // dec task count by 1
          tswtich();
      }
      sscanf(cmd, "%c%4d%5d", &opcode, &dev, &blk);
      if (opcode=='r')          // read (dev, blk)
          readBlk(dev, blk);
      if (opcode=='w')
          writeBlk(dev, blk);   // write (dev, blk)
   }
}
```

各任务的命令由 rand() 函数取模极限值来随机生成。每个命令都是三重形式

```
[r xxx yyyy] or [w xxx yyyy], where xxx=dev, yyyy=blkno;
```

对于 [r dev blk] 命令，任务调用

```
BUFFER *bread(dev, blk)
```

来获取包含有效数据的缓冲区（指针）。从缓冲区读取数据后，它会将缓冲区释放回缓冲区缓存以便重用。在 bread() 中，任务可以等待 getblk() 中的缓冲区，或者等待缓冲区数据生效。如果是这样，它会休眠（在 Unix 算法中）或者被阻塞（在 PV 算法中），将任务切换到运行 readyQueue 中的下一个任务。

对于 [w dev blk] 命令，任务试图获取用于（dev, blk）的缓冲区。如果必须要等待 getblk() 中的缓冲区，它将会休眠或被阻塞，并切换任务以运行下一个任务。在将数据写入缓冲区之后，它会将延迟写入的缓冲区标记为脏，并将该缓冲区释放到缓冲区缓存中。然后，它会试图执行下一个命令等。

12.5.3 缓冲区管理器

缓冲区管理器实现各缓冲区管理函数，包括：

```
BUFFER *bread(dev, blk)
        dwrite(BUFFER *bp)
        awrite(BUFFER *bp)
BUFFER *getblk(dev, blk)
        brelse(BUFFER *bp)
```

这些函数可能会调用磁盘驱动程序来发出物理磁盘 I/O。

12.5.4 磁盘驱动程序

磁盘驱动程序由两部分组成。

（1）start_io()：维护设备 I/O 队列，并对 I/O 队列中的缓冲区执行 I/O 操作。

（2）中断处理程序：在每次 I/O 操作结束时，磁盘控制器会中断 CPU。当接收到中断后，中断处理程序首先从 IntStatus 中读取中断状态，IntStatus 包含：

```
[dev   | R|W |  StatusCode]
|<----- e.g.10 chars---->|
```

其中 dev 会标识设备。中断处理程序从设备 I/O 队列的头部删除缓冲区，并为缓冲区处理中断。对于磁盘读取中断，中断处理程序会先将数据块从 DataIn 移动到缓冲区的数据区域。然后，它将缓冲区数据标记为有效，并唤醒或解除对正在缓冲区上等待有效数据的任务的阻塞。被唤醒的进程将会从缓冲区读取数据并将缓冲区释放回缓冲区缓存。对于磁盘写入中断，没有进程在缓冲区中等待 I/O 完成。此类异步写入缓冲区由中断处理程序处理。它会关闭缓冲区的异步写入标记并将缓冲区释放到缓冲区缓存中。然后，它会检查设备 I/O 队列。如果 I/O 队列为非空，它会对 I/O 队列中的第一个缓冲区发出 I/O 命令。在发出磁盘写入操作之前，它会将缓冲区的数据复制到 DataOut 以便磁盘控制器获取。最后，它会写一个 ACK 到 IntAck 来确认当前中断，允许磁盘控制器继续控制。当中断处理程序处理完当前中断后，会从最后一个中断点开始恢复正常任务执行。

12.5.5　磁盘控制器

（3）Box#3：这是磁盘控制器，它是**主进程**的一个**子进程**。因此，它与 CPU 端独立运行，除了它们之间的通信通道，通信通道是 CPU 和磁盘控制器之间的**接口**。通信通道由主进程和子进程之间的管道实现。

- 命令：从 CPU 到磁盘控制器的 I/O 命令。
- DataOut：在写操作中从 CPU 到磁盘控制器的数据输出。
- DataIn：在读操作中从磁盘控制器到 CPU 的数据。
- IntStatus：从磁盘控制器到 CPU 的中断状态。
- IntAck：从 CPU 到磁盘控制器的中断确认。

12.5.6　磁盘中断

从磁盘控制器到 CPU 的中断由 SIGUSR1（#10）信号实现。在每次 I/O 操作结束时，磁盘控制器会发出 kill(ppid, SIGUSR1) 系统调用，向父进程发送 SIGUSR1 信号，充当虚拟 CPU 中断。通常，虚拟 CPU 会在临界区屏蔽出 / 入磁盘中断（信号）。为防止竞态条件，磁盘控制器必须要从 CPU 接收一个中断确认，才能再次中断。

12.5.7　虚拟磁盘

（4）Box#4：这些是 Linux 文件模拟的虚拟磁盘。使用 Linux 系统调用 lseek()、read() 和 write()，我们可以支持虚拟磁盘上的任何块 I/O 操作。为了简单起见，将磁盘块大小设置为 16 字节。由于数据内容无关紧要，所以可以将它们设置为 16 个字符的固定序列。

12.5.8　项目要求

实现两个版本的缓冲区管理算法：

- 使用休眠 / 唤醒的 Unix 算法。
- 使用信号量的新算法。

比较两种算法在不同任务 / 缓冲区比率下的缓冲区缓存命中率、实际 I/O 操作数量、任务切换、重试和总运行时间等方面的性能。

12.5.9 基本代码示例

为了简单起见，我们只列出了 CPU 端的 main.c 文件和磁盘控制器的 controller.c 文件。仅对其他所含文件的函数做了简要解释。而且，并未说明用于收集性能统计信息的代码段，但是可以把它们轻松添加到基本代码中。

```
/************ CPU side: main.c file **********/
#include <stdio.h>
#include <stdlib.h>
#include <sys/time.h>
#include <signal.h>
#include <string.h>
#include "main.h"                  // constants, PROC struct pipes
struct itimerval itimer;           // for timer interrupt
struct timeval tv0, tv1, tv2;      // count running time
sigset_t sigmask;                  // signal mask for critical regions
int stack[NTASK][SSIZE];           // task stacks
int pd[5][2];                      // 5 pipes for communication
int cmdfile[NTASK];                // task command file descriptors
int record[NTASK+2][13];           // record each task use buff status
int task;                          // count alive task
int ntask;                         // number of tasks
int nbuffer;                       // number of buffers
int ndevice;                       // number of devices
int nblock;                        // number of blocks per device
int sample;                        // number of per task inputs
int timer_switch;                  // count time switch
char cmd[CMLEN];                   // command=[r|w dev blkno]
char buffer[BLOCK];                // save command to commuicate with device
char dev_stat[CMLEN];              // device status
PROC proc[NTASK], MainProc;        // NTASK tasks and main task
PROC *running;                     // pointer to current running task
PROC *readyQ, *freeQ, *waitQ;      // process link lists
FILE *fp;                          // record simulation results file
int body();                        // task body function
void catcher();                    // signal catcher (interrupt handler)

#include "proc.c"                  // PROC init, queue functions
#include "buf.c"                   // buffer struct, link lists
#include "utility.c"               // generate cmds, end_task()
#include "manager.c"               // getblk,brelse, bread,dwrite,awrite

int main(int argc, char *argv[ ])
{
   if(argc != 6){
     printf("usage: a.out ntask nbuffer ndevice nblock nsample\n");
     exit(0);
   }
```

```
   task = ntask  = atoi(argv[1]);
   nbuffer= atoi(argv[2]);
   ndevice= atoi(argv[3]);
   nblock = atoi(argv[4]);
   sample = atoi(argv[5]);
   if(ntask>NTASK||nbuffer>NBUFFER||ndevice>NDEVICE||nblock>NBLOCK)
      exit(1) ;
   fp = fopen("record", "a"); // record task activities
   printf("Welcome to UNIX Buffer Management System\n");
   printf("Initialize proc[] & buff[] & signal\n");
   initproc();            // build tasks to compete for buffers
   initbuf();             // build buffer link lists
   initsig();             // set sigmask & init itimer for timing
   printf("Install signal handler\n");
   signal(SIGUSR1, catcher); // install catcher for SIGUSR1 signal
   printf("Open pipes for communication\n");
   open_pipe();           // create communication pipes
   printf("fork(): parent as CPU child as Disk Controller\n");
   switch(fork()){
     case -1: perror("fork call"); exit(2); break;
     case  0: // child process as disk controller
        close_pipe(0);  // configure pipes at child side
        sprintf(buffer,"%d %d %d %d %d",\
                CMDIN, WRITEIN, READOUT, STATOUT, ACKIN);
        execl("controller","controller",buffer,argv[3],argv[4],0);
        break;
     default:           // parent process as virtual CPU
        close_pipe(1);  // configure pipes at parent side
        printf("MAIN: check device status\n");
        check_dev();    // wait for child has created devices
        printf("Generate command file for each task\n");
        gencmd();       // generate command files of tasks
        gettimeofday(&tv0,0);  // start record run time
        while(1){
           INTON              // enable interrupts
           while(tasks && readyQ == 0); // wait for ready tasks
           if(readyQ)
            kswitch();        // switch to run ready task
           else
             end_task();      // end processing
        }
     }
}

/******* task body function **********/
int body(int pid)
{
  char rw;            // opcode in command: read or write
  int dev, blk;       // dev, blk in command
  int n, count=0;     // task commands count
  while(1){
     n = read(cmdfile[running->pid], cmd, CMLEN);
     if (n==0){                    // if end of cmd file
        running->status = DEAD;    // mark task dead
        task--;                    // number of active task -1
```

```
      kswitch();                 // switch to next task
   }
   count++;                      // commands count++
   sscanf(cmd, "%c%5d%5d",&rw,&dev,&blk);
   if (rw == 'r')
      readBlk(dev, blk);         // READ (dev, blk)
   else
      writeBlk(dev, blk);        // WRITE (dev, blk)
  }
}
int kswitch()
{
  INTOFF       // disable interrupts
    tswitch(); // switch task in assembly code
  INTON        // enable interrupts

}
```

main.c 文件包含以下文件，对于这些文件仅作简要说明。

（1）main.h：该文件定义了系统常数、PROC 结构体类型和符号常数，还定义了以下宏：

```
#define INTON  sigprocmask(SIG_UNBLOCK, &sigmask, 0);
#define INTOFF sigprocmask(SIG_BLOCK,   &sigmask, 0);
```

用于屏蔽出 / 入中断，其中 sigmask 是一个 32 位的向量，其中 SIGUSR1 位（10）=1。

（2）proc.c：该文件可初始化多任务处理系统。它会创建 ntask 任务，通过缓冲区缓存来执行磁盘 I/O 操作。

（3）buf.c：该文件定义了信号量、缓冲区和设备结构。它可以初始化缓冲区和设备数据结构，并且包含用于缓冲区链表操作的代码。

（4）utility.c：该文件包含 CPU 端和磁盘控制器端通用的函数。例如管道的创建和配置、输入命令文件和模拟磁盘数据文件的生成等。它还可以实现 end_task() 函数，该函数在所有任务结束时收集和显示模拟统计数据。

（5）manager.c：该文件可以实现实际的缓冲区管理函数，如 readBlk(dev, blk)、writeBlk(dev, blk)、getblk() 和 brelse()，它们是缓冲区管理的核心函数。

在模拟器中，所有文件都是相同的，manager.c 文件除外，它有两个版本：一个版本可实现 Unix 算法，另一个版本可实现使用信号量的 PV 算法。

下面展示的是磁盘控制器源文件，它由主进程的子进程执行。子进程所需的信息通过 argv[] 中的命令行参数传递。初始化之后，磁盘控制器进程执行无限循环，直到它从 CPU 获得‘END’命令，这表示 CPU 端的所有任务都已经结束。然后，它会关闭所有虚拟磁盘文件并终止。

```
/******* Pseudo Code of Disk Controller *******/
initialization;
while(1){
   read command from CMDIN
   if  command = "END", break;
   decode and execute the command
   interrupt CPU;
   read ACK from ACKIN
}
```

```
   close all virtual disk files and exit

/******** Disk Controller: controller.c file: *******/
#include <stdio.h>
#include <fcntl.h>
#include <signal.h>
#include <string.h>

#define DEVICE  128      // max number of devices
#define BLOCK    16      // disk BLOCK size
#define CMLEN    10      // command record length
int main(int argc, char *argv[ ])
{
   char opcode, cmd[CMLEN], buf[BLOCK], status[CMLEN];
   int CMDIN,READIN,WRITEOUT,STATOUT,ACKIN;  // PIPEs
   int i, ndevice, dev, blk;
   int CPU = getppid();  // ppid for sending signal
   int fd[DEVICE];       // files simulate n devices
   char datafile[8];     // save file name
   // get command line parameters
   sscanf(argv[1], "%d %d %d %d %d",\
        &CMDIN, &READIN, &WRITEOUT, &STATOUT, &ACKIN);
   ndevice = atoi(argv[2]);
   printf("Controller: ndevice=%d\n", ndevice);
   gendata(ndevice);     // generate data files as virtual disks
   for (i=0; i<ndevice i++){  // open files to simulate devices
     sprintf(datafile,"data%d", i%128);
     if(( fd[i] = open(datafile, O_RDWR)) < 0)
       sprintf(status, "%dfail",i) ;
      else
       sprintf(status, " %d ok",i) ;
     write(STATOUT, status, CMLEN);  // send device status to CPU
   }
   // Disk Controller Processing loop
   while(1){
      read(CMDIN, cmd, CMLEN);             // read command from pipe
      if (!strcmp(cmd, "END"))             // end command from CPU, break
         break;
      sscanf(cmd,"%c%4d%5d", &opcode, &dev, &blk);
      if (opcoe == 'r'){
          read(fd[dev], buf, BLOCK);       // read data of (dev, blk)
          write(WRITEOUT, buf, BLOCK); // write data to WRITEOUT pipe
      }
      else if (opcode == 'w'){            // write cmd
          read(READIN, buf, BLOCK);        // read data from READIN pipe
          write(fd[dev], buf, BLOCK);      // write data to (dev, blk)
      }
      else
          strcpy(status, "IOerr");
      write(STATOUT, cmd, CMLEN);          // write INTstatus to STATOUT
      kill(CPU, SIGUSR1);                  // interrupt CPU by SIGUSR1
      read(ACKIN, buf, CMLEN);             // read ACK from CPU
   }
   // Disk controller end processing
   for (i=0; i < ndevice;i++)
```

```
      close(fd[i]);
}

int gendata(int ndev) // generate data file as virtual disks
{
  char cmd[2048], name[8];
  int fd[128], i, j;
  printf("generate data files\n");
  for(i=0; i<ndev; i++){
     sprintf(name,"data%d", i);
     fd[i]=creat(name, 0644);  // create data files
     for (j=0; j<2048; j++){   // 2 K bytes of '0'-'9' chars
         cmd[j]= i%10 +'0';
     }
     write(fd[i],cmd,2048);
     close(fd[i]);
  }
}
```

模拟器系统被编译并链接到两个二进制可执行文件中，a.out 用作 CPU 端，控制器用作磁盘控制器端。

```
cc main.c s.s                    # a.out as the CPU side
cc -o controller controller.c    # controller as Disk Controller
```

然后，运行以下模拟器系统：

```
a.out  ntask nbuffer ndevice nblock nsample
```

为了运行具有不同输入参数的模拟器，可以使用 sh 脚本，如：

```
# run: sh script to run the simulator systems
#! /bin/bash
for TASK in 4 8 16 64; do
  For BUF in 4 16 64 128; do
    a.out $TASK $BUF 16 64 1000    # ndev=16,nblk=64,sample=1000
  done
  echo -------------------------
done
```

12.5.10 示例解决方案

图 12.2 所示为使用 Unix 算法的模拟器输出：

每个设备有 4 个任务、4 个缓冲区、16 个块，每个任务有 100 个输入样本。图中的性能指标含义如下：

- run-time = 总运行时间
- rIO = 实际读操作的次数
- wIO = 实际写操作的次数
- intr = 磁盘中断的次数
- hits = 缓冲区缓存命中次数
- swtch = 任务切换次数
- dirty = 延迟写缓冲区的次数

- retry = getblk() 中的任务重试次数

输出的最后一行显示的是各种指标的百分比。最重要的性能指标是模拟的总运行时间和缓冲区缓存命中率。其他指标可以作为指导来证明算法的性能。

```
------------------------------------------------------------
ntask:4    nbuffer = 4      ndevice = 4      nblock = 16
run-time=29 msec
CMD  read  write  rIO   wIO  hits  intr swtch dirty retry
--0----1-----2-----3-----4----5-----6-----7-----8-----9--
100    52    48    51    44     3    93    57    43    48
100    51    49    50    55     4   106    59    55    63
100    44    56    44    50     1    94    51    50    56
100    55    45    54    44     3    98    59    44    48
-------------------- percentages ---------------------------
100    50    49    49    48     2    97    56    48    53
------------------------------------------------------------
```

图 12.2　Unix 算法的输出示例

图 12.3 所示为 PV 算法的模拟器输出，使用的输入参数与 Unix 算法的相同。从图中可以看出，运行时间（24 毫秒）比 Unix 算法（29 毫秒）短，缓冲区缓存命中率（6%）比 Unix 算法（2%）高。还可以看出，Unix 算法有大量的任务重试，但是 PV 算法中没有任务重试，这可能是两种算法性能差异的原因。

```
------------------------------------------------------------
ntask:4    nbuffer = 4      ndevice = 4      nblock = 16
run-time=24 msec
CMD  read  write  rIO   wIO   hits  intr swtch dirty retry
--0----1-----2-- ---3----4-----5-----6-----7-----8-----9--
100    52    48    49    47     5    91    52    46     0
100    51    49    47    53     5   100    52    53     0
100    44    56    42    55     5    99    47    55     0
100    55    45    49    40    10    85    57    40     0
-------------------- percentages ---------------------------
100    50    49    46    48     6    95    53    48     0
------------------------------------------------------------
```

图 12.3　PV 算法的输出示例

12.6　模拟系统的改进

可通过多种方式改进模拟系统，使其成为更好的真实文件系统模型。

（1）模拟系统可以扩展为支持多个磁盘控制器，而不是单独一个磁盘控制器，这样可通过一个数据信号来缓解 I/O 堵塞。

（2）可用非均匀分布生成输入命令，以改善实际系统中模型文件操作。例如，可以生成更多的读命令而不是写命令，以及一些设备上有更多的 I/O 需求等。

12.7　PV 算法的改进

PV 算法非常简单，易于实现，但是它有以下两个缺点。首先，它的缓存效果可能并非最佳。这是因为一旦没有空闲缓冲区，所有请求进程都将被阻塞在 getblk() 中的（1）处，即使它们所需的缓冲区可能已经存在于缓冲区缓存中了。其次，当进程从空闲列表信号量队列中唤醒时，它可能会发现所需的缓冲区已经存在，但处于繁忙状态，在这种情况下，它将在（4）处再次被阻塞。严格地说，进程被不必要地唤醒了，因为它被阻塞了两次。读者可参阅文献（Wang 2015）了解改进算法，改进算法在进程切换方面表现最佳，而且缓存性能也更好。

参考文献

Bach, M.J., "The Design of the Unix operating system", Prentice Hall, 1990
Lion, J., "Commentary on UNIX 6th Edition, with Source Code", Peer-To-Peer Communications, ISBN 1-57398-013-7, 1996
Ritchie, D.M., Thompson, K., "The UNIX Time-Sharing System", Bell System Technical Journal, Vol. 57, No. 6, Part 2, July, 1978
Wang, X., "Improved I/O Buffer Management Algorithms for Unix Operating System", M.S. thesis, EECS, WSU, 2002
Wang, K. C., Design and Implementation of the MTX Operating system, Springer A.G, 2015

第 13 章
Systems Programming in Unix/Linux

TCP/IP 和网络编程

摘要

本章论述了 TCP/IP 和网络编程，分为两个部分。第一部分论述了 TCP/IP 协议及其应用，具体包括 TCP/IP 栈、IP 地址、主机名、DNS、IP 数据包和路由器；介绍了 TCP/IP 网络中的 UDP 和 TCP 协议、端口号和数据流；阐述了服务器 - 客户机计算模型和套接字编程接口；通过使用 UDP 和 TCP 套接字的示例演示了网络编程。第一个编程项目可实现一对通过互联网执行文件操作的 TCP 服务器 - 客户机，可让用户定义其他通信协议来可靠地传输文件内容。

本章的第二部分介绍了 Web 和 CGI 编程，解释了 HTTP 编程模型、Web 页面和 Web 浏览器；展示了如何配置 Linux HTTPD 服务器来支持用户 Web 页面、PHP 和 CGI 编程；阐释了客户机和服务器端动态 Web 页面；演示了如何使用 PHP 和 CGI 创建服务器端动态 Web 页面。第二个编程项目可让读者在 Linux HTTPD 服务器上通过 CGI 编程实现服务器端动态 Web 页面。

13.1 网络编程简介

如今，上网已成为日常生活的需要。虽然大多数人可能只把互联网作为一种信息收集、网上购物和社交媒体等的工具，但计算机科学的学生必须对互联网技术有一定的了解，并掌握一定的网络编程的技能。在本章中，我们将介绍 TCP/IP 网络和网络编程的基础知识，包括 TCP/IP 协议、UDP 和 TCP 协议、服务器 - 客户机计算、HTTP 和 Web 页面、动态 Web 页面的 PHP 和 CGI 编程。

13.2 TCP/IP 协议

TCP/IP（Comer 1988, 2001; RFC1180 1991）是互联网的基础。TCP 代表传输控制协议。IP 代表互联网协议。目前有两个版本的 IP，即 IPv4 和 IPv6。IPv4 使用 32 位地址，IPv6 则使用 128 位地址。本节围绕 IPv4 进行讨论，它仍然是目前使用最多的 IP 版本。TCP/IP 的组织结构分为几个层级，通常称为 **TCP/IP 堆栈**。图 13.1 所示为 TCP/IP 的各个层级以及每一层级的代表性组件及其功能。

```
--------- Layer --------- Components --------- Functions ------------
| Application Layer  |  ssh    ping  | Application commands        |
--------------------------------------------------------------------
| Transport Layer    |  TCP    UCP   | Connection     Datagram     |
--------------------------------------------------------------------
| Internet Layer     |      IP       | send/receive  data frames   |
--------------------------------------------------------------------
| Link Layer         |   Ethernet    | send/receive  data frames   |
--------------------------------------------------------------------
```

图 13.1　TCP/IP 层

顶层是使用 TCP/IP 的应用程序。用于登录到远程主机的 ssh、用于交换电子邮件的

mail、用于 Web 页面的 http 等应用程序需要可靠的数据传输。通常，这类应用程序在传输层使用 TCP。另一方面，有些应用程序，例如用于查询其他主机的 ping 命令，则不需要可靠性。这类应用程序可以在传输层使用 UDP 来提高效率（RFC 768 1980; Comer 1988）。传输层负责以包的形式向 IP 主机发送 / 接收来自 IP 主机的应用程序数据。进程与主机之间的传输层或其上方的数据传输只是逻辑传输。实际数据传输发生在**互联网**（IP）和链路层，这些层将数据包分成数据帧，以便在物理网络之间传输。图 13.2 所示为 TCP/IP 网络中的数据流路径。

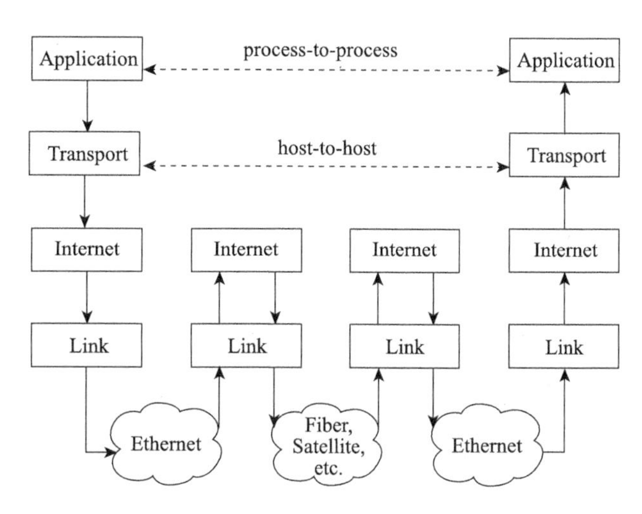

图 13.2　TCP/IP 网络中的数据流路径

13.3　IP 主机和 IP 地址

主机是支持 TCP/IP 协议的计算机或设备。每个主机由一个 32 位的 **IP 地址**来标识。为了方便起见，32 位的 IP 地址号通常用点记法表示，例如：134.121.64.1，其中各个字节用点号分开。主机也可以用**主机名**来表示，如 dns1.eec.wsu.edu。实际上，应用程序通常使用主机名而不是 IP 地址。在这个意义上说，主机名就等同于 IP 地址，因为给定其中一个，我们可以通过 DNS（域名系统）（RFC 134 1987; RFC 1035 1987）服务器找到另一个，它将 IP 地址转换为主机名，反之亦然。

IP 地址分为两部分，即 NetworkID 字段和 HostID 字段。根据划分，IP 地址分为 A~E 类。例如，一个 B 类 IP 地址被划分为一个 16 位 NetworkID，其中前 2 位是 10，然后是一个 16 位的 HostID 字段。发往 IP 地址的数据包首先被发送到具有相同 networkID 的**路由器**。路由器将通过 HostID 将数据包转发到网络中的特定主机。每个主机都有一个本地主机名 localhost，默认 IP 地址为 127.0.0.1。本地主机的链路层是一个回送虚拟设备，它将每个数据包路由回同一个 localhost。这个特性可以让我们在同一台计算机上运行 TCP/IP 应用程序，而不需要实际连接到互联网。

13.4　IP 协议

IP 协议用于在 IP 主机之间发送 / 接收数据包。IP 尽最大努力运行。IP 主机只向接收主机发送数据包，但它不能保证数据包会被发送到它们的目的地，也不能保证按顺序发送。这意味着 IP 并非可靠的协议。必要时，必须在 IP 层的上面实现可靠性。

13.5 IP数据包格式

IP数据包由IP头、发送方IP地址和接收方IP地址以及数据组成。每个IP数据包的大小最大为64KB。IP头包含有关数据包的更多信息，例如数据包的总长度、数据包使用TCP还是UDP、生存时间（TTL）计数、错误检测的校验和等。图13.3所示为IP头格式。

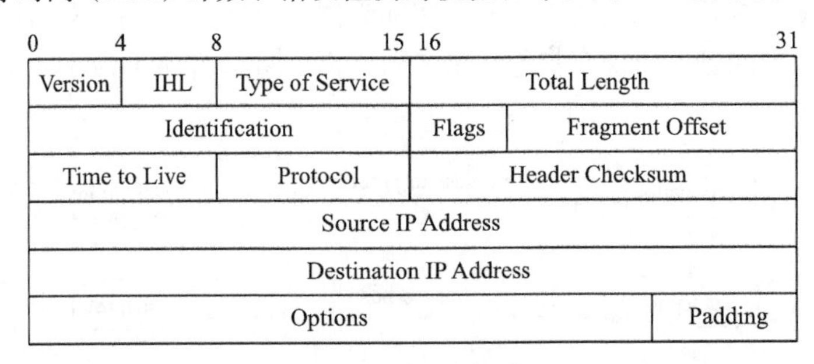

图13.3 IP头格式

13.6 路由器

IP主机之间可能相距很远。通常不可能从一个主机直接向另一个主机发送数据包。路由器是接收和转发数据包的特殊IP主机。如果有的话，一个IP数据包可能会经过许多路由器，或者跳跃到达某个目的地。图13.4显示了TCP/IP网络的拓扑结构。

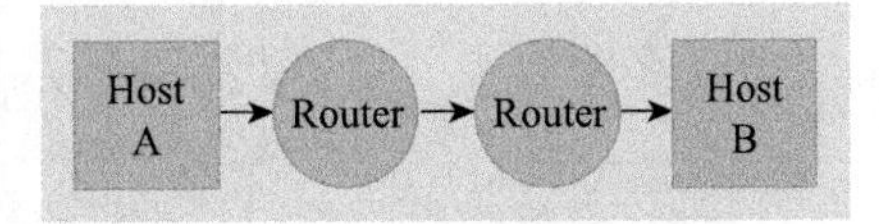

图13.4 TCP/IP网络拓扑结构

每个IP包在IP报头中都有一个8位生存时间（TTL）计数，其最大值为255。在每个路由器上，TTL会减小1。如果TTL减小到0，而包仍然没有到达目的地，则会直接丢弃它。这可以防止任何数据包在IP网络中无限循环。

13.7 UDP

UDP（用户数据报协议）（RFC 768 1980; Comer 1988）在IP上运行，用于发送/接收数据报。与IP类似，UDP不能保证可靠性，但是快速高效。它可用于可靠性不重要的情况。例如，用户可以使用ping命令探测目标主机，如

ping主机名或ping IP地址

ping是一个向目标主机发送带时间戳UDP包的应用程序。接收到一个pinging数据包后，目标主机将带有时间戳的UDP包回送给发送者，让发送者可以计算和显示往返时间。如果目标主机不存在或宕机，当TTL减小为0时，路由器将会丢弃pinging UDP数据包。在这种情况下，用户会发现目标主机没有任何响应。用户可以尝试再次ping，或者断定目标主机宕机。在这种情况下，最好使用UDP，因为不要求可靠性。

13.8 TCP

TCP（传输控制协议）是一种面向连接的协议，用于发送/接收数据流。TCP也可在IP上运行，但它保证了可靠的数据传输。通常，UDP类似于发送邮件的USPS，而TCP类似于电话连接。

13.9 端口编号

在各主机上，多个应用程序（进程）可同时使用TCP/UDP。每个应用程序由三个组成

部分唯一标识

应用程序 =（主机 IP，协议，端口号）

其中，协议是 TCP 或 UDP，端口号是分配给应用程序的唯一无符号短整数。要想使用 UDP 或 TCP，应用程序（进程）必须先选择或获取一个端口号。前 1024 个端口号已被预留。其他端口号可供一般使用。应用程序可以选择一个可用端口号，也可以让操作系统内核分配端口号。图 13.5 给出了在传输层中使用 TCP 的一些应用程序及其默认端口号。

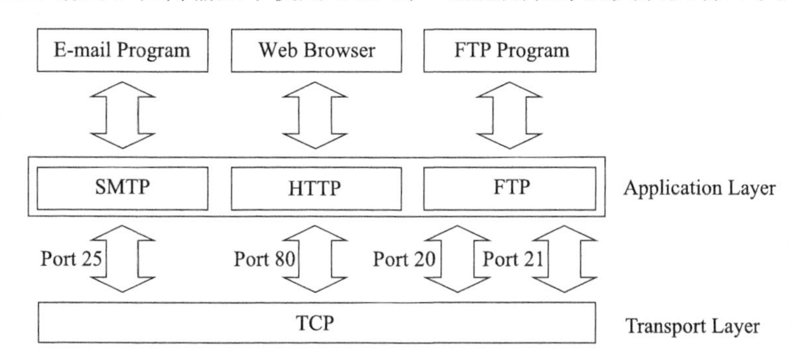

图 13.5 使用 TCP 的应用程序

13.10 网络和主机字节序

计算机可以使用大端字节序，也可以使用小端字节序。在互联网上，数据始终按网络序排列，这是大端。在小端机器上，例如基于 Intel x86 的 PC，htons()、htonl()、ntohs()、ntohl() 等库函数，可在主机序和网络序之间转换数据。例如，PC 中的端口号 1234 按主机字节序（小端）是无符号短整数。必须先通过 htons(1234) 把它转换成网络序，才能使用。相反，从互联网收到的端口号必须先通过 ntohs(port) 转换为主机序。

13.11 TCP/IP 网络中的数据流

图 13.6 给出了 TCP/IP 网络中的各层数据格式。它还给出了各层之间的数据流路径。

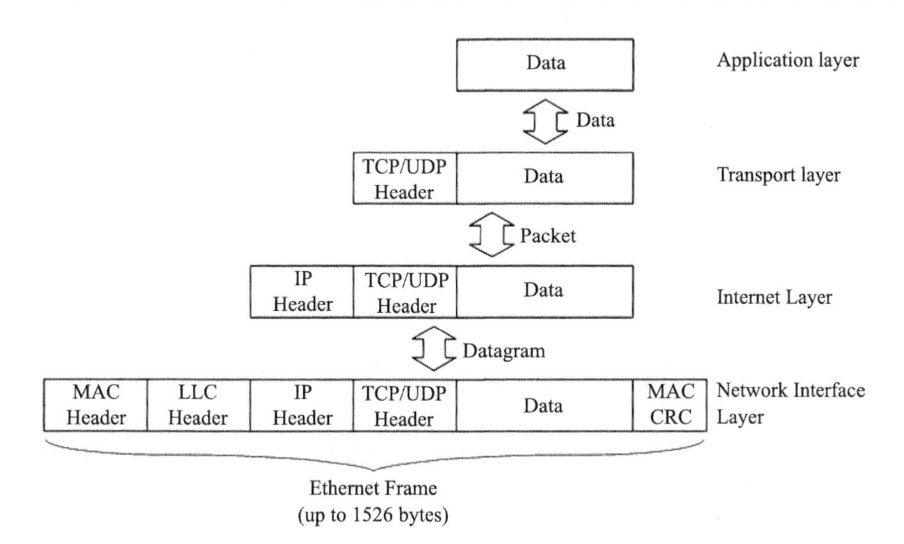

图 13.6 TCP/IP 层的数据格式

在图 13.6 中，应用程序层的数据被传递到传输层，传输层给数据添加一个 TCP 或 UDP 报头来标识使用的传输协议。合并后的数据被传递到 IP 网络层，添加一个包含 IP 地址的 IP 报头来标识发送和接收主机。然后，合并后的数据再被传递到网络链路层，网络链路层将数据分成多个帧，并添加发送和接收网络的地址，用于在物理网络之间传输。IP 地址到网络地址的映射由地址解析协议（ARP）执行（ARP 1982）。在接收端，数据编码过程是相反的。每一层通过剥离数据头来解包接收到的数据，重新组装数据并将数据传递到上一层。发送主机上的应用程序原始数据最终会被传递到接收主机上的相应应用程序。

13.12　网络编程

所有 Unix/Linux 系统都为网络编程提供 TCP/IP 支持。本节，我们将会阐释用于网络编程的平台和服务器 - 客户机计算模型。

13.12.1　网络编程平台

要进行网络编程，读者必须能够访问支持网络编程的平台。可通过下面几种方法访问这类平台。

（1）**服务器上的用户账户**：现在，几乎所有的教育机构都为它们的教职工和学生提供了网络接入，通常是以无线连接的形式。每位机构成员都要能够登录服务器以接入互联网。服务器是否允许一般的网络编程取决于本地网络管理策略。这里，我们要介绍作者所在机构（华盛顿州立大学电气工程与计算机科学系）的网络编程平台的设置。作者有一个专用服务器

cs360.eecs.wsu.edu

服务器运行 14.2 版 Slackware Linux，完全支持网络编程。这台服务器在华盛顿州立大学电气工程与计算机科学系的 DNS 服务器上注册。当服务器启动时，它会使用 DHCP（动态主机配置协议）从 DHCP 服务器上获取一个私有 IP 地址（RFC 2131 1997）。虽然它不是公共 IP 地址，但可以通过 NAT（网络地址转换）在互联网上访问它。然后，作者为 CS360 班级的学生创建用户账户，供他们登录。学生大多通过 WSU 无线网络将笔记本电脑连接到互联网上。一旦连上互联网，他们就可以登录 cs360 服务器了。

（2）**单独 PC 或笔记本电脑**：即便学生未接入服务器，仍然可以使用计算机的本地主机在单独计算机上进行网络编程。在这种情况下，学生需要下载安装一些网络部件。例如，Ubuntu Linux 用户可能需要安装和配置用于 HTTP 和 CGI 编程的 Apache 服务器，这将在后面的 13.17 节中讨论。

13.12.2　服务器 - 客户机计算模型

大多数网络编程任务都基于服务器 - 客户机计算模型。在**服务器 - 客户机**计算模型中，我们首先在服务器主机上运行服务器进程。然后，我们从客户机主机运行客户机。在 UDP 中，服务器等待来自客户机的数据报，处理数据报并生成对客户机的响应。在 TCP 中，服务器等待客户机连接。客户机首先连接到服务器，在客户机和服务器之间建立一个虚拟电路。建立连接后，服务器和客户机可以交换连续的数据流。下面，我们将展示如何使用 UDP 和 TCP 进行网络编程。

13.13　套接字编程

在网络编程中，TCP/IP 的用户界面是通过一系列 C 语言库函数和系统调用来实现的，这些函数和系统调用统称为**套接字 API**（(Rago 1993; Stevens 等 2004)。为了使用套接字 API，我们需要套接字地址结构，它用于标识服务器和客户机。netdb.h 和 sys/socket.h 中有套接字地址结构的定义。

13.13.1　套接字地址

```
struct sockaddr_in {
       sa_family_t sin_family;  // AF_INET for TCP/IP
       in_port_t   sin_port;    // port number
       struct in_addr sin_addr; // IP address
};
struct in_addr {                // internet address
       uint32_t    s_addr;      // IP address in network byte order
};
```

在套接字地址结构中，

- TCP/IP 网络的 sin_family 始终设置为 AF_INET。
- sin_port 包含按网络字节顺序排列的端口号。
- sin_addr 是按网络字节顺序排列的主机 IP 地址。

13.13.2　套接字 API

服务器必须创建一个套接字，并将其与包含服务器 IP 地址和端口号的套接字地址绑定。它可以使用一个固定端口号，或者让操作系统内核选择一个端口号（如果 sin_port 为 0）。为了与服务器通信，客户机必须创建一个套接字。对于 UPD 套接字，可以将套接字绑定到服务器地址。如果套接字没有绑定到任何特定的服务器，那么它必须在后续的 sendto()/recvfrom() 调用中提供一个包含服务器 IP 和端口号的套接字地址。下面给出了 socket() 系统调用，它创建一个套接字并返回一个文件描述符

1. int 套接字（int 域，int 类型，int 协议）

示例：

int udp_sock = socket(AF_INET, SOCK_DGRAM, 0);

将会创建一个用于发送 / 接收 UDP 数据报的套接字。

int tcp_sock = socket(AF_INET, SOCK_STREAM, 0);

将会创建一个用于发送 / 接收数据流的面向连接的 TCP 套接字。

新创建的套接字没有任何相联地址。它必须与主机地址和端口号绑定，以识别接收主机或发送主机。这通过 bind() 系统调用来完成。

2. int bind(int sockfd, **struct sockaddr** *addr, **socklen_t** addrlen)

bind() 系统调用将 addr 指定的地址分配给文件描述符 sockfd 所引用的套接字，addrlen 指定 addr 所指向地址结构的大小（以字节为单位）。对于用于联系其他 UDP 服务器主机的 UDP 套接字，必须绑定到客户机地址，允许服务器发回应答。对于用于接收客户机连接的

TCP 套接字，必须先将其绑定到服务器主机地址。

3. UDP 套接字

UDP 套接字使用 sendto()/recvfrom() 来发送 / 接收数据报。

```
ssize_t sendto(int sockfd, const void *buf, size_t len, int flags,
               const struct sockaddr *dest_addr, socklen_t addrlen);

ssize_t recvfrom(int sockfd, void *buf, size_t len, int flags,
                 struct sockaddr *src_addr, socklen_t *addrlen);
```

sendto() 将缓冲区中的 len 字节数据发送到由 dest_addr 标识的目标主机，该目标主机包含目标主机 IP 和端口号。recvfrom() 从客户机主机接收数据。除了数据之外，它还用客户机的 IP 和端口号填充 src_addr，从而允许服务器将应答发送回客户机。

4. TCP 套接字

在创建套接字并将其绑定到服务器地址之后，TCP 服务器使用 listen() 和 accept() 来接收来自客户机的连接

int listen(int sockfd, int backlog);

listen() 将 sockfd 引用的套接字标记为将用于接收连入连接的套接字。backlog 参数定义了等待连接的最大队列长度。

int accept(int sockfd, struct sockaddr *addr, socklen_t *addrlen);

accept() 系统调用与基于连接的套接字一起使用。它提取等待连接队列上的第一个连接请求用于监听套接字 sockfd，创建一个新的连接套接字，并返回一个引用该套接字的新文件描述符，与客户机主机连接。在执行 accept() 系统调用时，TCP 服务器阻塞，直到客户机通过 connect() 建立连接。

int connect(int sockfd, **const struct sockaddr** *addr, **socklen_t** addrlen**);**

connect() 系统调用将文件描述符 sockfd 引用的套接字连接到 addr 指定的地址，addrlen 参数指定 addr 的大小。addr 中的地址格式由套接字 sockfd 的地址空间决定。

如果套接字 sockfd 是 SOCK_DGRAM 类型，即 UDP 套接字，addr 是发送数据报的默认地址，也是接收数据报的唯一地址。这会限制 UDP 套接字与特定 UDP 主机的通信，但实际上很少使用。所以对于 UDP 套接字来说，连接是可选的或不必要的。如果套接字是 SOCK_STREAM 类型，即 TCP 套接字，connect() 调用尝试连接到绑定到 addr 指定地址的套接字。

5. send()/read() 以及 recv/write()

建立连接后，两个 TCP 主机都可以使用 send()/write() 发送数据，并使用 recv()/ read() 接收数据。它们唯一的区别是 send() 和 recv() 中的 flag 参数不同，通常情况下可以将其设置为 0。

```
ssize_t send(int sockfd, const void *buf, size_t len, int flags);
ssize_t write(sockfd, void *buf, size_t, len)

ssize_t recv(int sockfd, void *buf, size_t len, int flags);
ssize_t read(sockfd, void *buf, size_t len);
```

13.14 UDP 回显服务器 – 客户机程序

本节，我们将介绍一个使用 UDP 的简单回显服务器 / 客户机程序。为便于介绍，该程序用 C13.1 表示。下表列出了服务器和客户机的算法。

```
---------- UDP Server ---------------- UDP Client -------------
1. create a UDP socket          | 1. create a UDP socket
2. set addr = server [IP,port]| 2. set addr = server [IP,port]
3. bind socket to addr          |    while(1){
   while(1){                    | 3.   ask user for an input line
4.   recvfrom() from client     | 4.   sendto() line to server
5.   sendto() reply to client   | 5.   recvfrom() reply from server
   }                            |    }
----------------------------------------------------------------
```

为简单起见，我们假设服务器和客户机都在同一台计算机上运行。服务器在默认本地主机上运行（IP = 127.0.0.1），使用固定端口号 1234。这可以简化程序代码。读者可以在同一台计算机上以不同的 xterms 测试运行服务器和客户机程序。下面介绍使用 UDP 的服务器和客户机程序代码。

```
/*  C13.1.a: UDP server.c file */
#include <stdio.h>
#include <stdlib.h>
#include <string.h>
#include <sys/socket.h>
#include <netinet/ip.h>

#define BUFLEN  256  // max length of buffer
#define PORT   1234  // fixed server port number

char line[BUFLEN];
struct sockaddr_in me, client;
int sock, rlen, clen = sizeof(client);

int main()
{
  printf("1. create a UDP socket\n");
  sock = socket(AF_INET, SOCK_DGRAM, IPPROTO_UDP);

  printf("2. fill me with server address and port number\n");
  memset((char *)&me, 0, sizeof(me));
  me.sin_family = AF_INET;
  me.sin_port = htons(PORT);
  me.sin_addr.s_addr = htonl(INADDR_ANY); // use localhost

  printf("3. bind socket to server IP and port\n");
  bind(sock, (struct sockaddr*)&me, sizeof(me));

  printf("4. wait for datagram\n");
  while(1){
     memset(line, 0, BUFLEN);
     printf("UDP server: waiting for datagram\n");
```

```
    // recvfrom() gets client IP, port in sockaddr_in clinet
    rlen=recvfrom(sock,line,BUFLEN,0,(struct sockaddr *)&client,&clen);
    printf("received a datagram from [host:port] = [%s:%d]\n",
            inet_ntoa(client.sin_addr), ntohs(client.sin_port));
    printf("rlen=%d: line=%s\n", rlen, line);
    printf("send reply\n");
    sendto(sock, line, rlen, 0, (struct sockaddr*)&client, clen);
  }
}

/***** C13.1.b: UDP client.c file *****/
#include<stdio.h>
#include<stdlib.h>
#include<string.h>
#include<sys/socket.h>
#include <netinet/ip.h>

#define SERVER_HOST "127.0.0.1" // default server IP: localhost
#define SERVER_PORT  1234       // fixed server port number
#define BUFLEN       256        // max length of buffer

char line[BUFLEN];
struct sockaddr_in server;
int sock, rlen, slen=sizeof(server);

int main()
{
  printf("1. create a UDP socket\n");
  sock = socket(AF_INET, SOCK_DGRAM, IPPROTO_UDP);

  printf("2. fill in server address and port number\n");
  memset((char *) &server, 0, sizeof(server));
  other.sin_family = AF_INET;
  other.sin_port = htons(SERVER_PORT);
  inet_aton(SERVER_HOST, &server.sin_addr);

  while(1){
    printf("Enter a line : ");
    fgets(line, BUFLEN, stdin);
    line[strlen(line)-1] = 0;
    printf("send line to server\n");
    sendto(sock,line,strlen(line),0,(struct sockaddr *)&server,slen);
    memset(line, 0, BUFLEN);
    printf("try to receive a line from server\n");
    rlen=recvfrom(sock,line,BUFLEN,0,(struct sockaddr*)&server,&slen);
    printf("rlen=%d: line=%s\n", rlen, line);
  }
}
```

图 13.7 所示为运行 UDP 服务器 - 客户机程序 C13.1 的示例输出。

```
root@wang:~/NET# server
1. create a UDP socket
2. fill in server address with port number
3. bind socket to port
4. wait for datagram loop
UDP server: waiting for datagram
received a datagram from [host:port] = [127.0.0.1:46349]
rlen=9: line=test line
send reply
UDP server: waiting for datagram
```

```
root@wang:~/NET# client
1. create a UDP socket
2. fill in server address and port number
Enter a line : test line
send line to server
try to receive a line from server
rlen=9: line=test line
Enter a line :
```

图 13.7 UDP 服务器 - 客户机程序的输出

13.15 TCP 回显服务器 - 客户机程序

本节介绍了一个使用 TCP 的简单回显服务器 - 客户机程序。该程序用 C13.2 表示。为简单起见，我们假设服务器和客户机都在同一台计算机上运行，服务器端口号硬编码为1234。下图给出了 TCP 服务器和客户机的算法和操作顺序。

```
---------- TCP Server ----------- | ---------- TCP Client ----------
1. create a TCP socket            |  1. create a TCP socket sock
2. fill server_addr = [IP, port]  |  2. fill server_addr = [IP, port]
3. bind socket to server_addr     |
4. listen at socket by listen()   |
5. int csock = accept()        <==|= 3. connect() to server via sock
   -------------------------------|--------------------------------
                                     4. while(gets() line){
6. while(read()line from csock){<-|--5.   write() line   to  sock
     write() reply   to  csock  --|--->   read()  reply from sock
   }                              |      }
7. close newsock;                 |  6. exit
8. loop to 5 to accept new client |
-------------------------------------------------------------------
```

```
/******** C13.2.a: TCP server.c file ********/
#include <stdio.h>
#include <stdlib.h>
#include <string.h>
#include <sys/socket.h>
#include <netdb.h>

#define MAX          256
#define SERVER_HOST "localhost"
#define SERVER_IP   "127.0.0.1"
#define SERVER_PORT  1234

struct sockaddr_in  server_addr, client_addr;
```

```
int  mysock, csock;        // socket descriptors
int  r, len, n;            // help variables

int server_init()
{
  printf("================== server init =====================\n");
  //  create a TCP socket by socket() syscall

  printf("1 : create a TCP STREAM socket\n");
  mysock = socket(AF_INET, SOCK_STREAM, 0);
  if (mysock < 0){
     printf("socket call failed\n"); exit(1);
  }

  printf("2 : fill server_addr with host IP and PORT# info\n");
  // initialize the server_addr structure
  server_addr.sin_family = AF_INET;                 // for TCP/IP
  server_addr.sin_addr.s_addr = htonl(INADDR_ANY); // This HOST IP
  server_addr.sin_port = htons(SERVER_PORT);   // port number 1234

  printf("3 : bind socket to server address\n");
  r = bind(mysock,(struct sockaddr*)&server_addr,sizeof(server_addr));
  if (r < 0){
     printf("bind failed\n"); exit(3);
  }
  printf("    hostname = %s port = %d\n", SERVER_HOST, SERVER_PORT);
  printf("4 : server is listening ....\n");
  listen(mysock, 5); // queue length = 5
  printf("=================== init done =======================\n");
}

int main()
{
  char line[MAX];
  server_init();
  while(1){  // Try to accept a client request
    printf("server: accepting new connection ....\n");
    // Try to accept a client connection as descriptor newsock
    len = sizeof(client_addr);
    csock = accept(mysock, (struct sockaddr *)&client_addr, &len);
    if (csock < 0){
       printf("server: accept error\n"); exit(1);
    }
    printf("server: accepted a client connection from\n");
    printf("-----------------------------------------------\n");
    printf("Clinet: IP=%s  port=%d\n",
                inet_ntoa(client_addr.sin_addr.s_addr),
                ntohs(client_addr.sin_port));
    printf("-----------------------------------------------\n");
    // Processing loop: client_sock <== data ==> client
    while(1){
       n = read(csock, line, MAX);
       if (n==0){
           printf("server: client died, server loops\n");
           close(csock);
```

```
            break;
        }
        // show the line string
        printf("server: read  n=%d bytes; line=%s\n", n, line);
        // echo line to client
        n = write(csock, line, MAX);
        printf("server: wrote n=%d bytes; ECHO=%s\n", n, line);
        printf("server: ready for next request\n");
    }
  }
}

/******** C13.2.b: TCP client.c file TCP ********/
#include <stdio.h>
#include <stdlib.h>
#include <string.h>
#include <sys/socket.h>
#include <netdb.h>

#define MAX          256
#define SERVER_HOST "localhost"
#define SERVER_PORT  1234
struct sockaddr_in  server_addr;
int sock, r;

int client_init()
{
  printf("======= clinet init ==========\n");
  printf("1 : create a TCP socket\n");
  sock = socket(AF_INET, SOCK_STREAM, 0);
  if (sock<0){
     printf("socket call failed\n"); exit(1);
  }

  printf("2 : fill server_addr with server's IP and PORT#\n");
  server_addr.sin_family = AF_INET;
  server_addr.sin_addr.s_addr = htonl(INADDR_ANY); // localhost
  server_addr.sin_port = htons(SERVER_PORT); // server port number

  printf("3 : connecting to server ....\n");
  r = connect(sock,(struct sockaddr*)&server_addr, sizeof(server_addr));
  if (r < 0){
     printf("connect failed\n");  exit(3);
  }

  printf("4 : connected OK to\n");
  printf("-------------------------------------------------------\n");
  printf("Server hostname=%s PORT=%d\n", SERVER_HOST, SERVER_PORT);
  printf("-------------------------------------------------------\n");
  printf("========= init done ==========\n");
}

int main()
{
  int n;
```

```
  char line[MAX], ans[MAX];
  client_init();
  printf("********  processing loop  *********\n");
  while (1){
    printf("input a line : ");
    bzero(line, MAX);              // zero out line[ ]
    fgets(line, MAX, stdin);       // get a line from stdin
    line[strlen(line)-1] = 0;      // kill \n at end
    if (line[0]==0)                // exit if NULL line
       exit(0);
    // Send line to server
    n = write(sock, line, MAX);
    printf("client: wrote n=%d bytes; line=%s\n", n, line);
    // Read a line from sock and show it
    n = read(sock, ans, MAX);
    printf("client: read  n=%d bytes; echo=%s\n", n, ans);
  }
}
```

图 13.8 所示为运行 TCP 服务器 – 客户机程序 C13.2 的示例输出。

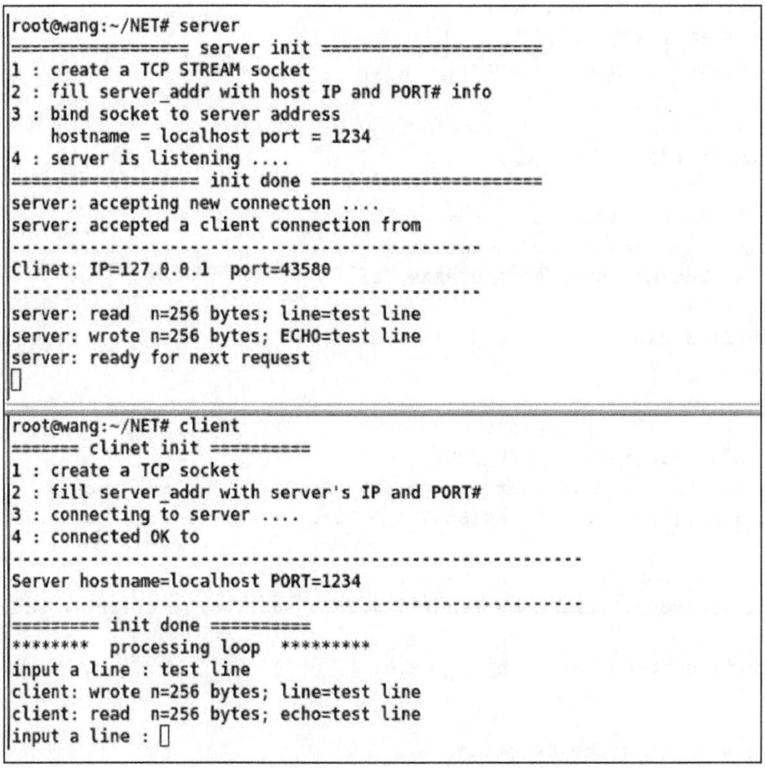

```
root@wang:~/NET# server
================== server init ======================
1 : create a TCP STREAM socket
2 : fill server_addr with host IP and PORT# info
3 : bind socket to server address
    hostname = localhost port = 1234
4 : server is listening ....
=================== init done =======================
server: accepting new connection ....
server: accepted a client connection from
-----------------------------------------------
Clinet: IP=127.0.0.1  port=43580
-----------------------------------------------
server: read  n=256 bytes; line=test line
server: wrote n=256 bytes; ECHO=test line
server: ready for next request
```

```
root@wang:~/NET# client
======= clinet init ==========
1 : create a TCP socket
2 : fill server_addr with server's IP and PORT#
3 : connecting to server ....
4 : connected OK to
---------------------------------------------------------
Server hostname=localhost PORT=1234
---------------------------------------------------------
========= init done ==========
********  processing loop  *********
input a line : test line
client: wrote n=256 bytes; line=test line
client: read  n=256 bytes; echo=test line
input a line : 
```

图 13.8　TCP 服务器 – 客户机程序的输出示例

13.16　主机名和 IP 地址

因此，我们假设服务器和客户机在同一台计算机上运行（使用本地主机或 IP=127.0.0.1），并且服务器使用固定端口号。如果读者打算在不同的主机上运行服务器和客户机，服务器端口号由操作系统内核分配，则需要知道服务器的主机名或 IP 地址及其端口号。如果某台计算机运行 TCP/IP，它的主机名通常记录在 /etc/hosts 文件中。库函数

gethostname(char *name, sizeof(name))

在 name 数组中返回计算机的主机名字符串，但它可能不是用点记法表示的完整正式名称，也不是其 IP 地址。库函数

struct hostent *gethostbyname(void *addr, socklen_t len, int typo)

可以用来获取计算机的全名及其 IP 地址。它会返回一个指向 <netdb.h> 中 hostent 结构体的指针

```
struct hostent {
   char  *h_name;          // official name of host
   char **h_aliases;       // alias list
   int    h_addrtype;      // host address type
   int    h_length;        // length of address
   char **h_addr_list;     // list of addresses
}
#define h_addr h_addr_list[0] // for backward compatibility
```

注意，h_addr 被定义为一个 char *，但它以网络字节序指向一个 4 字节的 IP 地址。h_addr 的内容可以存取为

- u32 NIP = *(u32 *)h_addr 是按网络字节序排列的主机 IP 地址。
- u32 HIP = ntohl(NIP) 是按主机字节序排列的 NIP。
- inet_ntoa(NIP) 将 NIP 转换为一个用点记法表示的字符串。

下面的代码段展示了如何使用 gethostbyname() 和 getsockname() 来获取服务器 IP 地址和端口号（若是动态分配）。服务器必须发布其主机名或 IP 地址和端口号，以便客户机连接。

```
/********* TCP server code *********/
char myname[64];
struct sockaddr_in server_addr, sock_addr;

// 1. gethostname(), gethostbyname()
  gethostname(myname,64);
  struct hostent *hp = gethostbyname(myname);
  if (hp == 0){
     printf("unknown host %s\n", myname); exit(1);
  }

// 2. initialize the server_addr structure
  server_addr.sin_family = AF_INET;    // for TCP/IP
  server_addr.sin_addr.s_addr = *(long *)hp->h_addr;
  server_addr.sin_port = 0; // let kernel assign port number

// 3. create a TCP socket
  int mysock = socket(AF_INET, SOCK_STREAM, 0);

// 4. bind socket with server_addr
  bind(mysock,(struct sockaddr *)&server_addr, sizeof(server_addr));

// 5. get socket addr to show port number assigned by kernel
  getsockname(mysock, (struct sockaddr *)&name_addr, &length);

// 6. show server host name and port number
```

```
  printf("hostname=%s IP=%s port=%d\n", hp->h_name,
         inet_ntoa(*(long *)hp->h_addr), ntohs(name_addr.sin_port));

/********* TCP client code *********/
// run as client server_name server_port
  struct sockaddr_in server_addr, sock_addr;
// 1.  get server IP by name
  struct hostent *hp = gethostbyname(argv[1]);
  SERVER_IP   = *(long *)hp->h_addr;
  SERVER_PORT = atoi(argv[2]);

// 2. create TCP socket
  int sock = socket(AF_INET, SOCK_STREAM, 0);

// 3. fill server_addr with server IP and PORT#
  server_addr.sin_family = AF_INET;
  server_addr.sin_addr.s_addr = SERVER_IP;
  server_addr.sin_port = htons(SERVER_PORT);

// 4. connect to server
  connect(sock,(struct sockaddr *)&server_addr, sizeof(server_addr));
```

将主机名和IP地址合并到TCP服务器 - 客户机程序中留作习题节的练习。

13.17 TCP编程项目：互联网上的文件服务器

上述TCP服务器 - 客户机程序可作为基于TCP网络编程的基础。只需更改数据内容及其处理数据的方式，就可以使其适应不同的应用程序。例如，客户机可能向服务器发送文件操作命令，而不是数字，服务器处理命令并将结果发送回客户机。本编程项目打算开发一对通过互联网执行文件操作的TCP服务器 - 客户机。项目规范如下。

13.17.1 项目规范

设计并实现一个TCP服务器和一个TCP客户机通过互联网来执行文件操作。下面的图表说明了服务器和客户机的算法。

-------------------------- 服务器 -----------------------------------
（1）将虚拟根目录设置为当前工作目录（CWD）
（2）通知服务器主机名和端口号
（3）接受来自客户机的连接
（4）从客户机获取：命令行 = cmd 路径名
（5）在路径名上执行 cmd
（6）将结果发送给客户机
（7）重复（4）直到客户机断开连接
（8）重复（3）以接受新的客户机连接

-------------------------- 客户机 --------------------------------
（1）按服务器主机名和端口号连接到服务器上
（2）提示用户输入：命令行 = cmd 路径名
（3）将命令行发送给服务器
（4）接收来自服务器的结果
（5）重复（2），直至命令行为空或退出命令

在命令行中，根据命令的不同，路径名可以是文件或目录。有效命令有

- mkdir：创建一个带路径名的目录。
- rmdir：删除名为路径名的目录。
- rm：删除名为路径名的文件。
- cd：将当前工作目录（CWD）更改为路径名。
- pwd：显示当前工作目录的绝对路径名。
- ls：按照与 Linux 的 ls –l 相同的格式列出当前工作目录或路径名。
- get：从服务器下载路径名文件。
- put：将路径名文件上传到服务器。

13.17.2 帮助和提示

（1）当在互联网主机上运行时，服务器必须发布其主机名或 IP 地址和端口号，以允许客户机连接。出于安全原因考虑，服务器应将虚拟根设置为服务器进程的当前工作目录，以防止客户机访问虚拟根上的文件。

（2）大多数命令都很容易实现。例如，前 5 个命令都只需要一个 Linux 系统调用，如下所示。

```
mkdir  pathname:   int r = mkdir(pathname, 0755);  // default permissions
rmdir  pathname:   int r = rmdir(pathname);
rm     pathname:   int r = unlink(pathname);
cd     pathname:   int r = chdir(pathname);
pwd :  char buf[SIZE]; char *getcwd(buf, SIZE);
```

对于 ls 命令，读者可参考第 8 章的 8.6.7 节中的 ls.c 程序。对于 get filename 命令，它实质上与分成两部分的文件复制程序相同。服务器打开文件进行读取，读取文件内容并将其发送给客户机。客户机打开一个文件进行写操作，从服务器接收数据并将接收到的数据写入文件。对于 put filename 命令，只需对调服务器和客户机的角色。

（3）当使用 TCP 时，数据是连续流。为确保客户机和服务器可以发送 / 接收命令和简单的回复，最好是双方都写 / 读固定大小的行，例如 256 字节。

（4）读取器必须设计用户级协议，以确保在服务器和客户机之间进行正确的数据传输。下文描述了需要用户级数据传输协议的命令。对于 ls 命令，服务器有两个选择。第一个选择是，对于目录中的每个条目，服务器生成一行表格

```
-rw-r--r--  link gid uid size date name
```

并将行保存到临时文件中。所有行累加完成后，服务器将整个临时文件发送给客户机。在这种情况下，服务器必须能够向客户机说明文件的开始和结束位置。因为 ls 命令只生成文本行，所以读者可以使用特殊 ASCII 字符作为文件的开始和结束标记。

第二个选择是，服务器可以生成一行并立即发送给客户机，然后再生成和发送下一行。在这种情况下，服务器必须能够在开始发送行以及没有更多行可发送时告诉客户机。

对于传输文件内容的 get/put 命令，不能使用特殊 ASCII 字符作为文件的开始和结束标记。这是因为二进制文件可能包含 ASCII 代码。这个问题的标准解决方案是位填充 [SDLC, IBM, 1979]。在该方案中，发送者在发送数据时，会在每个序列（5 个或更多连续的 1 位）后面插入一个外加 0 位，这样传输的数据就不可能包含 6 个或更多个连续的 1。在接收端，

接收者会去除每个序列（5 个连续 1）之后的外加 0 位。这样，双方都可以使用特殊标记位模式 01111110 作为文件的开始和结束标记。如果没有硬件支持，位填充会很低效。读者必须考虑其他方法来同步服务器和客户机。

提示：按文件大小。

（5）该项目适合 2 人团队合作。在开发期间，团队成员可以讨论如何设计用户级协议，并在团队成员之间分配实现工作。在测试期间，一个成员可在互联网主机上运行服务器，而另一个成员可在另一主机上运行客户机。

图 13.9 所示为运行项目程序的示例输出。

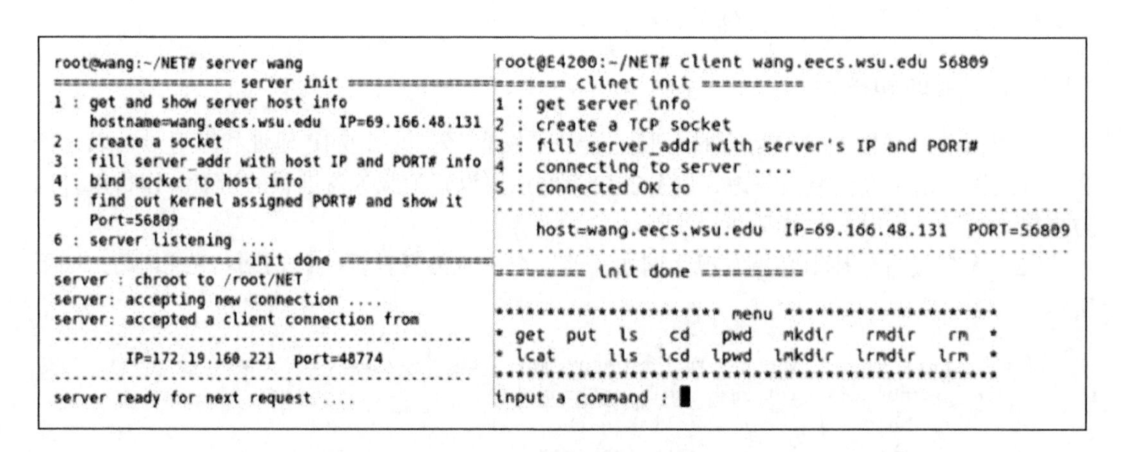

图 13.9　用于文件操作的 TCP 服务器 – 客户机

如图所示，服务器在主机 wang.eecs.wsu.edu 上运行，端口号为 56809。客户机在另一台主机上运行，IP=172.19.160.221，端口号为 48774。除了指定命令外，客户机端还可以实现本地文件操作命令，这些命令由客户机直接执行。

13.17.3　多线程 TCP 服务器

在 TCP 服务器 – 客户机编程项目中，服务器一次只服务一个客户机。在多线程服务器中，它可以接受来自多个客户机的连接，并同时或并发地为它们提供服务。这可以通过新服务器进程的 fork-exec 实现，也可以通过同一服务器进程中的多个线程实现。我们把这个扩展作为另一个可能的编程项目。

13.18　Web 和 CGI 编程

万维网（WWW）或 Web 是互联网上的资源和用户组合，它使用超文本传输协议（HTTP）（RFC 2616 1999）进行信息交换。自 20 世纪 90 年代初问世以来，随着互联网能力的不断扩展，Web 已经成为世界各地人们日常生活中不可或缺的一部分。因此，对于计算机科学的学生来说，了解这项技术非常重要。在本节中，我们将介绍 HTTP 和 Web 编程的基础知识。Web 编程通常包括 Web 开发中涉及的编写、标记和编码，其中包括 Web 内容、Web 客户机和服务器脚本以及网络安全。狭义上，Web 编程指的是创建和维护 Web 页面。Web 编程中最常用的语言是 HTML、XHTML、JavaScript、Perl 5 和 PHP。

13.18.1 HTTP 编程模型

HTTP 是一种基于服务器 - 客户机的协议，用于互联网上的应用程序。它在 TCP 上运行，因为它需要可靠的文件传输。图 13.10 所示为 HTTP 编程模型。

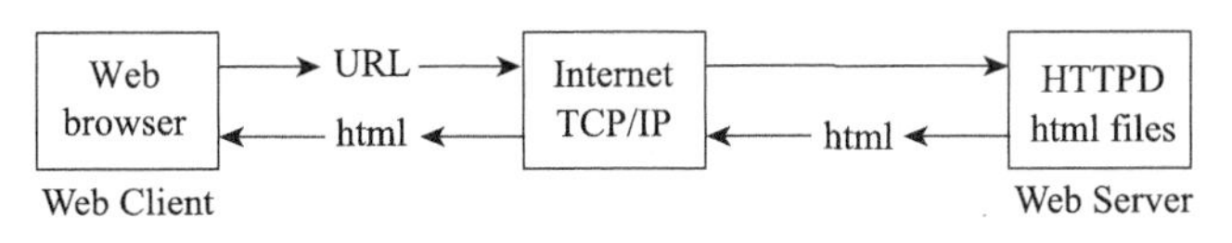

图 13.10 HTTP 编程模型

在 HTTP 编程模型中，HTTP 服务器在 Web 服务器主机上运行。它等待 HTTP 客户机（通常是 Web 浏览器）的请求。在 HTTP 客户机端，用户输入以下形式的 URL(统一资源定位符)：

http://hostname[/filename]

向 HTTP 服务器发送请求，请求文件。在 URL 中，http 标识 http 协议，hostname 是 http 服务器的主机名，filename 是请求的文件名。如果没有指定 filename，默认文件名是 index.html。客户机首先连接到服务器，以发送请求。服务器在接收到请求后，会将请求的文件发送回客户机。请求的文件通常是用 HTML 语言编写的 Web 页面文件，可在浏览器上解释和显示，但也可能是其他格式的文件，如视频、音频甚至是二进制文件。

在 HTTP 中，客户机可发出多个 URL，将请求发送到不同的 HTTP 服务器。客户机与特定服务器保持永久连接不但没有必要，也不可取。客户机连接到服务器只是为了发送请求，发送完毕后会关闭连接。同样，服务器连接到客户机也只是为了发送应答，发送完毕后会再次关闭连接。每个请求或应答都需要一个单独的连接。这意味着 HTTP 是一种无状态协议，因为在连续的请求或应答之间不需要维护任何信息。自然，这将导致大量系统开销和效率低下。为弥补这一缺乏状态信息的问题，HTTP 服务器和客户机可使用 cookie 来提供和维护它们之间的一些状态信息。

13.18.2 Web 页面

Web 页面是用 HTML 标记语言编写的文件。Web 文件通过一系列 HTML 元素指定 Web 页面的布局，可在 Web 浏览器上解释和显示。常用的 Web 浏览器有 Internet Explorer、Firefox、Google Chrome 等。创建 Web 页面相当于使用 HTML 元素作为构建块创建文本文件。与其说它是编程，不如说是文书类工作。因此，我们不讨论如何创建 Web 页面。相反，我们将只使用一个示例 HTML 文件来说明 Web 页面的本质。下面给出了一个简单的 HTML Web 文件。

```
1.  <html>
2.  <body>
3.  <h1>H1 heading: A Simple Web Page</h1>
4.  <P>This is a paragraph of text</P>
5.  <!---- this is a comment line ---->
6.  <P><img src="firefox.jpg" width=16></P>
7.  <a href="http://www.eecs.wsu.edu/~cs360">link to cs360 web page</a>
    <P>
8.  <font color="red">red</font>
9.  <font color="blue">blue</font>
10. <font color="green">green</font>
```

```
    </P>
    <!--- a table ---->
11. <table>
12.     <tr>
13.        <th>name</th>
14.        <th>ID</th>
15.     </tr>
16.     <tr>
17.         <th>kwang</th>
18.         <th>12345</th>
19.     </tr>
20. </table>
    <!---- a FORM ---->
21. <FORM>
22.   Enter command:  <INPUT NAME="command"><P>
23.   Submit command: <INPUT TYPE="submit" VALUE="Click to Submit">
24. </FORM>
25. </body>
26. </html>
```

HTML 文件内容说明

HTML 文件包含多个 HTML 元素。每个 HTML 元素由一对匹配的打开和关闭标记指定。

```
<tag>contents</tag>
```

实际上，HTML 文件本身可以看作是由一对匹配的 <html> 标记指定的 HTML 元素。

```
<html>HTML file</html>
```

第 1 至 26 行指定了一个 HTML 文件。一个 HTML 文件包含一个由一对匹配的 <body> 标记指定的主体

```
<body>body of HTML file</body>
```

第 2 至 25 行指定了 HTML 文件的主体。

HTML 文件可以使用标签 <H1> 至 <H7> 来显示不同字体大小的标题行。

第 3 行指定了一个标题行。

每对匹配的 <P> 标记指定一个段落，该段落显示在新行中。

第 4 行指定了一段文本。

第 5 行指定了一个注释行，浏览器将会忽略它。

第 6 行指定了一个图像文件，它将按每行的宽度像素显示。

第 7 行指定了一个链接元素

```
<a HREF="link_URL">link</a>
```

其中，属性 HREF 指定一个 link_URL 和一个描述链接的文本字符串。浏览器通常以深蓝色显示链接文本。如果用户单击链接，它将把请求定向到由 link_URL 标识的 Web 服务器。这可能是 Web 页面最强大的特性。它允许用户通过跟随链接导航到 Web 中的任何地方。

第 8 行至第 10 行使用 <font> 元素来显示不同颜色的文本。<font> 元素还可以指定不同

字体大小和样式的文本。

第 11 行至第 20 行指定一个表，<tr> 为行，<th> 为各行中的列。

第 21 至 24 行指定了一个表单，用于收集用户输入并将其提交给 Web 服务器进行处理。我们将在下一节的 CGI 编程中更详细地解释和演示 HTML 表单。

图 13.11 所示为上述 HTML 文件的 Web 页面。

图 13.11　来自 HTML 文件的 Web 页面

13.18.3　托管 Web 页面

现在我们有一个 HTML 文件。必须把它放在 Web 服务器中。当 Web 客户机通过 URL 请求 HTML 文件时，Web 服务器必须能够定位该文件并将其发送回客户机进行显示。有几种方法可以托管 Web 页面。

（1）与商业 Web 托管服务提供商签约，按月缴费。对于大多数普通用户来说，根本不会选择这个方法。

（2）机构或部门服务器上的用户账户。如果读者在运行 Linux 的服务器上有一个用户账户，那么通过以下步骤可在用户主目录中轻松创建一个私人网站

- 登录到服务器上的用户账户。
- 在用户主目录中，创建一个权限为 0755 的 public_html 目录。
- 在 public_html 目录中，创建 index.html 文件和其他 HTML 文件。

举个例子，从互联网上的 Web 浏览器，输入 URL

http://cs360.eecs.wsu.edu/~kcw

将会通过服务器 cs360.eecs.wsu.edu 访问作者的网站。

（3）独立 PC 或笔记本电脑：这里所述的步骤适用于运行标准 Linux 的独立 PC 或笔记本电脑，但是它也应该适用于其他 Unix 平台。出于某种原因，Ubuntu Linux 选择了不同的方式，偏离了 Linux 的标准设置。Ubuntu 用户可以查看官方 Ubuntu 文档的 HTTPD-Apache2 Web 服务器网页，以了解详细信息。

13.18.4　为 Web 页面配置 HTTPD

（3）.1 下载并安装 Apache Web 服务器。大多数 Linux 发行版，例如 Slackware Linux 14.2，都安装了 Apache Web 服务器，即大家熟知的 HTTPD。

（3）.2 输入 ps -x | grep httpd，查看 httpd 是否正在运行。如果没有运行，输入

```
sudo chmod  +x  /etc/rc.d/rc.httpd
```

使 rc.httpd 文件可执行。这将在下一次启动期间启动 httpd。或者，也可以通过输入以下内容来手动启动 httpd

```
sudo /usr/sbin/httpd -kstart.
```

（3）.3 配置 httpd.conf 文件：HTTPD 服务器的操作由 /etc/httpd/ 目录中的 httpd.conf 文件控制。要允许个人用户网站，可按以下方式编辑 httpd.conf 文件。

- 如果这些行被注释掉了，就取消注释

```
Loadmodule dir_module MODULE_PATH
Include /etc/httpd/extra/httpd-userdir.conf
```

- 在第一个目录块中

```
<Directory />
   Require all denied   # deny requests for all files in /
</Directory>
```

将“Require all denied”行更改为“Require all granted”。

- 所有用户主目录都在 /home 目录中。将行

```
DocumentRoot  /srv/httpd/htdocs 更改为 DocumentRoot  /home
```

- 所有 HTML 文件的默认目录是 htdoc。将行

```
<Directory /srv/httpd/htdocs> 更改为 <Directory /home>
```

编辑 httpd.conf 文件之后，重新启动 httpd 服务器或输入命令

```
ps -x | grep httpd    # to see httpd PID
sudo kill -s 1 httpdPID
```

kill 命令向 httpd 发送一个数字 1 信号，使它在不重新启动 httpd 服务器的情况下读取更新后的 httpd.conf 文件。

（3）.4 通过 adduser user_name 创建一个用户账户。通过以下用户名登录到用户账户：

```
ssh user_name@localhost
```

像以前一样创建 public_html 目录和 HTML 文件。

然后打开 Web 浏览器并输入 http://localhost/~user_name，以访问用户的 Web 网页。

13.18.5　动态 Web 页面

用标准 HTML 编写的 Web 页面都是静态的。当从服务器获取并用浏览器显示时，Web 页面的内容不会变化。要显示包含不同内容的 Web 页面，必须再次从服务器获取不同的 Web 页面文件。动态 Web 页面是内容可以变化的页面。动态 Web 页面有两种，分别称为客户机端动态 Web 页面和服务器端动态 Web 页面。客户机端动态 Web 页面文件包含用 JavaScript 写的代码，这些代码由 JavaScript 解释器在客户机上执行。它可以响应用户输入、时间事件等来对 Web 页面进行本地修改，而不需要与服务器进行任何交互。服务器端动态 Web 页面是真正的动态页面，因为它们是根据 URL 请求中的用户输入动态生成的。服务器端动态 Web 页面的核心在于服务器在 HTML 文件中执行 PHP 代码，或 CGI 程序通过用户输入生成 HTML 文件的能力。

13.18.6　PHP

PHP（超文本预处理器)(PHP 2017）是一种用于创建服务器端动态 Web 页面的脚本语言。PHP 文件用 .php 后缀标识。它们本质上是 HTML 文件，包含 Web 服务器要执行的 PHP 代码。当 Web 客户机请求 PHP 文件时，Web 服务器将首先处理 PHP 语句来生成一个 HTML 文件，然后将该文件发送给请求客户机。所有运行 Apache HTTPD 服务器的 Linux 系统都

支持 PHP，但是必须要启用它。要启用 PHP，只需对 httpd.conf 文件进行少量修改，如下所示。

（1）	DirectoryIndex index.php	# 默认 Web 页面是 index.php
（2）	AddType 应用程序 /x-httpd-php.php	# 添加 .php 扩展类型
（3）	包含 /etc/httpd/mod_php.conf	# 加载 php5 模块

在 httpd.conf 中启用 PHP 后，重启 httpd 服务器，它将把 PHP 模块加载到 Linux 内核中。当 Web 客户机请求 .php 文件时，httpd 服务器会复刻一个子进程，以执行 .php 文件中的 PHP 语句。由于子进程在映像中加载了 PHP 模块，因此可以快速高效地执行 PHP 代码。或者，httpd 服务器也可以配置为将 PHP 作为 CGI 执行，这会比较慢，因为它必须使用 fork-exec 来调用 PHP 解释器。为了提高效率，我们假设 .php 文件由 PHP 模块处理。下面，我们将通过示例展示基本的 PHP 编程。

（1）HTML 文件中的 PHP 语句：在 .php 文件中，PHP 语句包含在一对 PHP 标签中

```
<?php
  // PHP statements
?>
```

下面显示了一个简单的 PHP 文件 p1.php。

```
<html>
<body>
<?php
   echo "hello world<br>";      // hello world<br>
   print "see you later<br>";   // see you later<br>
?>
</body>
</html>
```

与 C 程序类似，每个 PHP 语句的末尾都必须使用分号。它可以在匹配的 /* 和 */ 对中包含注释块，或者单个注释行使用 // 和 #。对于输出，PHP 可以使用 echo 或 print。在 echo 或 print 语句中，多个项必须用点（字符串连接）运算符分开，而不是用空格分开，如

```
echo "hello world<br> . "see you later<br>";
```

当 Web 客户机请求 p1.php 文件时，httpd 服务器的 PHP 预处理器将先执行 PHP 语句来生成 HTML 行（显示在 PHP 行右侧），然后将生成的 HTML 文件发送到客户机。

（2）PHP 变量：在 PHP 中，变量以 $ 符号开头，其后是变量名。PHP 变量值可以是字符串、整数或浮点数。与 C 语言不同，PHP 是一种松散类型的语言。用户不需要使用类型定义变量。与 C 语言一样，PHP 也允许类型转换更改变量类型。对于大多数情况，PHP 还可以自动将变量转换为不同的类型。

```
<?php
  $PID = getmypid();          // return an integer
  echo "pid = $PID <br>";     // pid = php Process PID
  $STR = "hello world!";      // a string
  $A = 123; $B = "456";       // integer 123, string "456"
  $C = $A + $B;               // type conversion by PHP
```

```
echo "$STR Sum=$C<br>";    // hello world! Sum=579<br>
  ?>
```

与 C 语言或 sh 脚本中的变量一样，PHP 变量可以是局部变量、全局变量或静态变量。

（3）PHP 运算符：在 PHP 中，变量和值可以由以下运算符操作。

- 算术运算符
- 赋值运算符
- 比较运算符
- 递增 / 递减运算符
- 逻辑运算符
- 字符串运算符
- 数组运算符

大多数 PHP 运算符类似于 C 语言中的运算符。我们只在 PHP 中显示一些特殊的字符串和数组运算符。

（3）.1 字符串运算：大多数字符串运算，如 strlen()、strcmp() 等，与 C 语言相同。PHP 还支持许多其他字符串运算，它们的语法形式通常略有不同。例如，PHP 不使用 strcat()，而是使用点运算符进行字符串连接，如 “string1” . “string2”。

（3）.2 PHP 数组：PHP 数组用 array() 关键字定义。PHP 支持索引数组和多维数组。可通过数组索引单步遍历索引数组，如

```
<?php
 $name  = array('name0", "name1", "name2", "name3");
 $value = array(1,2,3,4);  // array of values
 $n = count($name);        // number of array elements
 for ($i=0; $i<n; $i++){   // print arrays by index
     echo $name[$i]; echo " = ";
     echo $value[$i];
 }
?>
```

此外，PHP 数组可由运算符进行集合运算，如并集（+）和比较，也可进行列表运算，可按不同的顺序排列。

关联数组：关联数组由名称 – 值对组成。

```
$A = array('name"=>1, "name1"=>2, "name2"=>3, "name"=>4);
```

关联数组允许按名称而不是索引访问元素值，如

```
echo "value of name1 = " . $A['name1'];
```

（4）PHP 条件语句：PHP 通过 if、if-else、if-elseif-else 和 switch-case 语句支持条件和测试条件，这些语句与 C 语言中的语句全相同，但语法上略有不同。

```
<?php
  if (123 < 456){           // test a condition
     echo "true<br>";       // in matched pair of {  }
  } else {
     echo "not true<br>";  // in matched pair of {  }
```

```
  }
?>
```

（5）PHP 循环语句：PHP 支持 while、do-while 和 for 循环语句，它们与 C 语言中的语句相同。foreach 语句可用于在没有显式索引变量的情况下遍历数组。

```
<?php
  $A = array(1,2,3,4);
  for ($i=0; $i<4; $i++){     // use an index variable
      echo "A[$i] = $A[$i]<br>";
  }
  foreach ($A as $value){     // step through array elements
      echo "$value<br>";
  }
?>
```

（6）PHP 函数：在 PHP 中，函数用函数关键字定义。函数的格式和用法与 C 语言中的函数类似。

```
<?php
  function nameValue($name, $value) {
    echo "$name . " has value " . $value <br>";}
  nameValue("abc", 123); // call function with 2 parameters
  nameValue("xyz", 456);
?>
```

（7）PHP 日期和时间函数：PHP 有许多内置函数，比如 date() 和 time()。

```
<?php
   echo date("y-m-d");  // time in year-month-day format
   echo date("h:i:sa);  // time in hh:mm:ss format
?>
```

（8）PHP 中的文件操作：PHP 的一大优点是它统一支持文件操作。PHP 中的文件操作包括系统调用函数，例如 mkdir()、link()、unlink()、stat() 等，以及 C 语言的标准 I/O 库函数，例如 fopen()、fread()、fwrite() 和 fclose() 等。这些函数的语法可能与 C 语言中的 I/O 库函数不同。大多数函数不需要特定的数据缓冲区，因为它们会直接接受字符串参数或返回字符串。通常，对于写操作，Apache 进程必须对用户目录具有写权限。读者可参阅 PHP 文件操作手册了解详细信息。下面的 PHP 代码段展示了如何显示文件内容，以及如何通过 fopen()、fread() 和 fwrite() 复制文件。

```
<?php
  readfile("filename");           // same as cat filename
  $fp = fopen("src.txt", "r");    // fopen() a file for READ
  $gp = fopen("dest.txt", "w");   // fopen a file for WRITE
  while(!feof($fp){               // feof() same as in C
     $s = fread($fp, 1024);       // read nbytes from $fp
     fwrite($gp, $s, strlen($s));// write string to  $gp
  }
?>
```

（9）PHP 中的表单：PHP 中的表单和表单提交与 HTML 中相同。通过一个包含 PHP 代码的 PHP 文件进行表单处理，PHP 代码由 PHP 预处理器执行。PHP 代码可以从提交的表单中获取输入，并以通常方式处理它们。我们用一个例子来说明 PHP 中的表单处理。

（9）.1 form.php 文件：该 .php 文件显示一个表单，收集用户输入，并向 httpd 服务器提交该表单，其中 METHOD=” post”，ACTION=” action.php”。图 13.12 所示为表单 .php 文件的 Web 页面。当用户单击 Submit 时，它将表单输入发送到 HPPTD 服务器进行处理。

```
<!------- form.php file ------->
<html><body>
<H1>Submit a Form</H1>
<form METHOD="post" ACTION="action.php">
  command:  <input type="text" name="command"><br>
  filename: <input type="text" name="filename"><br>
  parameter:<input type="text" name="parameter"><br>
  <input type="submit">
</form>
</body></html>
```

（9）.2 action.php 文件：action.php 文件包含处理用户输入的 PHP 代码。表单输入通过关键字从 global_POST 关联数组中提取。为简单起见，我们只回显用户提交的输入名 - 值对。图 13.13 显示返回的 action.php Web 页面。如图所示，它由服务器端的 Apache 进程执行，PID=30256。

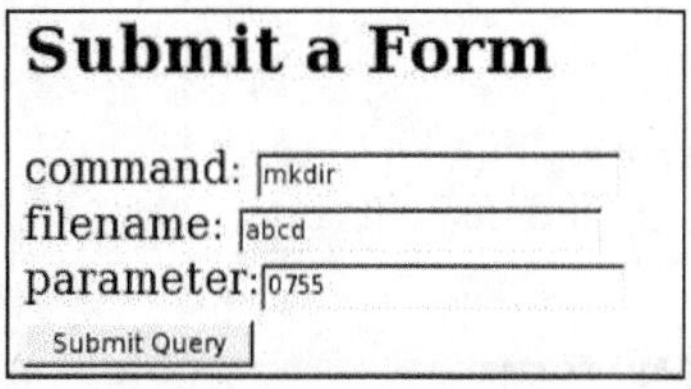

Submit a Form

command: mkdir
filename: abcd
parameter: 0755
Submit Query

图 13.12 PHP 提交表单

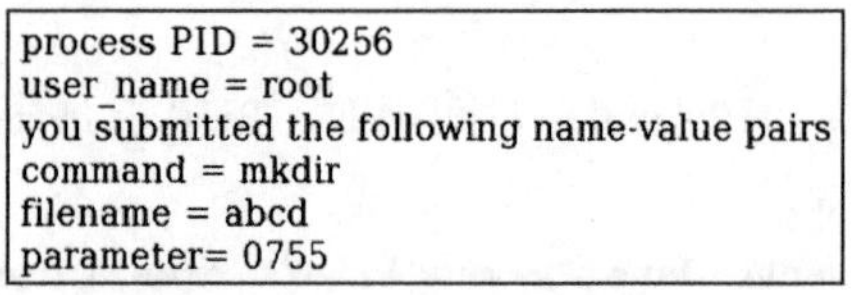

process PID = 30256
user_name = root
you submitted the following name-value pairs
command = mkdir
filename = abcd
parameter= 0755

图 13.13 PHP 表单处理

```
<!------- action.php file ------->
<html><body>
<?php
  echo "process PID = " . getmypid() . "<br>";
  echo "user_name = " . get_current_user() . "<br>";
  $command  = $_POST["command"];
  $filename = $_POST["filename"];
  $parameter= $_POST["parameter"];
  echo "you submitted the following name-value pairs<br>";
  echo "command   = " . $command .   "<br>";
  echo "filename = " . $filename .  "<br>";
  echo "parameter= " . $parameter . " <br>";
?>
</body></html>
```

PHP 概要

PHP 是一种用于在互联网上开发应用程序的通用脚本语言。从技术角度看，PHP 并无新颖之处，但它代表了 Web 编程数十年努力的发展。PHP 不是一种单一语言，而是多种其他语言的集成。它具有许多早期脚本语言（如 sh 和 Perl）的特性。它具有 C 语言的大部分标准特性和功能，还提供了 C 语言标准 I/O 库的文件操作。实际上，PHP 通常用作 Web 站点的前端，它与后端数据库引擎交互，通过动态 Web 页面在线存储和检索数据。PHP 与

MySQL 数据库的接口将在第 14 章中讨论。

13.18.7 CGI 编程

CGI 代表**通用网关接口**（RFC 3875 2004）。它是一种协议，允许 Web 服务器执行程序，根据用户输入动态生成 Web 页面。使用 CGI, Web 服务器不必维护数百万个静态 Web 页面文件来满足客户机请求。相反，它通过动态生成 Web 页面来满足客户机请求。图 13.14 显示了 CGI 编程模型。

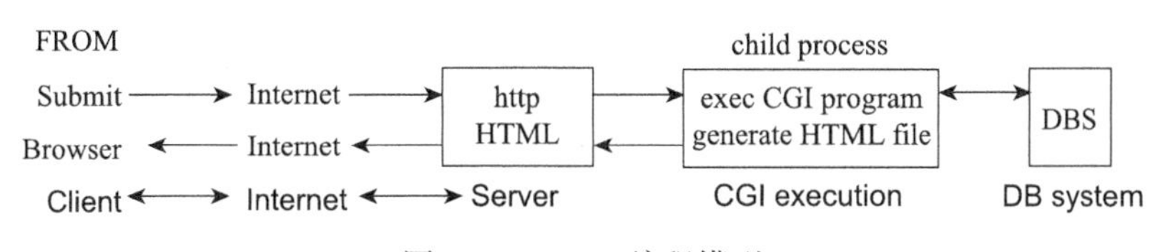

图 13.14　CGI 编程模型

在 CGI 编程模型中，客户机发送一个请求，该请求通常是一个 HTML 表单，包含供服务器执行的 CGI 程序的输入和名称。在接收到请求后，httpd 服务器会派生一个子进程来执行 CGI 程序。CGI 程序可以使用用户输入来查询数据库系统，如 MySQL，从而根据用户输入生成 HTML 文件。当子进程结束时，httpd 服务器将生成的 HTML 文件发送回客户机。CGI 程序可用任何编程语言编写，如 C 语言、sh 脚本和 Perl。

13.18.8 配置 CGI 的 HTTPD

在 HTTPD 中，CGI 程序的默认目录是 /srv/httpd/cgi-bin。这样网络管理员控制和监督可以执行 CGI 程序的用户。在许多机构中，出于安全原因，通常会禁用用户级 CGI 程序。为了允许用户级 CGI 编程，必须将 httpd 服务器配置为启用用户级 CGI。编辑 /etc/httpd/httpd.conf 文件，并将 CGI 目录设置更改为

```
<Directory "/home/*/public_html/cgi-bin">
    Options +ExecCGI
    AddHandler cgi-script .cgi .sh .bin .pl
    Order allow,deny
    Allow from all
</Directory>
```

修改后的 CGI 目录块将 CGI 目录设置为用户主目录中的 public_html/cgi-bin/。cgi-script 设置将后缀为 .cgi、.sh、.bin 和 .pl（用于 Perl 脚本）的文件指定为可执行的 CGI 程序。

13.19 CGI 编程项目：通过 CGI 实现动态 Web 页面

本编程项目可供读者练习 CGI 编程。它将远程文件操作、CGI 编程和服务器端动态 Web 页面组合成一个包。项目的组织结构如下。

（1）用户网站：在 cs360.eecs.wsu.edu 服务器上，每个用户都有一个登录账户。用户的主目录有一个 public_html 目录，其中包含一个 index.html 文件，可以通过 URL 从互联网上的 Web 浏览器访问该文件，URL 为

http://cs360.eecs.wsu.edu/~username

下面给出了用户 kcw 的 index.html 文件。

```
<!---------- index.html file ---------->
<html>
<body bgcolor="#00FFFF"
<H1>Welcome to KCW's Web Page</H1><P>
<img src="kcw.jpg" width=100><P>

<FORM METHOD="POST" ACTION=\
  "http://cs360.eecs.wsu.edu/~kcw/cgi-bin/mycgi.bin">
  Enter command: <INPUT NAME="command"> (mkdir|rmdir|rm|cat|cp|ls)<P>
  Enter filename1: <INPUT NAME="filename1"> <P>
  Enter filename2: <INPUT NAME="filename2"> <P>
  Submit command: <INPUT TYPE="submit" VALUE="Click to Submit"> <P>
</FORM>

</body>
</html>
```

图 13.15 所示为上述 index.html 文件对应的 Web 页面。

图 13.15　HTML 表单页面

（2）HTML 表单：index.html 文件包含一个 HTML 表单。

```
<FORM METHOD="POST" ACTION=\
  "http://cs360.eecs.wsu.edu/~kcw/cgi-bin/mycgi.bin">
  Enter command: <INPUT NAME="command"> (mkdir|rmdir|rm|cat|cp|ls)<P>
  Enter filename1: <INPUT NAME="filename1"> <P>
  Enter filename2: <INPUT NAME="filename2"> <P>
  Submit command: <INPUT TYPE="submit" VALUE="Click to Submit"> <P>
</FORM>
```

在 HTML 表单中，METHOD 指定如何提交表单输入，ACTION 指定 Web 服务器和 Web 服务器要执行的 CGI 程序。有两种表单提交方法。在 **GET 方法**中，用户输入包含在提交的 URL 中，这使得它们直接可见，而且输入数据的数量也受到了限制。由于这些原因，很少使用 GET 方法。在 **POST 方法**中，用户输入采用（URL）编码，并通过数据流进行传输，这更安全，而且输入数据的数量也不受限制。因此，大多数 HTML 表单使用 POST 方法。HTML 表单允许用户在提示框中输入。当用户单击 Submit 按钮时，输入将被发送到 Web 服务器进行处理。在 HTML 示例文件中，表单输入将被发送到 HTTP 服务器（cs360.eecs.wsu.edu），该服务器将在用户的 cgi-bin 目录中执行 CGI 程序 mycgi.bin。

（3）CGI 目录和 CGI 程序：HTTPD 服务器被配置为允许用户级 CGI。下图显示了用户级 CGI 设置。

```
/home/username:  public_html
                           | ----  index.html
                           | ----  cgi-bin
                                      | ---- mycgi.c
                                      | ---- util.o
                                      | ---- mycgi.bin, sample.bin
```

在 cgi-bin 目录中，mycgi.c 是一个 C 程序，它获取并显示提交的 HTML 表单中的用户输入。它回显用户输入并生成一个包含表单的 HTML 文件，表单被发送回 Web 客户机进行显示。下面给出了 mycgi.c 程序代码。

```
/************** mycgi.c file *************/
#include <stdio.h>
#include <stdlib.h>
#include <string.h>
#define MAX 1000
typedef struct{
    char *name;
    char *value;
}ENTRY;
ENTRY entry[MAX];
extern int getinputs(); // in util.o
int main(int argc, char *argv[])
{
  int i, n;
  char cwd[128];
  n = getinputs();      // get user inputs name=value into entry[ ]
  getcwd(cwd, 128);     // get CWD pathname
  // generate a HTML file containing a HTML FORM
  printf("Content-type: text/html\n\n"); // NOTE: 2 new line chars
  printf("<html>");
  printf("<body bgcolor=\"#FFFF00\"); // background color=YELLOW
  printf("<p>pid=%d uid=%d cwd=%s\n", getpid(), getuid(), cwd);
  printf("<H2>Echo Your Inputs</H2>");
  printf("You submitted the following name/value pairs:<p>");
  for(i=0; i<=n; i++)
     printf("%s = %s<P>", entry[i].name, entry[i].value);
  printf("<p>");
  // create a FORM webpage for user to submit again
  printf("---------- Send Back a Form Again ------------<P>");
  printf("<FORM METHOD=\"POST\"
   ACTION=\"http://cs360.eecs.wsu.edu/~kcw/cgi-bin/mycgi.bin\">");
   printf("<font color=\"RED\">");
   printf("Enter command : <INPUT NAME=\"command\"> <P>");
   printf("Enter filename1: <INPUT NAME=\"filename1\"> <P>");
   printf("Enter filename2: <INPUT NAME=\"filename2\"> <P>");
   printf("Submit command: <INPUT TYPE=\"submit\" VALUE=\"Click to \
           Submit\"> <P>");
  printf("</form>");
  printf("</font>");
  printf("----------------------------------------------<p>");
  printf("</body>");
  printf("</html>");
}
```

图 13.16 显示了上述 CGI 程序生成的 Web 页面。

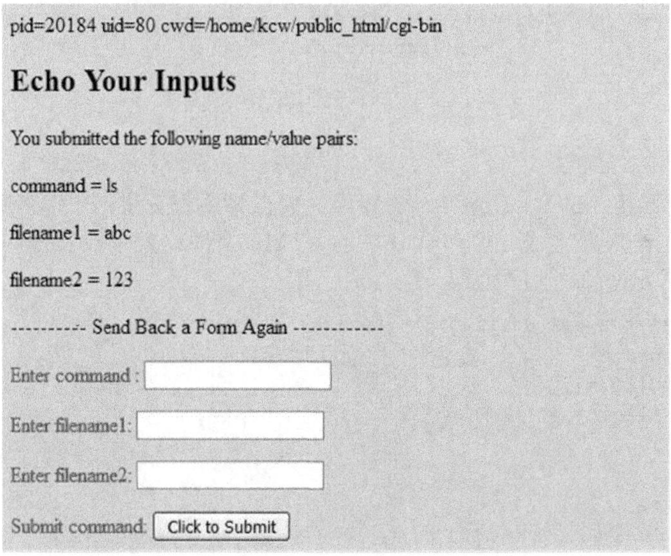

pid=20184 uid=80 cwd=/home/kcw/public_html/cgi-bin

Echo Your Inputs

You submitted the following name/value pairs:

command = ls

filename1 = abc

filename2 = 123

---------- Send Back a Form Again ------------

Enter command :

Enter filename1:

Enter filename2:

Submit command: Click to Submit

图 13.16　CGI 程序的 Web 页面

（4）CGI 程序：当 HTTPD 服务器接收到 CGI 请求时，它会复刻一个子进程（UID 80）来执行 CGI 程序。表单提交方法、输入编码和输入数据长度在进程的环境变量 REQUEST_METHOD, CONETENT_TYPE and CONTENT_LENGTH 中，输入数据在 stdin 中。输入数据通常采用 URL 编码。将输入解码为名称 – 值对很简单，但相当烦琐。因此，提供了一个预编译的 util.o 文件。它包含一个函数

```
int getinputs()
```

将用户输入解码到名称 – 值字符串对中。CGI 程序是由以下 Linux 命令生成的：

```
gcc -o mycgi.bin mycgi.c util.o
```

（5）动态 Web 页面：获取用户输入后，CGI 程序可以处理用户输入，以生成输出数据。在示例程序中，它只回显用户输入。然后，它将 HTML 语句作为行写入 stdout 来生成 HTML 文件。要使行成为 HTML 文件，第一行必须是

```
printf("Content-type: text/html\n\n");
```

有 2 个新行字符。其余各行可以是任何 HTML 语句。在示例程序中，它生成与提交的表单相同的表单，用于在下一次提交中获取新的用户输入。

（6）SETUID（设置用户标识符）程序：通常，CGI 程序仅使用用户输入从服务器端数据库读取数据来生成 HTML 文件。为了安全起见，它可能不允许 CGI 程序修改服务器端数据库。在编程项目中，我们允许使用请求来执行文件操作，例如 mkdir、rmdir、cp 文件等，这些操作需要将权限写入用户目录中。由于 CGI 进程的 UID=80，所以不应将它写入用户目录。CGI 进程可通过两种方法写入用户目录。在第一种方法中，用户可将 cgi-bin 目录权限设置为 0777，但这并不可取，因为这样任何人都可以在其目录中写入。第二种方法是通过以下操作使 CGI 程序成为 SETUID 程序：

```
chmod u+s mycgi.bin
```

当进程执行 SETUID 程序时，它会暂时假定程序所有者的 UID，允许它写入用户目录。

（7）文件操作用户请求：由于项目目标是 CGI 编程，我们只假设以下简单的文件操作是用户请求：

```
ls        [directory]           // list directory in ls -l form of Linux
mkdir dirname permission        // make a directory
rmdir  dirname                  // rm director y
unlink  filename                // rm file
cat       filename              // show file contents
cp        file1 file2           // copy file1 to file2
```

（8）示例解决方案：在 cgi-bin 目录中，sample.bin 是项目的示例解决方案。读者可将 index.html 文件中的 mycgi.bin 替换为 sample.bin，以测试用户请求并观察结果。

13.20 习题

1. 修改 C13.1 程序的 UDP 客户机代码，将 usec 中的时间戳附加到数据报。修改 UDP 服务器代码，将时间戳（也是在 usec 中）附加到接收到的数据报中，然后再将其发送回去。在客户机代码中，计算并显示 msec 中数据报的往返时间。
2. 修改程序 13.1 的 UDP 客户机代码，添加超时和重新发送。对于发送的每个数据报，启动一个带超时值的实时模式间隔定时器。如果客户机在超时到期之前收到服务器的回复，则取消定时器。如果超时过期时客户机未收到任何回复，则重新发送数据报。
3. 修改程序 C13.2 的 TCP 客户机代码，向服务器发送两个用空格分开的整数，如 1 2。修改 TCP 服务器代码，返回两个数字的和，如 1 2 sum=3。
4. 修改程序 13.2 的 TCP 服务器代码，使用 gethostbyname() 和内核分配的端口号。修改 TCP 客户机代码，以按名称查找服务器 IP。
5. POSIX 建议使用新的函数 getaddrinfo() 而不是 gethostbyname() 来获取主机名称和 IP 地址。请阅读关于 getaddrinfo() 的 Linux 手册页，并使用 getaddrinfo() 在程序 13.2 中重新实现 TCP 服务器和客户机。
6. 修改 13.7 节中的 TCP/IP 编程项目，在服务器和客户机实现位填充。
7. 在 13.7 节的 TCP/IP 编程项目中实现多线程服务器，以支持多个客户机。

参考文献

ARP, RFC 826, 1982
Apache, HTTP Server Project,https://httpd.apache.org, 2017
Comer, D., “Internetworking with TCP/IP Principles, Protocols, and Architecture”, Prentice Hall, 1988.
Comer, D., “Computer Networks and Internets with Internet Applications”, 3rd edition, Pretence Hall, 2001
IBM Synchronous Data Link Control, IBM, 1979
PHP: History of PHP and Related Projects, www, php.net, 2017
Rago, Stephen A, “Unix System V Network Programming”. Addision Wesley, 1993
RFC 134, Domain Names-Concepts and facilities, 1987
RFC 768, User Datagram Protocol, 1980
RFC 826, An Ethernet address Resolution protocol, 1982
RFC 1035, Domain Names-Implementation and Specification, 1987
RFC1180, A TCP/IP Tutorial, January 1991
RFC 2131, Dynamic Host Configuration Protocol, 1997
RFC 2616, Hypertext Transfer Protocol - HTTP/1.1, June 1999
RFC 3875, The Common Gateway Interface (CGI) Version 1.1, Oct. 2004
Stevens, Richard W., Fenner, Bill, Ruddof Andrew M., “UNIX Network Programming”, Volume 1, 3rd edition, Addison Wesley, 2004

MySQL 数据库系统

摘要

本章讨论了 MySQL 关系数据库系统；介绍了 MySQL 并指出了它的重要性；展示了如何在 Linux 机器上安装和运行 MySQL；演示了如何使用 MySQL 在命令模式和批处理模式下使用 SQL 脚本创建和管理数据库；说明了如何将 MySQL 与 C 编程相结合；演示了如何将 MySQL 与 PHP 集成，通过动态 Web 页面创建和管理数据库。

14.1 MySQL 简介

MySQL（MySQL 2018）是一个关系数据库系统（Codd 1970）。在关系数据库中，数据存储在表中。每个表由多个行和列组成。表中的数据相互关联。表也可能与其他表有关联。关系结构使得可在表上运行查询来检索信息并修改数据库中的数据。关系数据库系统的标准查询语言是 SQL（结构化查询语言），包括 MySQL。

MySQL 是一个开源数据库管理系统，由服务器和客户机组成。在将客户机连接到服务器后，用户可向服务器输入 SQL 命令，以便创建数据库，删除数据库，存储、组织和检索数据库中的数据。MySQL 有广泛的应用。除了提供标准的数据库系统服务外，MySQL 和 PHP（PHP 2018）已成为大多数数据管理和在线商务网站的主干网。本章介绍了 MySQL。我们将介绍 MySQL 的基础知识，包括如何在 Linux 中安装 / 配置 MySQL，如何使用 MySQL 创建和管理简单数据库，以及如何在 C 语言和 PHP 编程环境中与 MySQL 交互。

14.2 安装 MySQL

14.2.1 Ubuntu Linux

对于 Ubuntu 16.04 及以后版本，通过以下操作安装 MySQL：

sudo apt-get install mysql-server

mysql-server 包包括一个 MySQL 服务器和一个客户机。在安装 MySQL 时，它会询问根用户密码。用户可以使用与 Ubuntu 相同的登录密码。安装 MySQL 后，可通过运行脚本对其进行配置以获得更好的安全性：

mysql_secure_installation

要得到简单和标准的安全设置，读者可以按 Y，然后按 ENTER，以接受所有问题的默认值。

14.2.2 Slackware Linux

Slackware Linux 预装了 MySQL，但仍然需要配置。否则，Slackware 会在 MySQL 数据库启动时显示一条错误消息。在 Slackware 14.0 或更早版本中，可通过以下步骤配置 MySQL。

（1）设置 my.cnf：MySQL 在启动时加载一个名为 my.cnf 的配置文件。该文件要在首次设置 MySQL 时创建。在 /etc 目录中，有几个示例 my.cnf 文件，文件名分别是 my-small.cnf、my-large.cnf 等。选择所需的版本来创建 my.cnf 文件，如

cp /etc/my-small.cnf /etc/my.cnf

（2）安装所需数据库：MySQL 需要一个所需数据库集，用于用户识别等。要安装它们，可使用 mysql 用户作为超级用户，并使用以下命令安装所需的初始数据库。

mysql_install_db

（3）设置所需的系统权限：该步骤确保 mysql 用户拥有 mysql 系统的所有权。

chown -R mysql.mysql /var/lib/mysql

（4）通过以下操作使 /etc/rc.d/rc.mysqld 可执行：

chmod 755 /etc/rc.d/rc.mysqld

这将在后续系统引导上自动启动 MySQL 守护进程 mysqld。

Slackware 14.2 使用 MariaDB 代替 MySQL。除了未绑定到 Oracle 之外，MariaDB 与 MySQL 数据库基本相同。事实上，它仍然使用 mysqld 作为它的守护进程名。在 Slackware 14.2 中，MySQL 已经有一个默认 .cnf 文件，所以不再需要第 1 步。按照上面列出的第 2 步到第 4 步配置 MySQL。配置完成后，通过以下操作手动启动 MySQL 守护进程 mysqld:

/etc/rc.d/rc.mysqld -start

14.3 使用 MySQL

假设 MySQL 服务器已经设置好并在 Ubuntu 或 Slackware Linux 机器上运行。MySQL 服务器可以设置为支持不同的用户。为简单起见，我们将只假定根用户。为了使用 MySQL，用户必须运行一个 MySQL 客户机来连接到服务器。MySQL 支持来自远程 IP 主机的客户机连接。为了简单起见，我们将在同一台机器（即默认本地主机）上运行服务器和客户机。下面几节介绍如何使用 MySQL 管理数据库。

14.3.1 连接到 MySQL 服务器

使用 MySQL 的第一步是运行 MySQL 客户机程序。从 X-window 终端输入 MySQL 客户机命令 mysql，它连接到同一台计算机上默认本地主机上的 MySQL 服务器。

```
mysql -u root -p                 # specify the root user with password
Enter password:                  #  enter the MySQL root user password
mysql>                           # mysql prompt
```

连接到 MySQL 服务器后，即可访问 MySQL shell，如 mysql> 提示符所示。MySQL shell 类似于普通的 shell。它会显示一个 mysql> 提示符，要求用户输入可供 MySQL 服务器执行的 SQL 命令。与普通 sh 类似，它还维护一个命令历史记录，允许用户通过箭头键回忆和修改先前的命令。然而，它只接受 MySQL 命令或 MySQL 脚本，而不接受普通 sh 命令。在输入 MySQL 命令时，读者要注意以下几点。

- 所有的 MySQL 命令行末尾必须是分号。对于长命令，可在单独行中输入命令短语（按下 ENTER 键）。MySQL 将会通过 -> 符号继续提示更多的输入，直到它看到一个结束分号。
- MySQL 命令行不区分大小写。虽然不是强制要求，但为了清楚和更容易识别，通常使用大写编写 MySQL 命令，使用小写编写数据库、表、用户名或文本。

14.3.2 显示数据库

SHOW DATABASES 命令可显示 MySQL 中的当前数据库。

示例 14.1：

```
mysql> SHOW DATABASES;
+--------------------+
| Database           |
+--------------------+
| information_schema |
| mysql              |
| performance_schema |
| test               |
+--------------------+
4 rows in set (0.01 sec)
```

14.3.3 新建数据库

如果数据库 dbname 还不存在，那么 CREATE DATABASE dbname 命令将创建一个名为 dbname 的新数据库。如果数据库已经存在，则可以使用可选的 IF NOT EXISTS 子句对该命令进行限定，以避免出现错误消息。

示例 14.2：

```
mysql> CREATE DATABASE testdb
Query OK; 1 row affected (0.02 sec)  # mysql response
```

创建新数据库后，输入 SHOW DATABASES 以查看结果。

```
mysql> SHOW DATABASES;
+--------------------+
| Database           |
+--------------------+
| information_schema |
| mysql              |
| performance_schema |
| test               |
| testdb             |
+--------------------+
5 rows in set (0.00 sec)
```

14.3.4 删除数据库

DROP DATABASE dbname 会删除已存在的命名数据库。该命令可以用一个可选的 IF EXISTS 子句限定。注意，DROP 操作是不可逆的。一旦数据库被删除，就无法撤销或恢复。因此，须谨慎使用。

示例 14.3：

```
mysql> DROP DATABASE testdb;
Query OK; one row affected (0.04 sec)

mysql> SHOW DATABASES;
+--------------------+
| Database           |
+--------------------+
```

```
| information_schema |
| mysql              |
| performance_schema |
| test               |
+--------------------+
4 rows in set (0.01 sec)
```

14.3.5 选择数据库

假设 MySQL 已经有几个数据库。为了操作特定的数据库，用户必须通过 USE dbname 命令选择一个数据库。

示例 14.4：

```
mysql> CREATE DATABASE cs360;
Query OK, 1 row affected (0.00 sec)

mysql> use cs360;
Database changed
```

14.3.6 创建表

本节将介绍如何在 MySQL 数据库中创建表。假设 cs360 数据库包含以下学生记录。

```
struct students{
   int student_id;  # an integer ID number must exist
   char name[20];   # name string of 20 chars
   int score;       # an integer exam score, which may not exist
 }
```

CREATE TABLE table_name 命令会在当前数据库中创建一个表。命令语法如下：

```
CREATE TABLE [IF NOT EXISTS] tableName (
   columnName columnType columnAttribute, ...
   PRIMARY KEY(columnName),
   FOREIGN KEY (columnNmae) REFERENCES tableName (columnNmae)
)
```

对于单个表，不需要 FOREIGN KEY 子句。

示例 14.5：

```
mysql> CREATE TABLE students (
       student_id INT NOT NULL PRIMARY KEY AUTO_INCREMENT,
       name CHAR(20),
       score INT );
Query OK, 0 rows affected (0.00 sec)

mysql> SHOW TABLES;
+--------------------+
| Tables_in_cs360    |
+--------------------+
| students           |
+--------------------+
1 row in set (0.00 sec)
```

DESCRIBE 或 DESC 命令显示表格式和列属性。

```
mysql> DESCRIBE students
+------------+----------+------+-----+---------+----------------+
| Field      | Type     | Null | Key | Default | Extra          |
+------------+----------+------+-----+---------+----------------+
| student_id | int(11)  | NO   | PRI | NULL    | auto_increment |
| name       | char(20) | YES  |     | NULL    |                |
| score      | int(11)  | YES  |     | NULL    |                |
+------------+----------+------+-----+---------+----------------+
3 rows in set (0.00 sec)
```

可以看出，学生信息表有 3 个字段或列，其中：

- “student_id”是一个整数，不能为空，用作主键，当添加新行时，它的值将自动递增。
- “name”是一个固定长度的字符串，长度为 20 字符，**可以为空**。
- “score”是一个整数。

在表中，**主键**是一个列或一组列，可用于唯一地标识行。在默认情况下，主键是**唯一的**。从表中可以看出，name 列可以为空，可以生成一个具有有效的 student_id 但没有名称的行。为了避免这种情况，应该使用**非空**属性声明它。

14.3.7 删除表

DROP TABLE table_name 命令可删除表。

示例 14.6：

```
mysql> CREATE TABLE dummy (
       id INT PRIMARY KEY, name CHAR(20));
Query OK, 0 rows affected (0.00 sec)

mysql> SHOW TABLES;
```

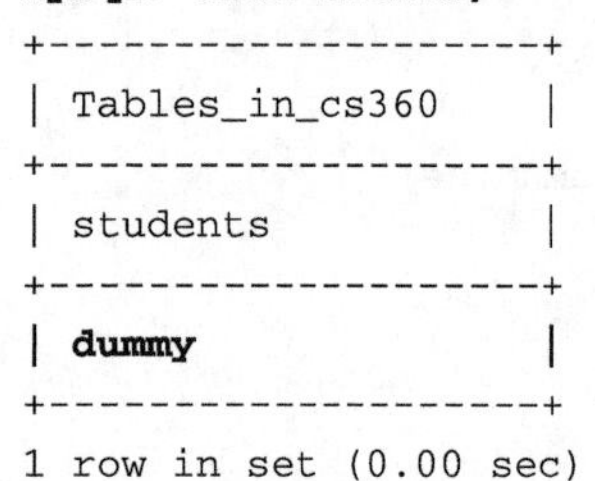

```
+--------------------+
| Tables_in_cs360    |
+--------------------+
| students           |
+--------------------+
| dummy              |
+--------------------+
1 row in set (0.00 sec)

mysql> DROP TABLE dummy;
mysql> SHOW TABLES;
```

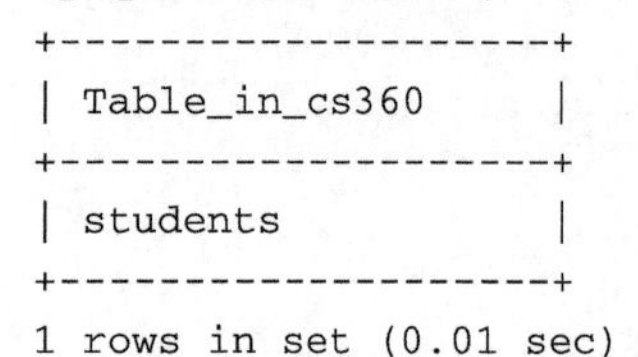

```
+--------------------+
| Table_in_cs360     |
+--------------------+
| students           |
+--------------------+
1 rows in set (0.01 sec)
```

14.3.8 MySQL 中的数据类型

在继续之前，要了解 MySQL 中使用的基本数据类型，主要有三类：数字、字符串、日期和时间。我们只显示每个类别中一些常用的数据类型。

数值类型：

- INT：整数（4 字节），TINYINT：（1 字节），SMALLINT：（2 字节）等。
- FLOAT：浮动指针数。

字符串类型：

- CHAR(size)：固定长度字符串，长度为 1～255 字符。
- VARCHAR(size)：可变长度字符串，但不能使用任何空格。
- TEXT：可变长度的字符串。

日期和时间类型：

- DATE：日期格式为 YYYY-MM-DD。
- TIME：以 HH:MM:SS 格式保存时间。

14.3.9 插入行

要在表中添加行，可使用 INSERT 命名，具有语法形式：

INSERT INTO table_name VLAUES(columnValue1, columnValue2, ...);

示例 14.7：

```
mysql> INSERT INTO students VALUES (1001, 'Baker', '50');
Query OK, 1 row affected (0.01 sec)
mysql> SELECT * FROM students;
+------------+-------+-------+
| student_id | name  | score |
+------------+-------+-------+
|       1001 | Baker |    50 |
+------------+-------+-------+
1 row in set (0.00 sec)
```

同样，我们可以向表中插入更多的学生记录。由于通过单个命令行手动插入多个条目非常烦琐，因此我们可以创建一个包含多个插入命令行的 MySQL 脚本文件，并将其用作 MySQL 的输入源文件。假设我们编辑了一个 .sql 脚本文件，如下所示。

```
/*********  insert.sql script file **********/
INSERT INTO students VALUES (NULL, "Miller", 65);
INSERT INTO students VALUES (2001, "Smith",  76);
INSERT INTO students VALUES (NULL, "Walton", 85);
```

输入 SOURCE 命令，让 mysql 使用 sql 脚本作为输入文件，如：

```
mysql> SOURCE insert.sql;   # use insert.sql as input source file
(Query OK, rows affected, etc)

mysql> SELECT * FROM students;
+------------+--------+-------+
| student_id | name   | score |
+------------+--------+-------+
|       1001 | Baker  |    50 |
|       1002 | Miller |    65 |
|       2001 | Smith  |    76 |
|       2002 | Walton |    85 |
+------------+--------+-------+
4 rows in set (0.00 sec)
```

注意，在插入新行时，如果 ID 号为空，那么 auto_increment 将为它分配下一个值（来自最后输入的 ID 值）。

14.3.10　删除行

使用 DELETE 命令从表中删除行，其语法形式如下：

```
DELETE FROM table_name;                  # delete all rows of a table
DELETE FROM table_name WHERE condition;  # delete row by condition
```

示例 14.8：要想演示删除操作，我们先添加一个新行，然后从表中删除它。

```
mysql> INSERT INTO students VALUES (NULL, 'Zach', '45');

mysql> SELECT * from students;
+------------+--------+-------+
| student_id | name   | score |
+------------+--------+-------+
|       1001 | Baker  |    50 |
|       1002 | Miller |    65 |
|       2001 | Smith  |    76 |
|       2002 | Walton |    85 |
|       2003 | Zach   |    45 |
+------------+--------+-------+
5 rows in set (0.00 sec)

mysql> DELETE FROM students WHERE name = 'Zach';
Query OK, 1 row affected (0.00 sec)

mysql> SELECT * FROM students;
+------------+--------+-------+
| student_id | name   | score |
+------------+--------+-------+
|       1001 | Baker  |    50 |
|       1002 | Miller |    65 |
|       2001 | Smith  |    76 |
|       2002 | Walton |    85 |
+------------+--------+-------+
4 rows in set (0.00 sec)
```

14.3.11　更新表

UPDATE 命令用于修改表中的现有记录（列）。它的语法形式如下：

```
UPDATE table_name SET col1 = value1, col2 = value2, ... WHERE condition;
```

示例 14.9：假设我们想把 Walton 的分数改成 92 分。

```
mysql> UPDATE students SET score = 92 WHERE name = 'Walton';

mysql> SELECT * FROM students;
+------------+--------+-------+
| student_id | name   | score |
+------------+--------+-------+
|       1001 | Baker  |    50 |
```

```
|        1002 | Miller |    65 |
|        2001 | Smith  |    76 |
|        2002 | Walton |    92 |
+------------+--------+-------+
4 rows in set (0.00 sec)
```

同样，我们可以通过修改 WHERE 条件来更新其他列的值。

14.3.12 修改表

ALTER TABLE 命令用于添加、删除或修改当前表中的列。它还用于添加和删除当前表中的各种约束条件。

1. 修改表名

如需修改表名，可使用以下命令：

```
ALTER TABLE table_name RENAME TO new_name;
```

2. 添加列

要在表中添加列，可使用以下命令：

```
ALTER TABLE table_name ADD column_name datatype;
```

3. 删除行

可使用以下命令删除列：

```
ALTER TABLE table_name DROP column_name datatype;
```

4. 更改 / 修改行

可使用以下命令修改表中某列的数据类型：

```
ALTER TABLE table_name ALTER COLUMN column_name datatype;
```

示例 14.10：假设我们要在 students 表中添加一个名为 grade 的列，然后根据学生的考试分数给他们分配相应的字母等级。

```
mysql> ALTER TABLE students ADD grade CHAR(2); // for 'A', 'B+', etc.
(Quert OK, 4 rows affected (0,01 sec)

mysql> SELECT * FROM students;
+------------+--------+-------+-------+
| student_id | name   | score | grade |
+------------+--------+-------+-------+
|       1001 | Baker  |    50 | NULL  |
|       1002 | Miller |    65 | NULL  |
|       2001 | Smith  |    76 | NULL  |
|       2002 | Walton |    92 | NULL  |
+------------+--------+-------+-------+
4 rows in set (0.00 sec)
```

在添加 grade 列后，所有的条目值最初都是 NULL。我们可以通过 UPDATE table_name 命令向学生分配字母等级。

示例 14.11：

```
mysql> UPDATE students SET grade = 'A' WHERE name = 'Walton';
Query OK, 1 row affected (0.00 sec)
Rows matched: 1  Changed: 1  Warnings: 0

mysql> SELECT * FROM students;
+------------+--------+-------+-------+
| student_id | name   | score | grade |
+------------+--------+-------+-------+
|       1001 | Baker  |    50 | NULL  |
|       1002 | Miller |    65 | NULL  |
|       2001 | Smith  |    76 | NULL  |
|       2002 | Walton |    92 | A     |
+------------+--------+-------+-------+
4 rows in set (0.00 sec)
```

或者，我们也可以使用 WHERE 条件句中的分数范围来分配字母等级。

示例 14.12：

```
mysql> UPDATE students SET grade = 'A' WHERE score > 80;

select * from students;
+------------+--------+-------+-------+
| student_id | name   | score | grade |
+------------+--------+-------+-------+
|       1001 | Baker  |    50 | NULL  |
|       1002 | Miller |    65 | NULL  |
|       2001 | Smith  |    76 | NULL  |
|       2002 | Walton |    92 | A     |
+------------+--------+-------+-------+
4 rows in set (0.00 sec)
```

假设我们已经执行了以下命令：

```
mysql> UPDATE students SET grade = 'B' WHERE score >= 70 AND score < 80;
mysql> UPDATE students SET grade = 'C' WHERE score >= 60 and score < 70;
mysql> UPDATE students SET grade = 'D' WHERE score < 60;

mysql> SELECT * FROM students;
+------------+--------+-------+-------+
| student_id | name   | score | grade |
+------------+--------+-------+-------+
|       1001 | Baker  |    50 | D     |
|       1002 | Miller |    65 | C     |
|       2001 | Smith  |    76 | B     |
|       2002 | Walton |    92 | A     |
+------------+--------+-------+-------+
4 rows in set (0.00 sec)
```

14.3.13 关联表

目前，我们已经展示了只包含一个表的数据库。一个真正的数据库可能包含多个相互关联的表。在 MySQL 中，使用**主键 - 外键**约束条件来定义表关系。在两个表之间创建链接，其中一个表的主键与另一个表的外键相关联。在 MySQL 中，表可能以几种方式相关联，包括：

1. 一对一（1-1）关系

一对一（1-1）关系是指两个表仅基于一个匹配行相互关联的关系。可以使用**主键－外键约束条件**创建这类关系。假设在 cs360 数据库中，每个学生都有一个唯一的电子邮箱地址。我们可在 students 表中添加每个学生的电子邮箱地址，但这需要在学生表中新增一列。相反，我们可以创建一个只包含学生电子邮箱地址的单独 email 表，并通过 email 表中的唯**一外键**定义两个表之间的 1-1 关系，该外键引用 students 表中的**主键**。

示例 14.13：

```
mysql> CREATE TABLE email (id INT PRIMARY KEY AUTO_INCREMENT,
       student_id INT UNIQUE NOT NULL, email CHAR (40),
       FOREIGN KEY (student_id) REFERENCES students(student_id));
Query OK, 0 rows affected (0.00 sec)

mysql> DESC email;
+------------+----------+------+-----+---------+----------------+
| Field      | Type     | Null | Key | Default | Extra          |
+------------+----------+------+-----+---------+----------------+
| id         | int(11)  | NO   | PRI | NULL    | auto_increment |
| student_id | int(11)  | NO   | UNI | NULL    |                |
| email      | char(40) | YES  |     | NULL    |                |
+------------+----------+------+-----+---------+----------------+
3 rows in set (0.00 sec)
```

接下来，我们将学生电子邮箱地址插入 email 表中。然后，我们查询 email 表，并显示表内容。

```
mysql> INSERT INTO email VALUES (NULL, 2002, 'walton@wsu.edu');
mysql> INSERT INTO email VALUES (NULL, 2001, 'smith@gmail.com');
mysql> INSERT INTO email VALUES (NULL, 1002, 'miller@hotmail.com');
mysql> INSERT INTO email VALUES (NULL, 1001, 'baker@gmail.com');

mysql> SELECT * FROM email;
+----+------------+--------------------+
| id | student_id | email              |
+----+------------+--------------------+
|  1 |       2002 | walton@wsu.edu     |
|  2 |       2001 | smith@gmail.com    |
|  3 |       1002 | miller@hotmail.com |
|  4 |       1001 | baker@gmail.com    |
+----+------------+--------------------+
4 rows in set (0.00 sec)
```

注意，在 email 表中，student_id 可以是任何独一无二的顺序。每个 student_id 都是一个引用学生表中主键的外键。

```
mysql> SELECT a.name, b.email FROM students a, email b WHERE
       a.student_id = b.student_id;
+--------+-------------------+
| name   | email             |
+--------+-------------------+
| Walton | walton@wsu.edu    |
| Smith  | smith@gmail.com   |
```

```
| Miller  | miller@hotmail.com |
| Baker   | baker@gmail.com    |
+--------+--------------------+
4 rows in set (0.00 sec)
```

2. 一对多（1-M）关系

在数据块中，一对多或1-M关系比1-1关系更常见，也更有用。一对多关系是指一个表中的一行在另一个表中有多个匹配行的关系。可以使用**主键－外键关系**来创建这种关系。假设cs360课程有一本必修教材和一本参考书。学生可以通过书店订购图书，书店使用book_order表来跟踪学生的图书订单。每个图书订单只属于一名学生，但每名学生可以订购一本或几本书，大多数学生甚至没有订购图书。因此，students表和book_order表之间的关系是一对多关系。在一对一关系的情况下，我们可以在book_order表中使用外键来引用学生表中的主键。使用这两个相互关联的表，我们就可以在表上运行查询来检查哪个学生没有订购所需的课本等。

示例 14.14：在本示例中，我们先创建一个book_order表，向其中添加一个外键，然后用学生的图书订单填充表。

```
mysql> CREATE TABLE book_order
       (order_no INT PRIMARY KEY AUTO_INCREMENT,
        student_id INT NOT NULL,
        book_name char(20),
        date DATE
       );
Query OK, 0 rows affected (0.01 sec)

mysql> DESC book_order;
+------------+----------+------+-----+---------+----------------+
| Field      | Type     | Null | Key | Default | Extra          |
+------------+----------+------+-----+---------+----------------+
| order_no   | int(11)  | NO   | PRI | NULL    | auto_increment |
| student_id | int(11)  | NO   |     | NULL    |                |
| book_name  | char(20) | YES  |     | NULL    |                |
| date       | date     | YES  |     | NULL    |                |
+------------+----------+------+-----+---------+----------------+
4 rows in set (0.00 sec)

mysql> ALTER TABLE book_order ADD FOREIGN KEY (student_id)
       REFERENCES students(student_id);
Query OK, 0 rows affected (0.03 sec)
Records: 0  Duplicates: 0  Warnings: 0

mysql> DESC book_order;
+------------+----------+------+-----+---------+----------------+
| Field      | Type     | Null | Key | Default | Extra          |
+------------+----------+------+-----+---------+----------------+
| order_no   | int(11)  | NO   | PRI | NULL    | auto_increment |
| student_id | int(11)  | NO   | MUL | NULL    |                |
| book_name  | char(20) | YES  |     | NULL    |                |
| date       | date     | YES  |     | NULL    |                |
+------------+----------+------+-----+---------+----------------+
4 rows in set (0.00 sec)
```

```
mysql> INSERT INTO book_order VALUES (   1, 1001, 'reference',
       '2018-04-16');
mysql> INSERT INTO book_order VALUES (NULL, 1001, 'textbook',
       '2018-04-16');
mysql> INSERT INTO book_order VALUES (NULL, 1002, 'textbook',
       '2018-04-18');
mysql> INSERT INTO book_order VALUES (NULL, 1002, 'reference',
       '2018-04-18');
mysql> INSERT INTO book_order VALUES (NULL, 2001, 'textbookd',
       '2018-04-20');
mysql> INSERT INTO book_order VALUES (NULL, 2002, 'reference',
       '2018-04-21');

mysql> SELECT * FROM book_order;
+----------+------------+-----------+------------+
| order_no | student_id | book_name | date       |
+----------+------------+-----------+------------+
|        1 |       1001 | reference | 2018-04-16 |
|        2 |       1001 | textbook  | 2018-04-16 |
|        3 |       1002 | textbook  | 2018-04-18 |
|        4 |       1002 | reference | 2018-04-18 |
|        5 |       2001 | textbookd | 2018-04-20 |
|        6 |       2002 | reference | 2018-04-21 |
+----------+------------+-----------+------------+
6 rows in set (0.00 sec)
```

3. 多对多（M-M）关系

如果一个表中的多条记录与另一个表中的多条记录相关，则两个表具有多对多（M-M）关系。例如，每名学生可以上几门课，每门课通常有多名学生。所以，students 表与课程注册表之间是 M-M 关系。处理 M-M 关系的标准方法是在两个表之间创建一个**连接表**。连接表使用外键来引用两个表中的主键，从而在两个表之间创建连接。

4. 自引用关系

表可以通过某些列自我关联。本书不讨论这种类型的自引用关系。

14.3.14 连接操作

在 MySQL 中，可使用连接操作在多个表中检索数据。连接操作有 4 种不同的类型。

- (INNER) JOIN table1, table2：检索两个表中共有的项。
- LEFT JOIN table1, table2：检索表 1 中的项以及两个表中共有的项。
- RIGHT JOIN table1, table2：检索表 2 中的项以及两个表中共有的项。
- OUTER JOIN tabel1, table2：检索两个表中**非共有以及没有用**的项。

对于正则集运算，MySQL 中的连接操作可以解释如下。+ 表示两个集合的**并集**，^ 表示两个集合的**交集**。则有

```
(INNER) JOIN t1, t2 = t1 ^ t2
LEFT    JOIN t1, t2 = t1 + (t1 ^ t2)
RIGHT   JOIN t1, t2 = t2 + (t1 ^ t2)
OUTER   JOIN t1, t2 = t1 + t2;
```

我们将通过示例演示 cs360 数据库中表的连接操作。

```
mysql> select * from students JOIN email ON
       students.student_id = email.student_id;
+------------+--------+-------+-------+----+------------+------------------+
| student_id | name   | score | grade | id | student_id | email            |
+------------+--------+-------+-------+----+------------+------------------+
|       1001 | Baker  |    50 | D     |  1 |       1001 | baker@wsu.edu    |
|       1002 | Miller |    65 | C     |  2 |       1002 | miller@gmail.com |
|       2002 | Walton |    92 | A     |  3 |       2002 | smith@yahoo.com  |
|       2002 | Walton |    92 | A     |  4 |       2002 | walton@wsu.edu   |
+------------+--------+-------+-------+----+------------+------------------+
4 rows in set (0.00 sec)

mysql> SELECT * FROM students JOIN book_order ON
       students.student_id = book_order.student_id;
+----------+------+-----+------+---------+----------+---------+----------+
|student_id|name  |score| grade| order_no |student_id| book_name | date       |
+----------+------+-----+------+---------+----------+---------+----------+
|      1001|Baker |   50| D    |        1 |      1001| reference | 2018-04-16 |
|      1001|Baker |   50| D    |        2 |      1001| textbook  | 2018-04-16 |
|      1002|Miller|   65| C    |        3 |      1002| textbook  | 2018-04-18 |
|      1002|Miller|   65| C    |        4 |      1002| reference | 2018-04-18 |
|      2001|Smith |   76| B    |        5 |      2001| textbookd | 2018-04-20 |
|      2002|Walton|   92| A    |        6 |      2002| reference | 2018-04-21 |
+----------+------+-----+------+---------+----------+---------+----------+
6 rows in set (0.00 sec)

mysql> SELECT * FROM students RIGHT JOIN book_order ON
       students.student_id = book_order.student_id;
+----------+------+-----+------+---------+----------+---------+----------+
|student_id|name  |score| grade| order_no |student_id| book_name | date       |
+----------+------+-----+------+---------+----------+---------+----------+
|      1001|Baker |   50| D    |        1 |      1001| reference | 2018-04-16 |
|      1001|Baker |   50| D    |        2 |      1001| textbook  | 2018-04-16 |
|      1002|Miller|   65| C    |        3 |      1002| textbook  | 2018-04-18 |
|      1002|Miller|   65| C    |        4 |      1002| reference | 2018-04-18 |
|      2001|Smith |   76| B    |        5 |      2001| textbookd | 2018-04-20 |
|      2002|Walton|   92| A    |        6 |      2002| reference | 2018-04-21 |
+----------+------+-----+------+---------+----------+---------+----------+
6 rows in set (0.00 sec)

mysql> SELECT * FROM students LEFT JOIN book_order ON
       students.student_id = book_order.student_id;
+----------+------+-----+------+---------+----------+---------+----------+
|student_id|name  |score| grade| order_no |student_id| book_name | date       |
+----------+------+-----+------+---------+----------+---------+----------+
|      1001|Baker |   50| D    |        1 |      1001| reference | 2018-04-16 |
|      1001|Baker |   50| D    |        2 |      1001| textbook  | 2018-04-16 |
|      1002|Miller|   65| C    |        3 |      1002| textbook  | 2018-04-18 |
|      1002|Miller|   65| C    |        4 |      1002| reference | 2018-04-18 |
|      2001|Smith |   76| B    |        5 |      2001| textbookd | 2018-04-20 |
|      2002|Walton|   92| A    |        6 |      2002| reference | 2018-04-21 |
+----------+------+-----+------+---------+----------+---------+----------+
6 rows in set (0.00 sec)

mysql> SELECT * FROM book_order JOIN students ON
```

```
        students.student_id = book_order.student_id;
+---------+-----------+---------+----------+----------+------+-----+-----+
| order_no | student_id | book_name | date | student_id|name | score | grade |
+---------+-----------+---------+----------+----------+------+-----+-----+
|      1 |      1001 | reference | 2018-04-16 |      1001|Baker |   50 | D    |
|      2 |      1001 | textbook  | 2018-04-16 |      1001|Baker |   50 | D    |
|      3 |      1002 | textbook  | 2018-04-18 |      1002|Miller|   65 | C    |
|      4 |      1002 | reference | 2018-04-18 |      1002|Miller|   65 | C    |
|      5 |      2001 | textbookd | 2018-04-20 |      2001|Smith |   76 | B    |
|      6 |      2002 | reference | 2018-04-21 |      2002|Walton|   92 | A    |
+---------+-----------+---------+----------+----------+------+-----+-----+
6 rows in set (0.00 sec)
```

14.3.15 MySQL 数据库关系图

在 MySQL 和所有关系数据库系统中，用数据库关系图来描述表之间的关系非常有用。这类关系图通常称为 ERD（实体关系图）或 EERD（增强 / 扩展 ERD）。它们可以直观地表示数据库中的各个组件及其关系。对于本章中使用的简单 cs360 数据库，可以用如下所示的数据库关系图表示。

```
     email                                    book_order
 ------------      students                  -------------
 |id         |     -------------             |order_no    |
 |student_id|<-->|student_id   |<========<<|student_id  |
 |email      |    |name          |          |book_name   |
 ------------     |score         |          |date        |
                  |grade         |          -------------
                  -------------
```

在数据库关系图中，箭头线通过将一个表中的**外键**与另一个表中引用的**主键**连接来描述表之间的关系。两端都有一个箭头标记的线表示 1-1 关系，一端有多个箭头标记的线表示 1-M 关系。

14.3.16 MySQL 脚本

与普通 Unix/Linux sh 一样，MySQL shell 也可以接受和执行脚本文件。MySQL 脚本文件的后缀是 .sql。它们包含 MySQL 服务器要执行的 MySQL 命令。我们可以使用 MySQL 脚本来创建数据库，在数据库中创建表，插入表条目和修改表内容，而不是手动输入命令行。本节介绍如何使用 MySQL 脚本。

首先，我们创建一个 do.sql 文件，如下所示。为方便起见，所有 MySQL 命令都以小写字母显示，但为了便于识别，全部加粗显示。

```
-- Comments: do.sql script file
drop database if exists cs360;

-- 1. create database cs360
create database if not exists cs360;
show databases;
use cs360;
show tables;
```

```
-- 2. create table students
create table students (id INT not null primary key auto_increment,
            name char (20) NOT NULL, score INT);
desc students;

-- 3. insert student records into students table
insert into students values (1001, 'Baker',  92);
insert into students values (NULL, 'Miller', 65);
insert into students values (2001, 'Smith',  76);
insert into students values (NULL, 'Walton', 87);
insert into students values (NULL, 'Zach',   50);

-- 4. show students table contents
select * from students;

-- 5 add grade column to students table
alter table students add grade char(2);
select * from students;

-- 6. Assign student grades by score ranges
update students set grade = 'A' where score >= 90;
update students set grade = 'B' where score >= 80 and score < 90;
update students set grade = 'C' where score >= 70 and score < 80;
update students set grade = 'D' where score >= 60 and score < 70;
update students set grade = 'F' where score < 60;

-- 7. Show students table contents
select * from students;
-- end of script file
```

使用sql脚本运行mysql客户机有两种方法。第一种方法是使用SOURCE命令让mysql接受脚本文件的输入。下面给出了带有添加备注的MySQL输出，用于标识脚本文件中的命令结果。

```
mysql> source do.sql;
Query OK, 0 rows affected, 1 warning (0.00 sec)
// create database cs360 and show it
Query OK, 1 row affected (0.00 sec)
+--------------------+
| Database           |
+--------------------+
| information_schema |
| cs360              |
| mysql              |
+--------------------+
3 rows in set (0.00 sec)

// use cs360
Database changed
Empty set (0.00 sec)

// create students table
Query OK, 0 rows affected (0.01 sec)
```

```
// describe students table
+-------+----------+------+-----+---------+----------------+
| Field | Type     | Null | Key | Default | Extra          |
+-------+----------+------+-----+---------+----------------+
| id    | int(11)  | NO   | PRI | NULL    | auto_increment |
| name  | char(20) | NO   |     | NULL    |                |
| score | int(11)  | YES  |     | NULL    |                |
+-------+----------+------+-----+---------+----------------+
3 rows in set (0.00 sec)

// insert student records
Query OK, 1 row affected (0.00 sec)
Query OK, 1 row affected (0.00 sec)
Query OK, 1 row affected (0.00 sec)
Query OK, 1 row affected (0.00 sec)
Query OK, 1 row affected (0.00 sec)

// select * from students
+------+--------+-------+
| id   | name   | score |
+------+--------+-------+
| 1001 | Baker  |    92 |
| 1002 | Miller |    65 |
| 2001 | Smith  |    76 |
| 2002 | Walton |    87 |
| 2003 | Zach   |    50 |
+------+--------+-------+
5 rows in set (0.00 sec)

// add grade column to table
Query OK, 5 rows affected (0.01 sec)
Records: 5  Duplicates: 0  Warnings: 0

// select * from students
+------+-------+-------+-------+
| id   | name  | score | grade |
+------+-------+-------+-------+
| 1001 | Baker |    92 | NULL  |
| 1002 | Miller|    65 | NULL  |
| 2001 | Smith |    76 | NULL  |
| 2002 | Walton|    87 | NULL  |
| 2003 | Zach  |    50 | NULL  |
+------+-------+-------+-------+
5 rows in set (0.00 sec)

// assign grades by scores
Query OK, 1 row affected (0.00 sec)
Rows matched: 1  Changed: 1  Warnings: 0
Query OK, 1 row affected (0.00 sec)
Rows matched: 1  Changed: 1  Warnings: 0
Query OK, 1 row affected (0.00 sec)
Rows matched: 1  Changed: 1  Warnings: 0
Query OK, 1 row affected (0.00 sec)
Rows matched: 1  Changed: 1  Warnings: 0
Query OK, 1 row affected (0.00 sec)
```

```
Rows matched: 1  Changed: 1  Warnings: 0

// show table with grades
+------+--------+-------+-------+
| id   | name   | score | grade |
+------+--------+-------+-------+
| 1001 | Baker  |    92 | A     |
| 1002 | Miller |    65 | D     |
| 2001 | Smith  |    76 | C     |
| 2002 | Walton |    87 | B     |
| 2003 | Zach   |    50 | F     |
+------+--------+-------+-------+
5 rows in set (0.00 sec)
```

第二种方法是使用 sql 脚本作为输入以批处理模式运行 mysql。

```
root@wang:~/SQL# mysql -u root -p < do.sql
Enter password:
Database
information_schema
cs360
mysql
// after creating students table and desc table
Field   Type     Null    Key     Default Extra
id      int(11) NO      PRI     NULL    auto_increment
name    char(20)         NO      NULL
score   int(11) YES             NULL
// after inserting rows and select * from students
id      name    score
1001    Baker   92
1002    Miller  65
2001    Smith   76
2002    Walton  87
2003    Zach    50
// after altering table with grade column
id      name    score   grade
1001    Baker   92      NULL
1002    Miller  65      NULL
2001    Smith   76      NULL
2002    Walton  87      NULL
2003    Zach    50      NULL
// after assigning grades
id      name    score   grade
1001    Baker   92      A
1002    Miller  65      D
2001    Smith   76      C
2002    Walton  87      B
2003    Zach    50      F
```

可以看出，批处理模式下的输出比交互或命令模式下的输出要简洁得多。

14.4 C 语言 MySQL 编程

MySQL 可以与多种编程语言配合使用，如 C 语言、Java 和 Python 等。本节介绍如何进行 C 语言 MySQL 编程。C 语言程序与 MySQL 之间的接口由 mysqlclient 库中的一系列 MySQL C API 函数（C API 2018a, b）支持。

14.4.1　使用 C 语言构建 MySQL 客户机程序

思考以下 C 程序，它可以打印 libmysqlclient 库版本。

```
// client.c file
#include <stdio.h>
#include <my_global.h>
#include <mysql.h>
int main(int argc, char *argc[])
{
  printf("MySQL client version is : %s\n", mysql_get_client_info());
}
```

若要编译程序，可输入

```
gcc client.c -I/usr/include/mysql/ -lmysqlclient
```

注意，–I 选项将包含文件路径指定为 /usr/include/mysql，–l 选项指定 mysqlclient 库。然后，运行 a.out。它会打印

```
MySQL client version is : version_number, e.g. 5.5.53
```

14.4.2　使用 C 语言连接到 MySQL 服务器

下面的 C 程序 C14.2 显示如何连接到 MySQL 服务器。

```
// c14.2.c file: connect to MySQL server
#include <stdio.h>
#include <stdlib.h>
#include <my_global.h>
#include <mysql.h>

int main(int argc, char *argv[ ])
{
   // 1. define a connection object
   MYSQL con;
   // 2. Initialize the connection object
    if (mysql_init(&con)) { // return object address printf("Connection handle
       initialized\n");
   } else {
       printf("Connection handle initialization failed\n");
       exit(1);
   }
   // 3. Connect to MySQL server on localhost
   if (mysql_real_connect(&con, "localhost", "root", "root_password",
                          "cs360", 3306, NULL, 0)) {
         printf("Connection to remote MySQL server OK\n");
   }
   else {
        printf("Connection to remote MySQL failed\n");
         exit(1);
   }
```

```
    // 4. Close the connection when done
    mysql_close(&con);
}
```

该项目包含4个步骤。

（1）将MYSQL对象con定义为连接句柄。它的作用类似于网络编程中的套接字。几乎所有的MySQL C API函数都需要这个对象指针作为参数。

（2）调用mysql_init(&con)来初始化con对象，这是规定操作。它会返回初始化对象的地址。大多数其他MySQL API函数会返回0，表示成功，返回非0，表示错误。如果出现错误，函数

```
unsigned int mysql_errno(&con) 返回错误编号。
constant char *mysql_error(&con) 返回描述错误的字符串。
```

（3）调用mysql_real_connect()，以连接到远程服务器。mysql_real_connect()的一般语法是

```
MYSQL *mysql_real_connect(MYSQL *mysql,
        const char *host,           // server hostname or IP
        const char *user,           // MySQL user name
        const char *passwd,         // user password
        const char *db,             // database name
        unsigned int port,          // 3306
        const char *unix_socket,    // can be NULL
        unsigned long client_flag); // 0
```

它可以用于连接到互联网上的任何MySQL服务器，前提是用户可以访问该MySQL服务器。为简单起见，我们假设MySQL客户机和服务器位于同一台本地主机上。连接到服务器之后，客户机程序就可以开始访问数据库了，下面将显示数据库。

（4）客户机程序要在退出之前关闭连接。

14.4.3　使用C语言构建MySQL数据库

本节将介绍如何在C语言程序中创建MySQL数据库。以下C语言程序C14.3在数据库cs360中构建students表，方法与前面几节使用MySQL命令或脚本完全相同。

```
// C14.1.c: build MySQL database in C
#include <stdio.h>
#include <stdlib.h>
#include <my_global.h>
#include <mysql.h>

MYSQL *con;        // connection object pointer
void error()
{
  printf("errno = %d %s\n", mysql_errno(con), mysql_error(con));
  mysql_close(con);
  exit(1);
}

int main(int argc, char *argv[ ])
{
```

```
  con =  mysql_init(NULL); // will allocate and initialize it
  if (con == NULL)
     error();
  printf("connect to mySQL server on localhost using database cs360\n");
  if (mysql_real_connect(con, "localhost", "root", NULL,
          "cs360", 0, NULL, 0) == NULL)
      error();
  printf("connection to server OK\n");

  printf("drop students table if exists\n");
  if (mysql_query(con, "DROP TABLE IF EXISTS students"))
     error();

  printf("create students tables in cs360\n");
  if (mysql_query(con, "CREATE TABLE students(Id INT NOT NULL
     PRIMARY KEY AUTO_INCREMENT, name CHAR(20) NOT NULL, score INT)"))
    error();
  printf("insert student records into students table\n");
  if (mysql_query(con, "INSERT INTO students VALUES(1001,'Baker',50)"))
     error();
  if (mysql_query(con, "INSERT INTO students VALUES(1002,'Miller',65)"))
     error();
  if (mysql_query(con, "INSERT INTO students VALUES(2001,'Miller',75)"))
     error();
  if (mysql_query(con, "INSERT INTO students VALUES(2002,'Smith',85)"))
     error();
  printf("all done\n");
  mysql_close(con);
}
```

连接到 MySQL 服务器后，程序使用 mysql_query() 函数对数据库进行操作。mysql_query() 函数的第二个参数与普通 MySQL 命令完全相同。它可以用于简单的 MySQL 命令，但如果参数包含二进制值或 null 字节，则不能用于命令。要执行可能包含二进制数据的命令，程序必须使用 mysql_real_query() 函数，如下一个程序所示。

```
// C14.2 Program: build MySQL database with mysql_real_query()
#include <stdio.h>
#include <stdlib.h>
#include <string.h>
#include <my_global.h>
#include <mysql.h>

#define N 5
int id[ ] =         { 1001,    1002,     2001,    2002,     2003};
char name[ ][20] = {"Baker", "Miller", "Smith", "Walton", "Zach"};
int score[ ] =      {  65,      50,       75,      95,       85};
MYSQL *con;
void error()
{
  printf("errno = %d %s\n", mysql_errno(con), mysql_error(con));
  mysql_close(con);
  exit(1);
}
int main(int argc, char *argv[ ])
```

```
{
  int i;
  char buf[1024];          // used to create commands for MySQL
  con =  mysql_init(NULL); // MySQL server will allocate and init con
  if (con == NULL)
     error();
  printf("connect to mySQL server on localhost using database cs360\n");
  if (mysql_real_connect(con, "localhost", "root", NULL,
          "cs360", 0, NULL, 0) == NULL)
      error();
  printf("connected to server OK\n");
  printf("drop students table if exists\n");
  if (mysql_query(con, "DROP TABLE IF EXISTS students"))
     error();
  printf("create students tables in cs360\n");
  if (mysql_query(con, "CREATE TABLE students(Id INT NOT NULL PRIMARY
      KEY AUTO_INCREMENT, name CHAR(20) NOT NULL, score INT)"))
     error();
  printf("insert student records into students table\n");
  for (i=0; i<N; i++){
    printf("id=%8d name=%10s, score=%4d\n",id[i],name[i],score[i]);
    sprintf(buf, "INSERT INTO students VALUES (%d, '%s', %d);",
             id[i], name[i], score[i]);
    if (mysql_real_query(con, buf, strlen(buf)))
      error();
  }
  printf("all done\n");
  mysql_close(con);
}
```

为了便于比较，我们重点介绍程序C14.3中的新特性。为了在students表中插入大量的行，最好在要插入的行数循环中发出INSERT命令。由于简单的mysql_query()函数不能接受二进制数据，因此必须使用mysql_real_query()函数。首先，我们将id[]、name[]和score[]定义为具有初始化学生数据的数组。对于每个学生记录，我们使用sprinf()在char buf[]中创建MySQL INSERT命令行，并调用mysql_real_query()，将buf[]中的字符串作为参数传递。实际上，学生记录通常来自一个数据文件。在这种情况下，C程序可以打开数据文件进行读取，读取每个学生记录，在向MySQL服务器发出命令之前对数据执行一些预处理。我们把这个问题留作14.6节的编程练习。

14.4.4 使用C语言检索MySQL查询结果

MySQL查询会返回结果，例如SELECT, DESCRIBE等。结果是一组列和行。MySQL C API提供了两种从MySQL服务器获取结果集的方法：一次性获取所有行或逐行获取。

函数mysql_store_result()用于一次性从MySQL服务器获取一组行并将其存储在本地内存中。它的用法是

```
MYSQL_RES *mysql_store_result(MYSQL *mysql);
```

函数mysql_use_result()用于启动逐行结果集检索。

```
MYSQL_RES *mysql_use_result(MYSQL *mysql);
```

每次使用返回结果的 MySQL 语句调用 mysql_query() 或 mysql_real_query() 之后，可调用其中一个函数来检索结果集。这两个函数都会返回用 MYSQL_RES 类型的对象表示的结果集。其他 API 函数使用返回的结果集对象来获取列和行集。当不再需要结果集时，使用 mysql_free_result() 函数来释放结果对象资源。下面的程序 C14.3 展示了如何检索结果集并显示表的列或行。

```
// C14.3 file: Retrieve MySQL query results
#include <my_global.h>
#include <mysql.h>
#include <string.h>

#define N 5
int id[ ] =         { 1001,    1002,     2001,     2002,     2003};
char name[ ][20] = {"Baker", "Miller", "Smith", "Walton", "Zach"};
int score[ ] =      {  65,      50,        75,       95,        85};
int grade[ ] =      { 'D',     'F',       'C',      'A',       'B'};

MYSQL *con;
void error()
{
  printf("errno = %d %s\n", mysql_errno(con), mysql_error(con));
  mysql_close(con);
  exit(1);
}

int main(int argc, char **argv)
{
  int i, ncols;
  MYSQL_ROW row;
  MYSQL_RES *result;
  MYSQL_FIELD *column;
  char buf[1024];

  con =  mysql_init(NULL);
  if (con == NULL)
     error();

  printf("connect to mySQL server on localhost using database cs360\n");
  if (mysql_real_connect(con, "localhost", "root", NULL,
          "cs360", 0, NULL, 0) == NULL)
     error();
  printf("connected to server OK\n");

  printf("drop students table if exists\n");
  if (mysql_query(con, "DROP TABLE IF EXISTS students"))
     error();

  printf("create students tables in cs360\n");
  if (mysql_query(con, "CREATE TABLE students(Id INT NOT NULL
                PRIMARY KEY AUTO_INCREMENT, name CHAR(20) NOT NULL,
                score INT, grade CHAR(2))"))
     error();
```

```
    printf("insert student records into students table\n");
    for (i=0; i<N; i++){
      printf("id =%4d name =%-8s score =%4d  %c\n",
               id[i], name[i], score[i], grade[i]);
      sprintf(buf, "INSERT INTO students VALUES (%d, '%s', %d, '%c');",
               id[i], name[i], score[i], grade[i]);
      if (mysql_real_query(con, buf, strlen(buf)))
         error();
    }
    printf("retrieve query results\n");
    mysql_query(con, "SELECT * FROM students");
    result = mysql_store_result(con); // retrieve result

    ncols = mysql_num_fields(result); // get number of columns in row
    printf("number of columns = %d\n", ncols);
    for (i=0; i<ncols; i++){
      column = mysql_fetch_field(result); // get each column
      printf("column no.%d name = %s\n", i+1, column->name);
    }

    mysql_query(con, "SELECT * FROM students");
    result = mysql_store_result(con);

    ncols = mysql_num_fields(result);
    printf("columns numbers = %d\n", ncols);
    while( row = mysql_fetch_row(result) ){
        for (i=0; i<ncols; i++)
          printf("%-8s ", row[i]);
        printf("\n");
    }
    printf("all done\n");
    mysql_close(con);
}
```

在程序 C14.3 中，为了便于识别，加粗显示专注于检索结果的代码行。除了 mysql_store_result() 函数，还会使用以下函数。

- MYSQL_FIELD *mysql_fetch_field(MYSQL_RES *result)：将结果集的一列定义为 MYSQL_FIELD 结构。重复调用该函数以检索结果集中所有列的信息。当没有其他字段时，它会返回 NULL。
- unsigned int mysql_num_fields(MYSQL_RES *result)：返回结果集中的列数。
- MYSQL_ROW mysql_fetch_row(MYSQL_RES *result)：检索结果集的下一行。当没有更多要检索的行时，会返回 NULL。

运行程序 C14.3 的输出如下。

```
connect to mySQL server on localhost using database cs360
connected to server OK
drop students table if exists
create students tables in cs360
insert student records into students table
id =1001 name =Baker    score =  65  D
id =1002 name =Miller   score =  50  F
id =2001 name =Smith    score =  75  C
```

```
id =2002 name =Walton   score =  95  A
id =2003 name =Zach     score =  85  B
retrieve query results
number of columns = 4
column no.1 name = Id
column no.2 name = name
column no.3 name = score
column no.4 name = grade
columns numbers = 4
1001      Baker    65       D
1002      Miller   50       F
2001      Smith    75       C
2002      Walton   95       A
2003      Zach     85       B
all done
```

总之，通过 C 语言编程访问 MySQL 非常简单。它只需要用户学习如何在 C 语言中使用一些 MySQL API 函数。一般来说，运行编译的二进制可执行 C 语言程序比运行解释性脚本程序更快、更有效。运行编译的 C 程序更适合进行大量的数据处理。或许，运行 C 语言程序的主要缺点是缺少 GUI 界面。因此，我们转向基于 GUI 的 PHP 编程环境。

14.5 PHP MySQL 编程

PHP（PHP 2018）通常用作 Web 站点的前端，它与后端数据库引擎（如 MySQL）交互，通过动态 Web 页面在线存储和检索数据。PHP 和 MySQL 相结合可提供创建所有类型 Web 页面的多种选择，从小型个人联系表到大型公司门户网站。本节介绍基本的 PHP MySQL 编程。

要想通过 PHP 进行 MySQL 编程，读者必须同时访问可与 MySQL 服务器交互的 HTTPD 服务器。对于 Linux 用户来说，这根本不是问题，因为 HTTPD 和 MySQL 可以位于同一台 Linux 机器上。虽然 HTTPD 服务器和 MySQL 服务器不一定在同一个 IP 主机上，但为了方便起见，我们假设它们在同一台计算机上，即同一个本地主机。以下讨论基于下面几点假设。

（1）读者的 Linux 机器同时运行 HTTPD 和 MySQL 服务器，并且 HTTPD 服务器被配置为支持 PHP。读者可参阅第 13 章，以了解如何配置可用于 PHP 的 HTTPD。

（2）读者有一个用户账号，可以托管网页，访问地址为 http://localhost/ ~user_name。

如果没有用户账号，则参考第 13 章，以了解如何设置个人 Web 页面。

（3）有一个 Web 浏览器，可访问互联网上的 Web 页面。

14.5.1 使用 PHP 连接到 MySQL 服务器

用 MySQL 编程 PHP 时，第一步是连接到 MySQL 服务器。为了方便起见，我们假设 MySQL 服务器在同一台 Linux 机器上。下面的程序 P14.1 显示了一个 PHP 脚本，它连接到 MySQL 服务器来创建一个新的数据库 cs362。

```
// P14.1: PHP script, connect to a MySQL Server
<html>
<head>
  <title>Creating MySQL Database</title>
```

```
</head>
<body>
  <?php
       $dbhost = 'localhost:3036';
       $dbuser = 'root';
       $dbpass = "; // replace with YOUR MySQL root password
       $con = mysql_connect($dbhost, $dbuser, $dbpass);
       if(! $con ) {
          die('Can not connect: ' . mysql_error());
       }
       echo 'Connected successfully<br />';

       $sql = 'DROP DATABASE IF EXISTS cs362';
       $retval = mysql_query( $sql, $con );
       if ($retval)
          echo "dropped database cs362 OK<br>";

       $sql = 'CREATE DATABASE cs362';
       $retval = mysql_query( $sql, $con );
       if(! $retval ) {
           die('Could not create database: ' . mysql_error());
       }
       echo "Database cs362 created successfully\n";
       mysql_close($con);
   ?>
   </body>
</html
```

在 C 语言中，PHP 和 MySQL 之间的接口由一系列 MySQL PHP API 函数支持（MySQL PHP API 2018）。PHP API 函数的语法通常比 C 函数简单。在 PHP 中调用 mysql 函数时，通常使用值定义 PHP 变量，而不是使用硬编码字符串或数字作为参数。如文中所示，该程序使用 mysql_connect() 连接到 MySQL 服务器。因为在 MySQL 中创建 / 删除数据库需要根用户权限，所以程序作为根用户连接到 MySQL 服务器。与 C 编程一样，标准 PHP 查询函数也是 mysql_query()。当程序结束时，它使用 mysql_close() 函数来关闭连接。

14.5.2 使用 PHP 创建数据库表

下一个程序将展示如何在现有数据库中创建表。为了便于识别，将加粗显示删除表（如果存在）并创建新表的 PHP 代码。

```
// P14.2 : PHP program Create Table
<html>
<head>
   <title>Creating MySQL Tables</title>
</head>
<body>
   <?php
        $dbhost = 'localhost:3036';
        $dbuser = 'root';
        $dbpass = ";
        $con = mysql_connect($dbhost, $dbuser, $dbpass);
        if(! $con ) {
            die('Could not connect: ' . mysql_error());
```

```
        }
        echo 'Connected to MySQL Server OK<br>';

        mysql_select_db( 'cs362' ); // use cs362 created earlier

        $sql = "DROP TABLE IF EXISTS students";
        $retval = mysql_query( $sql, $conn );

        echo "create table in cs362<br>";
        $sql = "CREATE TABLE students( ".
            "student_id INT NOT NULL PRIMARY KEY AUTO_INCREMENT, ".
            "name CHAR(20) NOT NULL, ".
            "score INT, ".
            "grade CHAR(2)); ";
        $retval = mysql_query( $sql, $conn );
        if(! $retval ) {
            die('Could not create table: ' . mysql_error());
        }
        echo "Table created successfully\n";
        mysql_close($conn);
    ?>
</body>
</html
```

图 14.1 所示了从 Web 浏览器运行程序 P14.1 的结果。

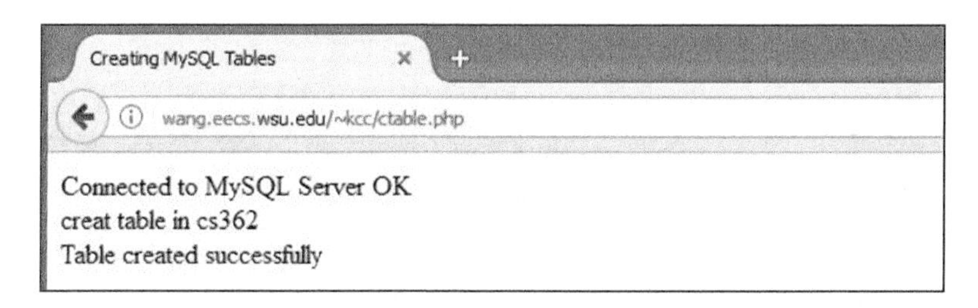

图 14.1 使用 PHP 连接到 MySQL 服务器

14.5.3 使用 PHP 将记录插入表中

下面的 P14.2 程序演示了如何使用 PHP 将记录插入数据库的表中。

```
// P14.2: Insert records into table
<html>
<head>
  <title>Insert Record to MySQL Database</title>
</head>
<body>
<?php
 if ( isset($_POST['submit']) ){ // if user has submitted data
    $dbhost = 'localhost:3036';
    $dbuser = 'root';
    $dbpass = '';
    $con = mysql_connect($dbhost, $dbuser, $dbpass);

    if(! $con ) {
      die('Could not connect: ' . mysql_error());
```

```
    }
    echo "Connected to server OK<br>";
    mysql_select_db('cs362');      // use cs362 database

    $student_id = $_POST['student_id'];
    $name       = $_POST['name'];
    $score      = $_POST['score'];
    echo "ID=" . $student_id . " name=" . $name .
         " score=".$score."<br>";
    if ($student_id == NULL || $name == NULL || $score == NULL){
       echo "ID or name or score can not be NULL<br>";
    }
    else{
       $sql = "INSERT INTO students ".
              "(student_id, name, score) "."VALUES ".
              "('$student_id','$name','$score')";
       $retval = mysql_query( $sql, $con );

       if(! $retval ) {
         echo "Error " . mysql_error() . "<br>";
       }
       else{
         echo "Entered data OK\n";
         mysql_close($con);
       }
     }
   }
 ?>

<br>
// a FORM for user to enter and submit data
<form method = "post" action = "<?php $_PHP_SELF ?>">
  ID number : <input name="student_id" type="text" id=student_id"><br>
  name    : <input name="name" type="text" id="name"><br>
  score   : <input name="score" type="text"
                    id="score"><br><br>
  <input name="submit" type="submit" id="submit" value="Submit">
</form>
</body>
</html>
```

图 14.2 所示为 PHP 生成的可供用户输入数据和提交请求的表单。

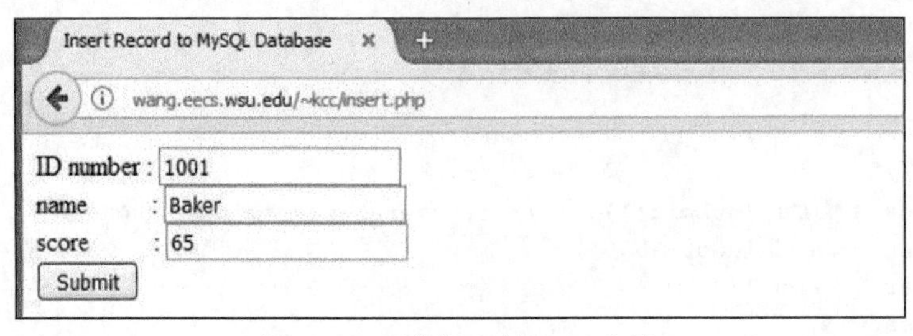

图 14.2　可供插入的 PHP 表单

图 14.3 所示为插入操作的结果。如果操作成功，它将返回另一个表单，供用户输入和提交更多的插入请求。

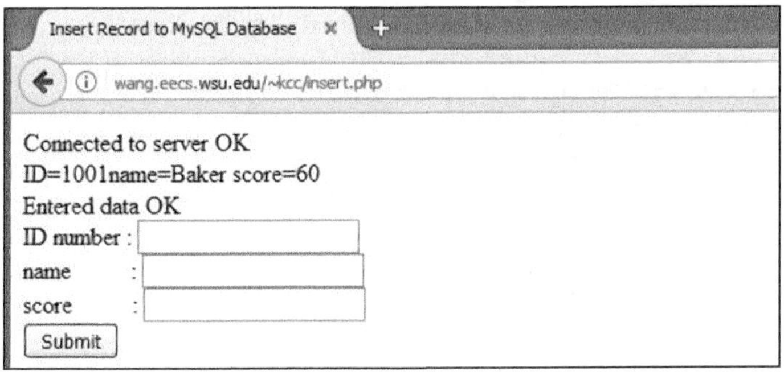

图 14.3 插入操作的结果

14.5.4 在 PHP 中检索 MySQL 查询结果

下面的程序 P14.3 展示了如何使用 PHP 检索和显示查询结果。

```
// P14.3: Select query and display results
<html>
<head>
 <title>Select Records from MySQL Database</title>
</head>
<body>
<?php
   $dbhost = 'localhost:3036';
   $dbuser = 'root';
   $dbpass = '';
   $conn = mysql_connect($dbhost, $dbuser, $dbpass);

   if(! $con ) {
      die('Could not connect: ' . mysql_error());
   }

   mysql_select_db('cs362');

   echo "select 1 row<br>";
   $sql = 'SELECT student_id, name, score FROM students
           WHERE name="Baker"';
   $retval = mysql_query( $sql, $con );

   if(! $retval ) {
      die('Could not retrieve data: ' . mysql_error());
   }
   while($row = mysql_fetch_array($retval, MYSQL_ASSOC)) {
      echo "student_id = {$row['student_id']}  ".
         "name = {$row['name']}  ".
         "score = {$row['score']} <br> ".
         "--------------------------------<br>";
   }

   echo "select all rows<br>";
   $sql = 'SELECT * FROM students';
   $retval = mysql_query( $sql, $con );

   if(! $retval ) {
```

```
        die('Could not retrieve data: ' . mysql_error());
    }
    while($row = mysql_fetch_assoc($retval)) {
        echo "student_id = {$row['student_id']}  ".
            "name = {$row['name']}  ".
            "score = {$row['score']}<br>  ";
    }

    echo "--------------------------------<br>";

    echo "selected data successfully\n";
    mysql_close($con);
?>
</body>
</html>
```

程序 P14.3 展示了如何选择数据库中的行。它通过 mysql_fetch_assoc() 函数以关联数组 $row 的形式检索行，并按名称显示 $row 的内容。它还展示了如何选择所有行，通过相同的 mysql_fetch_assoc() 函数检索这些行，并将每一行的内容显示为关联数组。关联数组是 PHP 和 Perl 中的标准特性，但在 C 语言中不可用。

14.5.5　使用 PHP 进行更新操作

下一个程序 P14.4 展示了如何更新 PHP 中的 MySQL 表。

```
// P14.4: Update operations in PHP
<html>
<head> <title>Update tables in NySQL Database</title></head>
<body>
<?php
     $dbhost = 'localhost:3036';
     $dbuser = 'root';
     $dbpass = '';
     $conn = mysql_connect($dbhost, $dbuser, $dbpass);
     if(! $con ) {
       die('Could not connect: ' . mysql_error());
     }
     echo 'connected to Server OK<br>';
     mysql_select_db('cs362');
     $sql = 'UPDATE students SET grade = \'A\'
             WHERE score >= 90';
     $retval = mysql_query( $sql, $conn );
     if(! $retval ) {
        die('Could not update data: ' . mysql_error());
     }
     echo "Updated data successfully\n";
     mysql_close($conn);
?>
</body>
</html>
```

练习：修改程序 C14.4，根据学生分数范围分配相应的字母成绩等级“B”到“F”。更新成绩等级后，运行 PHP 程序 P14.3 来验证结果。

14.5.6 使用 PHP 删除行

以下 PHP 程序 P14.5 可删除 MySQL 表中的行。

```
<?php
   $dbhost = 'localhost:3036';
   $dbuser = 'root';
   $dbpass = '';
   $conn = mysql_connect($dbhost, $dbuser, $dbpass);

   if(! $con ) {
      die('Could not connect: ' . mysql_error());
   }
   mysql_select_db('cs362');

   $sql = 'DELETE FROM students WHERE name=\'Zach\'';
   $retval = mysql_query( $sql, $con );

   if(! $retval ) {
      die('Could not delete data: ' . mysql_error());
   }
   echo "Deleted data OK\n";
   mysql_close($conn);
?>
```

14.6 习题

1. 假设 CS360 班级列表是一个文件，每行包含一个学生记录。每个学生的记录都是 C 结构

```
struct student_record{
       int  ID;
       char name[20];
       int  score;
       char email[40];   // email address
       struct book_order{
         char [20];       // "text" or "reference"
         date [10];       // yyyy-mm-dd
}
```

编写一个 C 程序，为学生建立一个 cs360 班级数据库。该数据库应有 3 个表：students、email 和 book_order。students 表应包含一个字母成绩等级列。email 表包含学生的电子邮箱地址以及引用 students 表中学生 ID 的外键。book_order 表的外键也引用了 Students 表中的 ID。

在构建 cs360 班级数据库之后，运行 UPDATE 命令来分配字母成绩等级，在表上运行 JOIN 查询并显示查询结果。

2. 假设和要求与习题 1 中相同。编写 PHP 脚本，以用 PHP 构建 cs360 数据库。通过 Web 页面测试结果数据库。
3. 为简单起见，本章中的所有示例数据库表之间只使用 1-1 和 1-M 关系。设计具有 M-M 关系的其他学生信息表，例如学生选修的课程和课程注册表。然后创建一个数据库来支持 M-M 关联表。

参考文献

C API Functions, MySQL 5.7 Reference Manual::27.8.6 C API Function OverView, https://dev.mysql.com/doc/refman/

5.7/en/c-api-function-overview.html, 2018a
C API Functions, MySQL 5.7 Reference Manual:: 27.8.7 C API Function Descriptions, https://dev.mysql.com/doc/refman/5.7/en/c-api-functions.html, 2018b
Codd, E.F. A relational model for data for large shared data banks, communications of the ACM, 13, 1970
MySQL, https://dev.mysql.com/doc/refman/8.0/en/history.html, 2018
MySQL PHP API, https://dev.mysql.com/doc/apis-php/en/, 2018
MySQL Functions, https://dev.mysql.com/doc/apis-php/en/apis-php-ref.mysql.html, 2018
PHP: History of PHP and Related Projects, www.php.net, 2018

推荐阅读

深入理解计算机系统（英文版·第3版）

作者：（美）兰德尔 E.布莱恩特 大卫 R. 奥哈拉伦 ISBN：978-7-111-56127-9 定价：239.00元

本书是一本将计算机软件和硬件理论结合讲述的经典教材，内容涵盖计算机导论、体系结构和处理器设计等多门课程。本书最大的特点是为程序员描述计算机系统的实现细节，通过描述程序是如何映射到系统上，以及程序是如何执行的，使读者更好地理解程序的行为，找到程序效率低下的原因。

编译原理（英文版·第2版）

作者：（美）Alfred V. Aho 等 ISBN：978-7-111-32674-8 定价：78.00元

本书是编译领域无可替代的经典著作，被广大计算机专业人士誉为“龙书”。本书上一版自1986年出版以来，被世界各地的著名高等院校和研究机构（包括美国哥伦比亚大学、斯坦福大学、哈佛大学、普林斯顿大学、贝尔实验室）作为本科生和研究生的编译原理课程的教材。该书对我国高等计算机教育领域也产生了重大影响。

第2版对每一章都进行了全面的修订，以反映自上一版出版二十多年来软件工程、程序设计语言和计算机体系结构方面的发展对编译技术的影响。

推荐阅读

现代操作系统（原书第4版）

书号：978-7-111-57369-2 作者：[荷]安德鲁 S. 塔嫩鲍姆 赫伯特·博斯 定价：89.00元

本书是操作系统的经典教材。在这一版中，Tanenbaum教授力邀来自谷歌和微软的技术专家撰写关于Android和Windows 8的新章节，此外，还添加了云、虚拟化和安全等新技术的介绍。书中处处融会着作者对于设计与实现操作系统的各种技术的思考，他们的深刻洞察与清晰阐释使得本书脱颖而出且经久不衰。

第4版重要更新

- 新增一章讨论虚拟化和云，新增一节讲解Android操作系统，新增研究实例Windows 8。此外，安全方面还引入了攻击和防御技术的新知识。
- 习题更加丰富和灵活，这些题目不仅能考查读者对基本原理的理解，提高动手能力，更重要的是启发思考，在问题中挖掘操作系统的精髓。
- 每章的相关研究一节全部重写，参考文献收录了上一版推出后的233篇新论文，这些对于在该领域进行深入探索的读者而言非常有益。

作者简介

安德鲁 S. 塔嫩鲍姆（Andrew S. Tanenbaum） 阿姆斯特丹自由大学教授，荷兰皇家艺术与科学院教授。他撰写的计算机教材享誉全球，被翻译为20种语言在各国大学中使用。他开发的MINIX操作系统是一个开源项目，专注于高可靠性、灵活性及安全性。他曾赢得享有盛名的欧洲研究理事会卓越贡献奖，以及ACM和IEEE的诸多奖项。

赫伯特·博斯（Herbert Bos） 阿姆斯特丹自由大学教授。他是一名全方位的系统专家，尤其是在安全和UNIX方面。目前致力于系统与网络安全领域的研究，2011年因在恶意软件逆向工程方面的贡献而获得ERC奖。